Women Working Wonders

Women Working Wonders

Small-scale Farming and the Role of Women
in Vihiga District, Kenya

A Case Study of North Maragoli

Basilida Anyona Mutoro

THELA PUBLISHERS AMSTERDAM
1997

CERES Series, no. 6

CIP-DATA KONINKLIJKE BIBLIOTHEEK, DEN HAAG

This publication has been prepared as a part of CERES Research School for Resource Studies for Development. Utrecht University, Wageningen Agricultural University, as well as the Netherlands Foundation for the Advancement of Tropical Studies (grant number W 52-494) have participated in funding the research.

© Basilida Anyona Mutoro - 1997

Cover design: Mirjam Bode
Layout: KartLab, FRW/UU, Gérard van Betlehem & Margot Stoete
Drawings: KartLab, FRW/UU, Margot Stoete & Lilian van de Zande
Photographs: Basilida Anyona Mutoro

All rights reserved. Save exceptions stated by the law no part of this publication may be reproduced, stored in a retrieval system of any nature, or transmitted in any form or by means, electronic, mechanical, photocopying, recording or otherwise, included a complete or partial transcription, without pior written permission in writing of the publisher, application for which should be addressed to the publisher. Thesis Publishers/Thela Publishers, Prinseneiland 305, 1013 LP Amsterdam, the Netherlands.

ISBN 90 5538 019 9
NUGI 661/652

Preface

Preface and Acknowledgements

Due to high population increase, rapid urbanization, low technology and haphazard industrialization in Kenya and many other developing countries, the problems of poverty, inequality, environmental degradation and poor resource management are acute. This explains the urgent and rapidly growing need for environmental specialists to carry out research, provide advice on environmental management and resource use, and to train environmental personnel at various levels.

The establishment of the School of Environmental Studies at Moi University in Eldoret, Kenya was part of the national policy towards the rational management of natural resources and the environment for sustainable development. The School of Environmental Studies was charged with the responsibility of training specialized personnel. However, the School was handicapped in its duties because it lacked staff adequately qualified in environmental disciplines, equipment, documentation and capital.

The School of Environmental Studies approached the University of Amsterdam, The Netherlands in the hope of a possible co-operation. This co-operation was realized in 1991 with financial and material support from the Dutch Government through NUFFIC. In this way, the School of Environmental Studies is being provided with financial resources and equipment and opportunities for staff development are being created for the School.

This study has been made possible because of the co-operation that exists between the Faculty of Environmental Sciences, University of Amsterdam, The Netherlands and the School of Environmental Studies, Moi University. I feel myself privileged and am grateful to both the Dutch Government through NUFFIC, and the Faculty of Environmental Sciences, University of Amsterdam, who have made it possible for me to be among the first Ph.D candidates to be trained in The Netherlands in the framework of this co-operation.

This study would never have taken place without the initiative of several people. In a research project of this scale it is impossible to mention every one who assisted either directly or indirectly with data collection and analysis. I am grateful to Prof. Dr. C.O Okidi and Prof. Dr. J.A. Dietz, 'the brains' behind the co-operation between the School of Environmental Studies, Moi University and the Faculty of Environmental Sciences, University of Amsterdam. Financial support was only approved because of the excellent project proposals written by these two distinguished scholars. I would like to thank the Vice Chancellor of Moi University for granting me study leave. Gratitude as well to my colleagues in the Department of Geography who readily accepted taking over my teaching responsibilities.

In the Netherlands I would like to express my sincere gratitude to my chief promotor Prof. Dr. J.Hinderink for his dedication and attachment to this study from its inception in 1991 to its successful conclusion in 1995. He provided continual inspiration and excellent supervision both in Kenya, and in The Netherlands where we spent long working hours, days and months even when his health was in jeopardy. Thank you for your resolute, unreserved, uncompromising but sympathetic guidance and academic advice.

This study had a strong team of co-promotors to whom I am greatly indebted. They were always open and pleasant to me as well as being prepared to read endless drafts of my manuscript. They have showed great interest in my work giving it priority in their schedules and provided me with very close criticism and supervision. Our discussions were crucial to me and their contributions towards improving the book are visible in each chapter. I therefore wish to thank Prof. Dr. A.J. Dietz for his invaluable hints, suggestions and limitless supervision. Prof. Dr. H.A Luning with his long experience of land evaluation and farming systems helped greatly with this study. I acknowledge and thank Prof. Dr. R.S Odingo for his commitment, interest, and firm supervision while I was in the field in Kenya and for inspiring me towards agricultural and rural-based research.

Much gratitude also goes to Prof. Dr. L.O. Fresco of the Department of Agronomy, the Agricultural University of Wageningen, Prof. Dr. J. Hoorweg of the African Studies Centre, University of Leiden, Prof. Dr. J.T. Schrijvers of Institute of Women and Development, University of Amsterdam, and Dr. W. Van Beek of the Department of Cultural Anthropology, University of Utrecht for reading the manuscript and acting as external examiners. A special word of thanks here to the Cartographic Department of the Faculty of Geographical Sciences, University of Utrecht. I particularly appreciate the work of Simone Haller-Albers, Margot Stoete, Gérard van Betlehem and Lilian van de Zande in drawing all the maps and diagrams, scanning the pictures and making camera ready text for publication. I acknowledge the support of Rob van der Sanden, Albert Visser, Andri Harianja, Marcel Heemskerk, and Erik Smeerdijk of the computer Department of the University of Amsterdam who have patiently shown me how to use the computer and the necessary programmes. The secretariat staff of the Department of Human Geography, University of Amsterdam: Lotty Jansen Saan, Saskia Peletier and Joyce Van Galen Last have been very kind to me and provided me with all the necessary working materials.

This work would be incomplete without a word of thanks to my colleagues with whom I have shared the joys and hardships of writing a Ph.D dissertation in a foreign country away from our families. I would like to thank Dr. Godfrey Anyumba, Carey Ombura and Jockey B. Nyakaana for their support especially when the going seemed very tough. They have always been there to listen and give advice. A special word of thanks to Simone Sas who has been a dear friend for all the years I have lived in The Netherlands. Much thanks too to Drs J.M. van Haastrecht for her efficient administrative and financial co-ordination and for making my stay in Amsterdam so very comfortable. The field research kept to schedule because of Prof. Dr. G. Linden's efficiency in the financial administration of the research grants and I am grateful to him for the support he gave me during my fieldwork by his visit, academic support and advice. I would also like to acknowledge Mark Lamers who was the first project-co-ordinator. I convey my appreciation to all my colleagues in the Department of Human Geography, University of Amsterdam for their friendliness and support particularly Gonda de Haan, and Maike Kromhout and Antje Van Driel who accepted and were my assistants (paranimfen) during the defence and graduation. Much gratitude to Ms Beatrice Achiamaa and her daughters Gifty Omusu and Catherine Achiamaa for the hospitality and support I have received at their house during my last few months in Amsterdam.

I would like to record my gratitude to all the small-scale farmers who, over the years, opened their doors to me and gave me such friendship that it made this research project a most pleasant experience. It is important to mention that since the 1970s a new preoccupation with the position of women emerged. This began in the developed world and spread into the developing countries. However, a preoccupation with women issues in workshops, seminars and conferences is far removed from the rural women who have not benefitted from the calls for womens' liberation, women and development, and womens' empowerment. Rural women are only preoccupied with the survival of their households. It is encouraging to note that despite rural womens' constraints, hardships and poverty, they are still welcoming and ready to share the little they have. I very much appreciate the way in which all the women of North Maragoli and Vihiga District gave me their friendship and provided me with all the necessary and relevant information for this study. I thank them for allowing me to be part of their households for the two years I was in the field. I appreciate too the women groups who were always ready to entertain, sing and dance any time I arrived, apart from providing the required information. Knowing how busy these women are, I am indebted to them for giving up time to be with me on their farms and elsewhere. The women did indeed expect me to put forward their problems. I hope that those who read this book will help in devising strategies to improve the rural life of millions of Kenyan women who labour daily to sustain the livelihood of their households. I also wish to thank the male heads of households who participated in this study for their patience during our long interview sessions and farm visits. Without the valuable information and the co-operation of the women and men of Maragoli, this study could not have been a success. I thank them once more for their willingness, trust and tolerance. This study was inspired by the small-scale farmers especially the women of North Maragoli. To protect their privacy, I have used fiction names in our case studies analyses.

I also wish to express my sincere gratitude to all government officials (especially those from the Ministry of Agriculture, Livestock Development and Marketing) and organizations participating in small-scale farming, such as the NGOs whether in their headquarters in Nairobi, or in Kakamega, Vihiga and Sabatia who kindly and willingly assisted me in my research. I acknowledge the help, material support and advice I received from the staff of the Departments of Geography at Nairobi University and at Moi University. My thanks goes to Isaiah Angiro Nyandega of the Department of Geography, Nairobi University for the preliminary data analysis and Raphael Kareri of the Department of Geography, Moi University for his inspirational academic discourses. A special word of gratitude to Mr T. Okuku, former DDO and the staff of the district planning office of Kakamega, who allocated me an office, allowed me to use the departmental computer and provided me with transport now and again.
Last, but not least, I am grateful to all my research assistants: Margriet de Haan previously of the Department of Human Geography, University of Amsterdam, and Florence Ngoya, Japheth Agwusioma, Judith Imali, Mary Mideva, Charles Kibisu and Grace Agufa for doing a commendable job and for being there for me all the time during the field work. I would also like to mention here Mr Alfred Nyairo and Dr. D'lima who agreed to be my guarantors.

Finally, the support and interest shown by my friends and relatives was deeply highly appreciated. I am indebted to my friends and colleagues at Moi University, particularly Mrs Anne Nangulu-Ayuku and Dr. Titus Sisenda who have been very instrumental in the administration of my household during my long absence. A special word of appreciation as well as compliments to my dear friend Ms Aurelia Osogo who is like a sister to me and who took on the responsibilities of providing and caring for my children in my absence. I am deeply grateful for her constant reassurance. I would also like to thank my brothers and sisters: Andrew Maina Andieri, Lawrence Tabu, Teresia Nechesa, Anne Malenya, Maurice Lumumba, Sylvester Juma, Imelda Nina, Isaac Kilo and all my other relatives who I cannot mention individually here for their support, encouragement and for being on my side always. I extend my gratitude and appreciation to my mother Leonida Anyanga and my late father Alfred Andieri for their unfailing love, support, encouragement, prayers and their belief that all children are the same and should be provided with equal educational opportunities regardless of gender. I acknowledge too the inspiration and support provided by my husband Mr Benedict Mayende Mutoro. Finally, I am more than indebted to my children Karen Anne Mutoro, Nancy Nekesa Mutoro and Immaculate Anyanga Mutoro who have endured all the pains of being without parents and all the bad times during this study. Though still of tender age, they have provided me with perpetual encouragement, love, prayers and support throughout this study. I apologize to them and believe that we shall catch up on what we have missed in the next phase. I dedicate this book to the people I love most and who suffered the most during its realization: Karen Anne Mutoro, Nancy Nekesa Mutoro and Immaculate Anyanga Mutoro.

Table of Contents

Preface and acknowledgements   VII

Contents   XI

1. Introduction

1.1	Subject of the Study	1
1.2	The Importance of Agriculture to Kenya	2
1.3	Justification for the Emphasis on the Role of Women	4
1.4	Research Objective and Research Questions	6
1.5	Research Methodology	8
1.6	Organization of the Thesis	9

2. A Framework of the Major Components of Small-scale Farming and the Role of Women in Farm Management

2.1	Introduction	11
2.2	Views on African Agriculture	12
2.3	Characteristics of Small-Scale Farming in Kenya	15
2.3.1	Land Tenure	16
2.3.2	Labour Input and Labour Constraints	19
2.3.3	Agro-Support Services in Small-Scale Farming	22
2.4	Gender and Farm Management	27
2.4.1	Gender Relations and Farming in Tribal Society	28
2.4.2	The Impact of Colonialism on Gender Relations	29
2.4.3	Gender and Agricultural Extension and Credit	31
2.4.4	The Household as a Unit of Production and Consumption?	32
2.4.5	Women Groups and Agricultural Production	34
2.5	Conclusions and Research Hypotheses	35

3. Vihiga District: Background Information

3.1	Introduction	39
3.2	The Physical Resource Base	42
3.2.1	Topography and Geology	42
3.2.2	Soil Characteristics and Agro-Ecological Zones	42
3.2.3	Climatic Characteristics and Agro-Ecological Zones	45
3.2.4	Forestry Resources	47

3.3	The Human Resource Base	51
3.3.1	Population: Settlement and Distribution	51
3.3.2	Out-Migration and Wage Employment Opportunities	56
3.4	The Economic Situation of Vihiga District	58
3.4.1	The Agricultural Resource Base	58
3.4.2	Other Primary Production Activities	65
3.4.3	Commerce, Trade, Industry and Services	65
3.5	Physical and Social Infrastructure	67
3.6	Recapitulation of the Main Problems in Vihiga District	74

4. Small-scale Farming in North Maragoli

4.1	Introduction	79
4.2	Characteristics of North Maragoli	79
4.3	The North Maragoli Farming System	82
4.4	The Food Crop Sub-system	85
4.4.1	Maize	86
4.4.2	Reasons for Low Production and Low Yields of Maize	87
4.4.3	Other Food Crops	91
4.5	The Commercial Crop Sub-system	92
4.5.1	Tea Cultivation	92
4.5.2	Problems in Tea Cultivation	95
4.5.3	Coffee Production	98
4.5.4	French Beans	101
4.6	The Livestock Farming Sub-system	102
4.7	The Tree Production Sub-system	104
4.8	Agricultural Labour	105
4.9	Agro-Support Institutions in North Maragoli	108
4.9.1	Agricultural Credit	108
4.9.2	Co-operatives in Production and Marketing	111
4.9.3	Agricultural Extension	112
4.9.4	Types of Information Transmitted and Target Groups	113
4.10	Conclusions	117

5. Farm Management Situations and Types of Households in Small-scale Farming in North Maragoli

5.1	Introduction	119
5.2	The Selection of Respondents	120
5.3	Characteristics of the Selected Households	121
5.4	Land Tenure and Farm Size in Relation to Gender and Types of Households	124
5.5	The Household Type and its Effect on the Maize Farming Sub-system	126
5.6	The Impact of Household Type on the Tea Farming Sub-system	132
5.7	Livestock Sub-system and Milk Production in the Three Types of Households	139
5.8	The Tree Production Sub-system	142

5.9	Gender and Labour Distribution and their Effect on the Farm-Household System	143
5.10	Fertilizer Use in the Three Types of Households	150
5.11	Farm and Household Incomes and Farming Investments in Relation to Gender	152
5.12	Gender Differentiation and Access to Agro-Support Institutions	154
5.13	Coping Strategies and Non-Agricultural Activities in North Maragoli	156
5.14	Conclusions	159

6. Inter- and Intra-household Farm Management Situations: Case Studies

6.1	Introduction	163
6.2	Male-Headed Households	164
6.3	The Example of an Independent Female in a Male-Headed Household	183
6.4	Female-Managed Households	187
6.5	Female-Headed Households	196
6.6	Conclusions	204

7. The Impact of Selected Agricultural Policies and Programmes on Farm Management and Agricultural Production in Relation to Gender in North Maragoli

7.1	Introduction	207
7.2	Land Tenure Policy	209
7.3	Agricultural Extension Policy	215
7.4	Agricultural Credit Policy	221
7.5	Policies Affecting Cash Crops, Inputs and Seeds	226
7.6	Food Policy	231
7.7	Livestock Development Policy	234
7.8	Recent Changes in Agricultural Policy	236
7.9	Policy and Gender	238
7.10	Conclusions	240

8. Women Groups as a Strategy for Agricultural and Rural Development

8.1	Introduction	243
8.2	A Review of the Debate on Women Organizations	244
8.3	A History of Women Organizations in Kenya	247
8.4	Formation and Objectives of Women Groups in North Maragoli	250
8.5	Activities and/or Projects of Women Groups in North Maragoli	253
8.5.1	Crop Production and Income-Generation	254
8.5.2	Livestock Production and Income-Generation	259
8.5.3	Tree Production	265
8.5.4	Secondary Sector Activities	266
8.5.5	Tertiary Sector Activities	267

8.6	The Role of Women Groups in Agro-forestry and Environmental Conservation in North Maragoli	271
8.7	The Role of Women Groups in Farm Management	276
8.8	Women Groups and Agricultural Extension Services and Credit	278
8.9	Views of Women Groups on the General Socio-Economic Life of Women in the Rural Areas	280
8.10	Problems and Aspirations of Women Groups	284
8.11	Conclusions	288

9. Conclusion

9.1	Introduction	299
9.2	The Major Problems in Small-scale Farming	299
9.3	The Role of Women in Farm Management	301
9.4	Research Hypotheses Revisited	302

References	305
Abbreviations and glossary	326
Appendices	328
Curriculum Vitae	343

Figures

1.1	Sectoral National Growth Rates and Contribution to GDP 1964-1992	3
1.2	Coffee and Tea Production in Kenya 1979-1991	3
1.3	The Conceptual Research Model	7
1.4	Interactions Between the Institutional Support Systems and Farm Management	7
3.1	Location of Vihiga District in Kenya	40
3.2	Administrative Boundaries of Vihiga District	41
3.3	Administrative Divisions of Vihiga District	41
3.4	Drainage and Relief of Vihiga District	43
3.5	Soil Map of Vihiga District	45
3.6	Vihiga District Annual Average Rainfall and Mean Temperatures	48
3.7	Vihiga District Annual Rainfall and Temperatures	48
3.8	Mean Annual Rainfall Map of Vihiga District	49
3.9	Agro-Climatic Zones Map of Vihiga District	50
3.10	Kenya Population Pyramid by Sex and Age	54
3.11	Vihiga Population Pyramid by Sex and Age	54
3.12	Vihiga District Population Growth 1962-1989	55
3.13	Communication Network of Vihiga District	74
3.14	Health Facilities in Vihiga District	75
3.15	Location of Cattle Dips in Vihiga District	76
4.1	North Maragoli Population Pyramid by Sex and Age	80
4.2	Farming System Hierarchy	83

4.3	The Farm Household System	83
4.4	Simplified North Maragoli Farming System	84
4.5	Structure of Agricultural Extension in North Maragoli	114
5.1	Population Pyramid by Sex and Age	122
5.2	Total Maize Production, 1993	129
5.3	Annual Average Maize Yield per Hectare	130
5.4	Mean Annual Tea Yield per Hectare, 1990-1993	135
5.5	Monthly Tea Production, 1993	136
5.6	Monthly Tea Income, 1993	137
5.7	Mean Annual Tea Income per Hectare	138
5.8	Total Milk Production, 1993	141
5.9	Monthly Milk Income, 1993	141
5.10	Monthly Investments in Farming, 1993	154
6.1	Symbols used in Farm Sketches	168
6.2	Agesa Farm	169
6.3	Eboso Farm	173
6.4	Manase Farm	178
6.5	Iminza Farm	184
6.6	Chagenda Farm	188
6.7	Mmbone Farm	191
6.8	Muhonja Farm	195
6.9	Ajema Farm	197
6.10	Kaveza Farm	200
6.11	Keyonzo Farm	203

Tables

3.1	Agro-Ecological Zones in Vihiga District	44
3.2	Agro-Ecological Zones by Division	45
3.3	Climatic Data of Vihiga District	47
3.4	Population Size and Density per Division, Vihiga District 1989	53
3.5	The Population of North Maragoli in 1989	55
3.6	The Population of Urban Centres in Vihiga District	56
3.7	Production Figures for Major Crops in Vihiga District 1991/92	59
3.8	Land Suitable and Under Tea in Vihiga District	62
3.9	Green Leaf Production in Vihiga District	62
3.10	Comparison of Green Leaf Production in Vihiga and Other Tea-Growing Districts in Kenya	63
3.11	Zebu and Grade Cattle Population, 1993	63
3.12	Livestock Production in Vihiga District	64
3.13	Distribution of Business Enterprises by Division, 1993	65
3.14	Employment Profile Vihiga District in 1994	66
3.15	Distribution of Educational Institutions in Vihiga District	72

4.1	Population Distribution by Sex and Age	80
4.2	Levels of Education of Heads of Households	81
4.3	Frequency Distribution of Farm Sizes	82
4.4	Cropping Patterns for Food and Cash Crops, and Livestock Production	85
4.5	Maize Production 1991	87
4.6	The Purchase of Certified Hybrid Maize Seeds in 1991	88
4.7	Land Allocated to Tea by Sample Farmers	92
4.8	Total Tea Production by Sample Farmers 1991	93
4.9	Annual Tea Income to Farm Households	93
4.10	The Economics of Tea Production in Vihiga District	94
4.11	Purchase of Fertilizer for Tea Cultivation in 1991	96
4.12	Land under Coffee Cultivation in 1991	99
4.13	Coffee Production 1991 among Sample Farmers	99
4.14	Coffee Income in 1991 of the among Sample Farmers	99
4.15	Livestock Population among the Sample Households 1992	103
4.16	Expenditure on Hired Agricultural Labour	107
4.17	Sources of Loans taken by Farmers 1991	110
5.1	Case Studies Population Distribution by Age and Sex	121
5.2	Levels of Education for Household Heads	123
5.3	Farm Sizes Among the Case Study Samples	125
5.4	Farm Size Under Maize Production	127
5.5	General Summary Table on Maize Production in North Maragoli	129
5.6	Distribution of Plot Size Under Tea by Household Types in 1993	132
5.7	Population Density of Tea Bushes	133
5.8	Average Tea Yield in the Three Types of Households in 1993	134
5.9	Summary of Tea Production and Incomes in 1991	135
5.10	1993 End of Year Tea Incomes in the Three Types of Households in 1993	137
5.11	Land Under Tree Cultivation in the Three Types of Households	142
5.12	Money Spent on Hired Labour in 1991	145
5.13	Average Time Devoted to Tea Production	146
5.14	Labour Time Allocation to Food Production	146
5.15	Weighted Farm Labour in Hours	147
5.16	Gender Division of Working Hours	148
5.17	Amount of Fertilizer taken on Credit from KTDA by Household Type	150
5.18	Total and Estimated Net Annual Farm and Household Incomes by Household Type, 1993	152
5.19	Farm Investments by Household Type	153
5.20	Visits by Extension Agents in 1993	155
6.1	Summary of Annual Tea Production, Tea Incomes and Total Household Incomes	337
6.2	Summary Maize Production	338
6.3	Range of Activities in Relation to Agricultural Production and Household Incomes	339

Preface xvii

7.1	Land Tenure System and Ownership in North Maragoli	212
8.1	Number of Women Groups Assisted by the Kenya Government	249
8.2	The Most Prevalent Activities among Women Groups in North Maragoli	254
8.3	Financial Assistance to Women Groups in North Maragoli	270
8.4	A List of Women Groups in North Maragoli Location 1992	341

Table 1   330

Plates

1	Mother and Children Weeding their Maize Field	290
2	Types of Food Crops: Yams and Onions in the Foreground, 'Sukuma Wiki' in the Middle, Maize and Bananas in the Background	290
3	Healthy Maize	291
4	Maize Badly Affected by Hailstorm	291
5	Tea Fields	292
6	Plucking Tea	292
7	Coffe Trees with Food crops	293
8	Organic Farming of Vegetable Plots	293
9	Zero-Grazing Unit with Grade Cow	294
10	Feeding a Cross-Bred on the Homestead	294
11	Eucalyptus Groves in Marshy Valley and Tea on Slopes	295
12	Women Group Engaged in Tailoring	295
13	Women Group Displaying Home-made Pottery	296
14	Women Choir Practising	296
15	Future Farmers	297

Exchange Rates between 1991 and 1995:

1 Dutch Guilder = between 25 - 38 Kenya Shillings

1 U.S.A Dollar = between 45 - 75 Kenya Shillings

1 Sterling Pound = between 55 - 98 Kenya Shillings

1

Introduction

1.1 Subject of the Study

"There is virtually no land which produces economically useful products such as crops, livestock or trees which cannot be managed to maintain yields indefinitely. Even for the least resilient ecosystems, there are techniques of land management providing protection from degradation." (Blaikie, P. 1989).

Many scholars writing on developing countries have tended to associate poor agricultural development - defined in terms of low yields, low farm incomes and land degradation - with physical constraints and the increasing pressure on land caused by population growth. It is indeed true that the vagaries of weather may be too much for farmers in certain ecological zones. Population pressure is also very important because it determines individual farm size and therefore the intensity of land use and overall production. However, it is the contention of this study that farm management practices play just as an important role in explaining poor agricultural development. Farm management is heavily influenced by the nature of the policies pursued in boosting agricultural development, and by the functioning of institutions and their programmes in relation to farm managers.

In most rural areas in Kenya, women are actively involved in almost all smallholder farming operations. Therefore, to understand the problems of farm management in Kenya, a thorough analysis of the role and perceptions of women is crucial. Without taking a gender approach to the improvement of the Kenyan farming situation, it will be difficult to promote good farm management and, thus, sustainable agriculture. In Kenya, as in all countries, the political economy determines the policies pursued and the institutional programmes designed. These policies and programmes may influence agricultural development negatively or positively, depending on their practical relevance - i.e. their physical and economic feasibility, acceptability and remunerativeness in the eyes of the farmers concerned. It is important to realize that many of these farmers are women, and that their perceptions and constraints should be fully taken into consideration.

The area selected for this study is Vihiga District in densely populated Western Kenya. This rural district belongs to one of the high potential areas of Kenya because of its deep fertile soils and abundant precipitation. Yet, paradoxically, the district's agricultural situation is so bad that many rural households have to buy food from other districts. Food insecurity (a situation where food for a healthy life is not accessible to all people at all times) is both a chronic and permanent phenomenon in Vihiga District. How can this apparent paradox be explained?
In Vihiga District, many farm households are managed and headed by women because their husbands have migrated to other places, outside the district, to seek remunerative off-farm employment. These women experience all types of difficulties when farming including problems

with the land tenure system, the growing of cash crops, the use of fertilizers, access to credits and useful agricultural information, membership of co-operatives, and other factors essential to optimal farm management. Therefore, to understand the problematic agricultural situation in Vihiga District it is extremely important to analyze the problems women face in farming, their perceptions of proper farm management and the relevance to them of policies pursued and programmes designed to boost agricultural development. Such an analysis is the subject of this study.

1.2 The Importance of Agriculture to Kenya

Kenya, a country with an area of approximately 582,000 sq.kms and a population of approximately 24 million people, has virtually no mineral wealth or oil deposits. The agricultural sector is the economic mainstay, providing the country both with substantial and reasonably stable export earnings and a basis for industrial and commercial growth. At Independence in 1963, agriculture accounted for 80 to 90% of rural employment and approximately 80% of the foreign exchange earnings (Kenya, 1965).More than thirty years later, despite considerable expansion in the industrial and manufacturing sectors, the agricultural sector remains the single most important sector in the Kenyan economy, providing 80% of the working population with a living and accounting for over 70% of Kenya's foreign exchange earnings (Kenya Development Plan 1994-96, Muchena and Kiome 1995).

After Independence, much emphasis was given to the development of cash crops in the small farm sector. The intention was to increase rural incomes and provide employment for the majority of the rural population. The emphasis was necessary because the colonial government had largely concentrated its efforts on the large-scale farming areas because these areas were seen as crucial to the economy. The main cash crops today include coffee, tea and horticultural products, such as cut flowers, French beans, fruits and vegetables. Moreover, livestock farming has also developed, including dairy farming, beef ranching, and the keeping of small livestock. In recent years agricultural policy has paid attention to food production in the interest of national food self-security, and there has been particular emphasis on maize production.

Since Independence the Kenyan economy has experienced a very fluctuating development (see Figure 1.1). Between 1963 and 1989 the Kenyan GDP grew at an average rate of 5% per annum. Agriculture provided the main engine for growth. During the period in question there was rapid expansion in the areas devoted to the cultivation of cash crops. The heavy dependence on agricultural exports to provide the main source of foreign exchange implies that the country was very vulnerable to the whims of world markets. This was dramatically proven by the collapse of the coffee market after the boom years of 1976-82, an event that had severe repercussions for the Kenyan economy (Kenya 1994).

Figure 1.1 makes clear that since 1990 in particular, the contribution of agriculture to GDP has deteriorated. Agriculture actually shows a negative growth of minus 4.2%. This is contrary to what the government has tried to achieve with its interventions and agricultural policies. Coffee production has been steadily declining which calls for a thorough analysis of the underlying problems. In contrast to coffee, however, tea has performed better in recent years (Figure 1.2).

Introduction

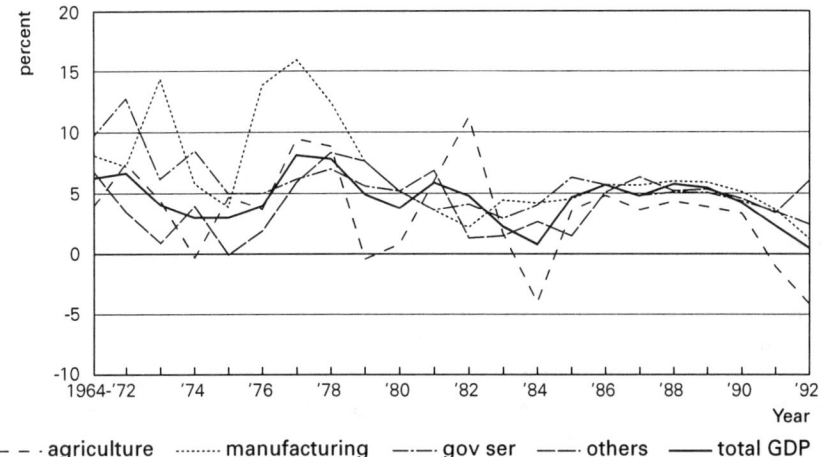

Figure 1.1
Sectorial National Growth Rates and Contribution to GDP, 1964-1992 (source: Kenya (1994), Kenya Development Plan 1994-1996).

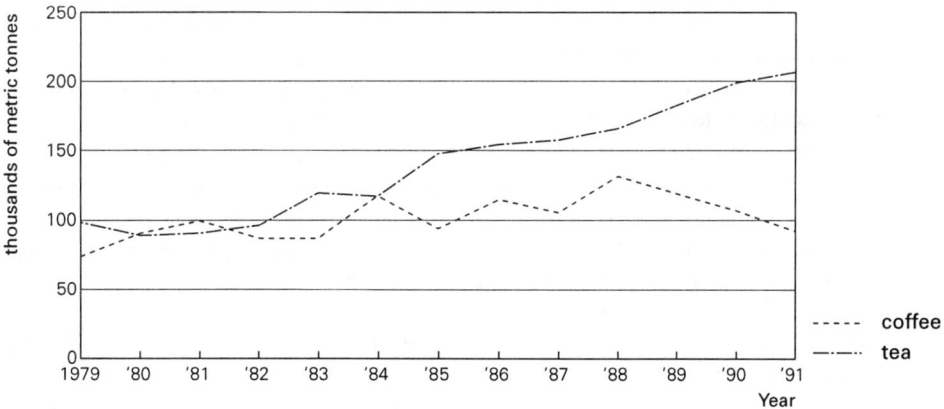

Figure 1.2
Coffee and Tea Production in Kenya 1979-1991 (source: Statistic Yearbook 1990-1991).

The slowdown in the growth of GDP since 1990 can be explained in terms of an actual decline in real output and in value added in agriculture. This is attributed to below average amounts of rainfall, sluggish growth of aggregate private domestic demand and foreign exchange shortages, and to the impact of the suspension of donor aid. All of these factors led to reduced imports of intermediate goods (Kenya 1994). However, the major causes of poor economic growth in recent years have been the social and political tensions in many parts of the country. These tensions culminated in tribal clashes and ethnic purges which resulted in the displacement of many farmers and the destruction of crops, livestock and produce, and have led to a low level agricultural activity in the main farming areas. In addition, the excitement of multipartism and the elections in 1992 also contributed to an economic turmoil that is still affecting the country.

Because of donor dissatisfaction with the negative political situation in Kenya, the international community suspended foreign aid and quick disbursement grants. The result of all these factors was the slowdown of development in all sectors of the economy, low imports of inputs and materials, skyrocketing prices and the increasing poverty of many Kenyans.

Apart from the problems mentioned above, there is the constraint that there are limited areas of high potential land on which agricultural production can be increased and maintained. Of Kenya's 44.6 million hectares of land, only about 8.6 million hectares (19%) are medium to high potential agricultural land. Of this, about 60% or 5.2 million hectares can be described as good in the sense that it has adequate and reliable rainfall and good soils and is not too hilly (Heyer 1976). The best land is used for crop and milk production. Much of the rest is used for extensive grazing or is taken up by forest reserves. It has been estimated that some 500,000 hectares of additional land could be brought into production by using irrigation, flood control and other engineering techniques. The medium and high potential areas support two-thirds of the country's population which explains the pressure on and the acute shortage of arable land. This alarming situation calls for far-reaching measures to solve the continuing slowdown of agricultural production and to promote the sustainable management of this limited area of good agricultural land.

According to the figures available, a total of 1,172,000 hectares of land is farmed by large-scale farmers with holdings varying from 20 to over 20,000 hectares, whereas some 2,519,055 hectares of land is farmed by small-scale farmers with holdings varying from less than 1 hectare to 20 hectares. The majority of small-scale farmers have holdings of less than 2 hectares, and as the population continues to increase and farms decrease in size, landlessness becomes a growing problem. Despite the problems of small farms, more than half of the cash crops grown for export and most of the food crops are produced by the small-scale farmers. Therefore, policy makers have to find ways of making farming efficient and sustainable enough to go on providing a living for the large majority of rural people who are unable to find alternative employment away from agriculture and family holdings (Kenya Statistical Abstract 1990).

Agriculture will remain essential for Kenya's development for the following reasons: (1) the creation in Kenya of productive employment every year for almost 200,000 new entrants to the labour market can only be sustained by a rapid growth in agriculture, and to a lesser extent by revitalizing industry and small-scale enterprises; (2) in the context of the current government policy of food self-sufficiency agriculture is expected to feed the ever increasing population; (3) agriculture is central to national efforts to find funds for servicing the national debt, and repaying loans owed to the World Bank and other bilateral and multilateral donors; and (4) the prosperity of industry, commerce and other sectors of the economy are largely dependent on the prosperity of the agricultural sector which has great potential for inter-sectoral economic linkages.

1.3 Justification for the Emphasis on the Role of Women

In spite of the lip-service paid in development plans to the crucial role which could be played by women, the position of women in Kenyan agriculture has been very neglected as far as the implementation of policy and programmes is concerned.

Introduction

"To see the development of any society is to see the position of women in that society." (Muro, A. in Kiros (ed), 1985:61). Women have historically been producers and reproducers, particularly in the subsistence agricultural sector. Ninety percent of Third World women depend on their land for their survival (Dankelman and Davidson 1988). On the whole, women are the world's farmers: they grow the crops, gather firewood, tend animals and bring in water. Women, particularly those living in the Third World, are therefore seen as playing a major role in managing natural resources (Dankelman et al. 1988). Because women are at the centre of world food production, an analysis of the utilization of land resources must include an appreciation of their pivotal role.

Africa is the region of female farming par-excellence. In many African societies, nearly all the tasks connected with food production continue to be carried out by women (Boserup 1970, Davison 1988, Parpart et al 1989, and Staudt 1987). Nearly 70% of the stable food in the continent is provided by female farmers (World Bank 1989a, Saito et al.1990). Women are also important in other agricultural activities, including food processing and marketing, cash cropping and animal husbandry. The role of women in managing farms and in heading households on a day - to - day basis is increasing as more men migrate to cities and other countries for work (Saito et al. 1990).

In Kenya, about four-fifths of the female population lives in rural areas and most of them are engaged in agricultural production as a major economic activity (Kenya 1988, 1990, 1994). Women provide three-quarters of the labour used on small holdings and 95% of women small holders work on their own farms (Kenya 1990, 1994). Kenyan rural women have two major agricultural responsibilities. The first is household food crop production, for which they take the major responsibility and contribute more than 80% of the necessary labour. The second is production of cash crops for the market. Here they contribute more than 50% of the labour input. In addition, rural women generate incomes through non-agricultural activities and contribute 95% of the labour needed for family and household maintenance (Kenya 1994). Indeed, this implies that women should be involved in all programmes geared to the improvement of agricultural management and production in the country.

Declining per capita food production in many areas during the past two decades has led to a closer examination of indigenous farming systems and the factors that may impede efforts to improve agricultural productivity. However, the gender aspect has received little attention up till now despite the growing awareness of its importance in policy documents.

Whereas the Kenya Development Plan 1989-1993 acknowledges the central role of women in economic/agricultural development, little or no consideration is given to them when measures to improve agriculture are taken. All changes that pertain to land entitlement, agricultural technology, extension services, the co-operative movement, credit and capital acquisition, and conservational technology appear to favour men alone. How can women manage the farms efficiently without the above mentioned resources?

Therefore, apart from the problems which have been advanced as explanations for declining agricultural production, the gender issue in farm management and agricultural production should explicitly be taken into consideration when studying problems of agriculture in a country like Kenya. Recognition of research on women has been slow, but worse still is the fact that available research findings on women issues, especially in farming, have scarcely entered the mainstream of policy formulation. Yet, understanding how women cope under the strenuous and marginalizing conditions of the rural areas and within their households, knowing what women think is their role in farm management and agricultural production, and what their perceptions and aspirations are with regard to proper farm management happens to be of vital importance for the formulation of policies geared to the social and economic improvement of Kenya's rural areas.

1.4 Research Objective and Research Questions

The research objective of this study is to examine the problems of small-scale farming with an emphasis on the role of women in farm management and agricultural production. Physical factors and population pressure have taken the blame for many years for being the main causes of poor agricultural development in terms of low yields and low incomes in many parts of Africa, and Kenya in particular. This study goes further to investigate other dynamics that may play an important role, viz. the actual practices of farm management; the role of the farm manager; and the inter- and intra-household relations and interactions in farm management and agricultural production.

This study is based on the following assumptions:
(1) Good farm management is a crucial factor for redressing low agricultural production in terms of low investments, poor yields, and low farm incomes;
(2) Government policies towards agriculture influence the quality of farm management and the resulting production to a great extent; and
(3) In small-scale farming women are crucial for good farm management, and government policies influence greatly the way women perform their role in managing their farms.

Based on the above three assumptions, the researcher argues that the role and perception of women are often ignored in government policies which contributes to problems of poor farm management and low agricultural production in many rural small-scale farming areas of Kenya. This results in household impoverishment creating a vicious cycle of marginalization of agriculture, rural poverty and the continued out-migration of males.
The research objective stated above can be presented in the following conceptual models (see Figures 1.3 and 1.4). The idea underlying these models is that agriculture is a system that interacts with many other systems, and that the farm manager's position in these interactions is very crucial for good farm management and agricultural production. In this study, the farm manager is either a man or a woman who manages the farm on a daily basis at the household level. Our aim is to understand how the different types of farm households interact with the different sectors/systems and how these interactions with the different systems influence farm management and agricultural production.
As already mentioned, despite fertile soils and adequate rainfall throughout the year, our research area Vihiga District is unable to feed itself because most farmers obtain low yields from most of their crops. The district enjoys almost similar ecological and demographic conditions in all its divisions. To analyze the role of women in farm management and agricultural production, differences must be considered in farming situations between the various types of households

Introduction

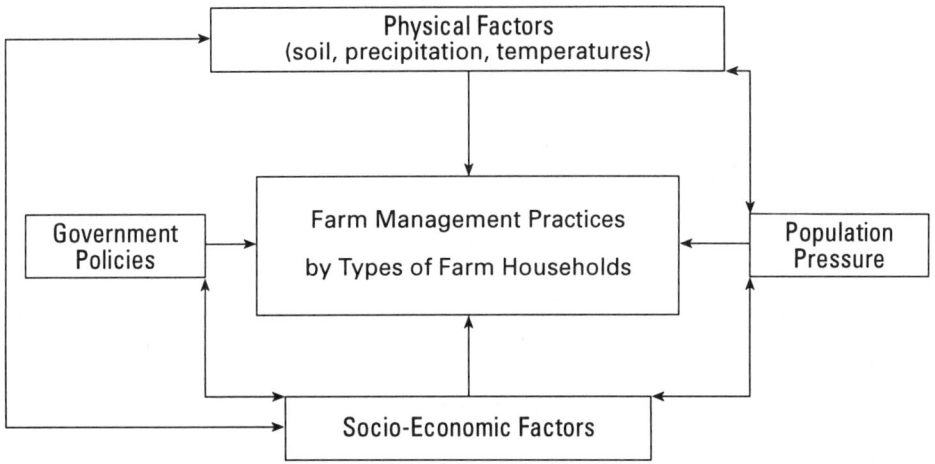

Figure 1.3
The Conceptual Research Model.

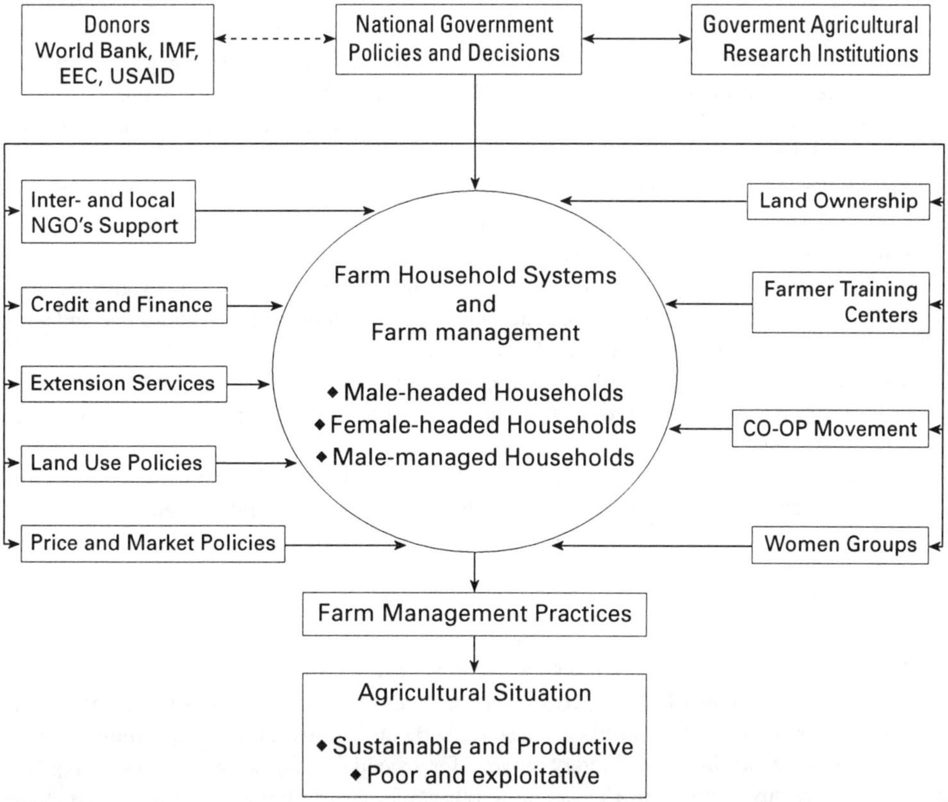

Figure 1.4
Interactions Between the Institutional Support Systems and Farm Management.

found in our research area. Therefore, three types of households common in Vihiga District were investigated as to their farm management practices, their specific problems and the strategies developed to cope with these. Three types of farm households were distinguished:
1. Male-headed households (dual hh).
2. Female-managed households (defacto fem.hh).
3. Female-headed households (dejure fem.hh).

These three types of farm households will bring out the role of women in farm management practices in each type of farm and farming system. To explain observed differences, a gender analysis was made of the content and implementation of government and other institutional policies affecting farm management and agricultural production. Special attention was paid to: (a) land ownership and control over the means of production; (b) access to extension services; (c) membership of co-operatives; (d) access to credit and other capital assets; (e) conservation measures; and (f) crop-related gender-specific information.

Research Questions
From the research objective formulated above, the following general research questions were derived:
1. What differences can be observed between and among the various types of rural households and farm-household systems with respect to:
   •the quality of farm management in terms of labour allocation, farming techniques and practices, farm investments, crop yields, farm incomes and other sources of income.
   •the role women play in farm management (quality of farm operations for cash and food crop and livestock production).
2. What are the main factors that negatively affect the adoption of modern farming techniques, household labour use and allocation, farming investments, cash crop production and yields, and farm incomes as perceived by the heads of the various types of households?
3. What are the main problems women farmers experience in overcoming these negative factors?
4. What are the strategies farm households develop to cope with the problems observed and what differences can be found in this respect between the various types of households?
5. What are Government's policies towards agriculture? How do these policies influence the farm manager in his/her farm management and agricultural production? Is there a gender bias in agricultural policy?
6. To what extent are women groups successful as a strategy for agriculture and rural development and as a means of improving the living conditions of rural women?

1.5 Research Methodology

The research material for this study was collected in various ways in the period 1991 to 1994. The findings here are based on a large variety of sources of information and methods of investigation. Initial secondary sources were consulted in The Netherlands in 1991 through a literature survey in a number of university libraries. Library surveys for secondary data were also carried out in a number of university libraries in Kenya and at other relevant institutions such as the Food and Agriculture Organization, agricultural research institutions and government ministries. Archival research on Vihiga District was done at the Kakamega District archival centre.

The primary data on farm management, agricultural production and the role of women in overall rural development was collected during field studies and came mainly from three sources, viz. (a) household heads, (b) key informants such as government officials and organizations in the field of farming and (c) women groups. The respondents in the households and women groups were subjected to structured individual and group questionnaires, interviews and in-depth case study check lists. A more detailed explanation of the methods used and the definitions of operational terms can be found in the Appendices 1 and 2.

1.6 Organization of the Thesis

Chapter 1, the introduction, outlines the subject of the study and the importance of agriculture to the economy of Kenya. It provides a justification for the emphasis on the role of women in farm management and agricultural production, the research objective and research questions, and the research methodology. **Chapter 2** analyses the literature on small-scale farming and the role of women in farm management. Emphasis is put on policy issues and government support services relevant to the small-scale farming sector. Attention is given to the household as a unit of production and the role of women groups. The chapter concludes with four research hypotheses. **Chapter 3** describes Vihiga District in detail, presenting the physical resource base, population characteristics, the economic situation, the physical and social infrastructure and the main problems of the district. **Chapter 4** gives a picture of small-scale farming in North Maragoli, the study area. A farming systems analysis is carried out detailing the four farming sub-systems found at the farm household level, viz. food crop, cash crop, livestock and tree production sub-systems. Farm management problems are discussed to explain the paradox of low agricultural production in a high potential agricultural area. **Chapter 5** outlines farm management situations and types of households in small-scale farming in North Maragoli. This chapter attempts to demonstrate how gender differentiation plays an important role in access to agricultural production resources and how this differentiation in turn negatively influences farm management and agricultural production in the four farming sub-systems. **Chapter 6** gives an in-depth analysis of inter- and intra-household farm management situations on the bases of case studies and discusses the coping strategies developed by the different types of households to overcome poverty and deteriorating agricultural production. **Chapter 7** reviews the impact of selected agricultural policies and programmes of farm management and agricultural production in relation to gender. Land tenure policy, agricultural extension policy, agricultural credit policy, policies affecting cash crops, food policy and livestock policy are analyzed with a particularly critical eye to the role of women. **Chapter 8** analyses women groups as a strategy for agricultural and rural development. It reviews the debate on women organizations, gives a history of women organizations in Kenya, and details the income generating projects of women groups, the aspirations and problems encountered and the views of women groups on the general socio-economic life of women in the rural areas of Kenya. Finally, **Chapter 9** summarizes the problems in small-scale farming and the role of women in farm management, while bringing the discussion back to the research hypotheses.

Notes

1 Citation of Kenya refers to government documents authored by Republic of Kenya

2 In theory it is now realized that land inheritance patterns favour males over female offsprings. It is claimed that measures will have to be taken that encourage joint decision-making between spouses on land utility and its accruing benefits (Kenya 1986, 1989, 1990). In order to increase the role of women in the agricultural sector, the Government of Kenya has committed itself to: (a) allocating more human and financial resources to research, production, storage, and processing of nutritionally important household food crops; (b) improving the productive capacity of women farmers, especially in strengthening extension service and also providing inputs in appropriate quantities and qualities; (c) improving agricultural planning and project implementation, so as to mobilize more support for women' multiple responsibilities such as food and cash producers and as household maintainers; (d) increase women's access to land rights and extension in order to improve their participation in agricultural and rural development; (e) collecting gender-disaggregated data by all agricultural institutions and agencies, in order to monitor the degree of women's participation and benefits; (f) ensure that female staff in the agricultural sector are considered for further training and promotion on an equal basis with their male colleagues; (g) allocate more human and financial resources to research and development of appropriate time and labour-saving technologies to alleviate women's work burdens on the farm and in the household; and (h) ensure household food security through a balanced approach to cash and subsistence crops coupled with improved agricultural extension services (Kenya 1980, 1981, 1986, 1989, 1994). It remains to be seen how much of these policy objectives have been implemented and whether rural women are placed in the perspective of agricultural development as mentioned in these objectives.

3 Reijntjes (1992) and others write that agricultural research tends to be the preserve of women. Western-based researchers are usually imbued with western models of division of labour between men and women, in which the men dominate the external economic domain and the women the household domain. This scenario, inappropriate as it is even for industrial societies, has blinded researchers to the fact that women play a significant role in agriculture. As a result, agricultural research has paid little attention to solving the problems of female farmers. The design of new technologies, often disregards important questions concerning women's influence on decision-making and labour allocation.

2

A Framework of the Major Components of Small-scale Farming and the Role of Women in Farm Management

2.1 Introduction

Kenya, like other independent African states, has been faced with the challenge of developing the rural areas. It is estimated that about 80% of the Kenyan population lives and supports itself in these areas (Pala, 1974, 1978, 1984). Policy makers, planners and researchers have the task of developing programmes to provide the rural population with the technical skills and capital, sufficient to create a strong economic base. Such programmes would improve farm management practices, increase agricultural production and, by raising the per capita output of the rural population, ameliorate rural living conditions and stimulate and expand employment opportunities at the local level. In this way out-migration of people to the large urban centres can perhaps be reduced. The success of such programmes will be enhanced if the problems of both male and female farmers in the small-scale sector are properly identified and addressed.

Intensification of land use, rather than any structural change in the distribution of land or the implementation of large-scale irrigation and drainage schemes is considered the key strategy for increasing agricultural productivity by both the government (1984, 1994) and the World Bank (1983). Intensification is to be facilitated through the extension of an individual freehold land tenure system, the provision of extension services financed by the World Bank, a realistic pricing policy, improvements to the system of marketing and credit, improving access to and the availability of inputs, and through agricultural research. Such measures are specifically directed at small-scale farmers. However, research has made clear that the strategy is not specific enough given the different realities experienced by women vis a vis men in the process of agricultural change. This perception is now becoming evidenced by the growing attention being given to women farmers in national agricultural policy (Ellis 1992, Gladwin 1991, Mackenzie 1986, Moock 1986, and Pala 1984).

A number of studies have shown that the small-scale agricultural sector in Africa and particularly in Kenya is run by women (60-80%). Yet, their role in agricultural production does not seem to be recognized (Gladwin 1991, Davison 1988, Bifani 1985, and Pala 1984). If Kenya has to continue relying on the small-scale sector to produce enough food to feed the rapidly growing population and to absorb more labour, the government has to formulate policies that are gender-sensitive and -specific and that address the problems facing female small-scale farmers. Whereas the conventional wisdom held that Kenya's increasing population is one of the major causes of poor agricultural performance in both cash crop, food crop and animal production, there is now a growing feeling that the policy negligence of women and the consequent continuance of poor farm management practices may be equally important causative factors. However, very little is known of the prevailing conditions of work and the decision-making of female farm managers and the effect the policies formulated and the measures taken to intensify land use and to increase agricultural output has on them. A gender-sensitive analysis is therefore crucial to understanding farm management practices and agricultural production in small-scale farming in Kenya.

2.2 Views on African Agriculture

Agriculture is an important economic activity in Sub-Saharan Africa: it employs a majority of the working population and accounts for about 20% of the value of total Gross Domestic Product (World Development Report, 1994). This implies that economic development and the welfare of the majority of the African people are heavily dependent on the performance of the agricultural sector. However, this performance leaves very much to be desired. A most disturbing aspect of Sub-Saharan agriculture is the steady decline in food production per head of population, which has taken place over the last fifteen to twenty years. The growth in food production of about 1.3% per year for Sub-Saharan Africa has been insufficient to keep pace with the population growth of over 2.5% which characterizes this region (World Development Report 1994, Upton 1987). Many governments in Africa have resorted to large imports of food. At the same time the production of cash crops for export has stagnated or even declined. In fact, the post-independence agricultural production performance in Africa south of the Sahara is said to be rather bleak. This is evidenced in the numerous publications of international organizations and individual research workers who sketch a picture of decreasing agricultural output, stagnating exports, food shortages, increasing food imports, soil erosion, famines and even mass starvation as pointed out by Hinderink and Sterkenburg (Hinderink and Sterkenburg 1982 & 1987; Lofchie and Commins, 1982; and World Bank, 1978, 1980 and 1981). Therefore, most African governments are increasingly concerned with bringing about a change which will reverse these trends and help to increase the rate of growth of agricultural production. But although governments may facilitate, stimulate and promote the growth of agricultural production through policy instruments and programmes, agricultural development must occur at the farm/household level. Success ultimately depends on the decisions of the multitude of farm households involved in agricultural production (Singh 1988 and Upton 1987).

Farm management decisions are influenced by the context in which farm managers operate. In areas without large inequalities in land ownership, the density of the population engaged in agriculture has an important influence on the types of farming adopted, since population density controls the average amount of land available to the farm families and hence the intensity of farming (Upton, 1987). Population increase and high population density have been associated with poor agricultural performance by some scholars. There are a number of views on the issue of population pressure, land degradation, and poor agricultural performance in Third World countries, mainly Africa. Boserup (1965, 1970, 1975) was amongst the first economists to argue strongly that the growth of population is the main force by which agricultural development is brought about since intensive farming always involves more advanced techniques and more capital per hectare and per family than extensive farming. This also implies more intensive labour use. It is also noted that while arable land in some places is going out of cultivation because of erosion and other destructive forces including the important aspect of population pressure, both Simon (1981:81 in Blaikie, 1985:12) and Boserup (1970) are quite positive that population pressure acts as a catalyst to invent new technologies which create new resources and make it possible to use the existing ones more efficiently. Therefore, the more technically advanced agriculture becomes, the less it becomes dependent on natural resources per se. In fact, population becomes the independent variable while the dependent variable becomes agro-technology. This effectively means that the socio-economic characteristics of the peasant farming system will be subject to change. The problem in developing countries is the scarcity of capital and this presents a major constraint to improvements in intensifying land use, and the management and adoption of advanced and relevant technology.

Apart from population pressure on resources, a number of scholars claim that political and economic factors may also lead to degradation and a decline in agricultural production. Blaikie (1985), and Blaikie and Brookfield (1987:27), in their various studies argue that soil/land degradation and therefore a deteriorating agricultural situation is caused by the political economy of developing countries. They claim that soil degradation, which may eventually lead to poor agricultural production, results from cumulative land use decisions and policies through time, and these decisions and policies are part of a wider political and economic system. The above authors are of the opinion that the whole complex of relations between land managers with land-crop rotations, fuel-wood use, stocking densities and capital investments are affected by the social stratification and the ensuing, conflicting interests of the countries concerned and by conditioning global factors. They are, therefore, emphasizing that the political and economic systems may carry the explanation behind deteriorating agricultural production and land degradation. Most of the African countries south of the Sahara suffer from what can be referred to as political and policy instability and unpredictability. This instability leads to poor planning or no planning at all as the political systems are only interested in the consolidation of their powers (Wanjiku,1993).

The chronic lack of progress for most Sub-Saharan African countries throughout the 1980s and 1990s has been blamed by many on external factors such as poor prices for primary commodities, debt and the lack of Western investments. But a report by the London based Overseas Development Institute says that although these factors are indeed important, far greater emphasis should be put on domestic factors, in particular the economic policies carried out by those in power. According to the Overseas Development Institute's Report (ODI 1993, Daily Nation, 24 August 1993), the main factors hindering development across Sub-Saharan Africa are the policies of leaders or regimes that came to power at the end of the colonial period. These policies, the report says, are based on negative short-term thinking and a need to maintain the regime in power by rewarding supporters (tribal alignments). This approach is referred to in the report as 'the personal rule' model and it is distinguished by (a) a proliferation of generally over-manned public enterprises, (b) in-ward looking import substitution policies and the neglect of the politically unimportant peasant farmers, (c) financial repression and (d) politicized credit allocation mechanism, in order to have cheap credit to offer to supporters. The governments in Sub-Saharan Africa are criticized for their intervention in the market place and their inability to rectify mistakes once policy clearly fails because of the notion that 'the Government is always right'.

These far reaching criticisms from the Overseas Development Institute (ODI) have already been pointed out by a number of authors on a number of occasions. For instance Harrison (1987) and

Timberlake (1985) claimed that Africa is suffering because of ill-planned, and ill-advised attempts to modernize. The two authors further emphasized that African governments are taking too much from the farmers (in terms of labour) and from their land in order to feed the urban centres and to earn foreign exchange. This results both in environmental deterioration and explains the present crisis in a number of sectors. Many countries in Africa have shown themselves to lack political will power and commitment when it comes to the plight of the large populations in small-scale farming and the unemployed. Conditions in the small-scale sector have been deteriorating because of the high prices of farm inputs, the low and irregular incomes from agricultural products, a lack of adequate incentives and, more recently, because of the structural adjustment programmes that most African governments have rather belatedly implemented, therefore negatively affecting their economies. Broadly speaking, the policy instruments of the World Bank and International Monetary Fund's SAPs emphasize economic liberalization, the deregulation of the prices of all commodities and services, large-scale privatization of public enterprises, and cutbacks in government expenditure on social services and employment (Gladwin 1991, Nzomo 1992). The SAPs are a way of invigorating the stagnating economies of these countries. However, for small-scale agricultural producers, especially women, a decrease in government spending means higher school fees and no access to medicine; wage freezes and early retirements which mean no remittances from working husbands,sons and relatives; the removal of subsidies which may mean fertilizers become so expensive that they are beyond their reach; and an emphasis on exportables which may mean a decrease in the resources of land, labour, and capital/credit available for their food crops (Gladwin, 1991).

There is a growing literature, therefore, that puts the blame for Africa's economic problems on African governments themselves for pursuing policies that are biased against the agricultural sector in general and small farmers in particular and which have had detrimental effects on farm management (Nzomo 1992, Bevan et al.1989, Mellor et al.1987, Hinderink and Sterkenburg, 1987, Bates, 1981, World Bank, 1978 & 1981, Lipton 1977, Lamb 1974). It is noted by these authors that when Independence was achieved, most young African governments expressed their intentions of stimulating economic development, which for the agricultural sector implied an acceleration of the commercialization and commoditization process. The post independence period was, therefore, characterized by an intensification of intervention and control by national governments and parastatal institutions over the agricultural sector. Despite the differences in professed development pathways, in practice, there has been little difference between the countries of Sub-Saharan Africa as far as the general aspects of agricultural policy is concerned. In most of these countries, many features of general development policy emerge which influence the agricultural sector. These features have been summarized as follows:
1. the considerable intensity of government interference in all the sectors of the economy;
2. a wide gap between the proposed development plans and implementation as measured by budgetary commitments and concrete efforts;
3. unfavourable trade and exchange rate policies with an inherent bias against agriculture;
4. and finally, the overvalued exchange rates as a result of inflation at home leading to import restrictions which force farmers to purchase high cost local implements, and causing a downward effect on the prices farmers receive for their export crops (Hinderink and Sterkenburg,1987).

These background features of development policy are reflected in the general characteristics of policies directed towards agriculture and in turn such policies influence farm management practices in Africa.

This general overview shows that it would be too simple to link poor agricultural development to population pressure as the principle causative factor. Other factors may, in fact, turn out to provide a more satisfactory explanation for poor agricultural performance in Sub-Saharan Africa. Good and efficient farm management in small-scale agriculture depends on the possibilities available to farmers that allow them to benefit from opportunities to improve farming. These opportunities are strongly conditioned by the content and modes of implementing government policies. However, farmers are not a homogeneous mass, but differ greatly in their resource position. A gender perspective is important for a better understanding of these differences in resource positions that affect farm management in different household types, especially in the African context.

2.3 Characteristics of Small-Scale Farming in Kenya

The problem of high population growth and population pressure on resources and the consequent negative effects this has on economic development and agricultural production in Kenya have been discussed by a number of authors (Rutten, 1992, Collier, et.al.1989;, Livingstone, 1986;, Hunt, 1985;, Bernard, 1982;, Stichter, 1982, and I.L.O, 1972). Kenya's demographic situation has been characterized by them as one of the most alarming in Africa. The country's population has grown over the past decades from 3.0, and 3.4% per annum to 3.8% and increased to 4% in the late 1980s and early 1990s (Livingstone, 1986, World Development Report 1992 and 1994, Kenya Development Plans 1989-1993 and 1994-1996). It is predicted that Kenya's population will increase to a total of 30 million people by the year 2000 and approximately 50 million by 2025 (World Development Report, 1994). The labour force is expected to rise from 9 million people in 1989 to an estimated 14 million by the year 2000.

This rapidly increasing population cannot easily be absorbed into the labour market because the Kenyan economy is one of the world's low income economies, ranking nineteenth from the bottom on a list of the world's poorest countries with an average per capita income of $310 in 1992 (World Development Report 1994). In view of the limited employment opportunities in the non-agricultural sector, the agricultural sector will probably have to provide employment for about 8 million people by the year 2000, and most of them in small-scale farming. On the other hand, Kenya has limited amounts of high- and medium-potential land to support the increasing population. The present population growth rate in high-potential areas has already led to situations where resources are being depleted faster than they are being generated. The reason for this is that small-scale farmers have been trying to produce enough food for their families with very low inputs, and this is to the detriment of soil fertility.

But despite increasing population pressure on land, the small-scale agricultural sector might still support a large population, absorb more labour, produce sufficient food for the Kenyan population and cash crops for the world market if there was to be an improvement in farming conditions in terms of efficient management.

In the following sub-sections attention will be paid to some important characteristics of small-scale farming in Kenya

2.3.1 Land Tenure

In many parts of Africa, a burning issue has been, and still is, whether land may properly be bought and sold, or, if not, exchanged in a more gradual, conditional, and subtle kind of transfer: pledged or mortgaged against a loan. Many African governments have a policy to title farmland as private property in the hands of individuals or group owners (Parker, 1992). They are aware that small-scale farmers need seasonal credit to adopt new crops, inputs, and techniques, and therefore, that farmers' access to credit depends on possessing individual land titles that can act as collateral. The reasons put briefly for the policy on titlement of land are as follows:- (i) to raise agricultural productivity, farmers need more inputs and new technology. (ii) Farmers, especially in small-scale farming, are too poor to be able to afford or to save for the inputs and new technology themselves. (iii) They therefore need loans from endowed institutions or people until they can afford to finance their own needs. (iv) Loans require collateral as security, and this will ensure their repayment. (v) The best collateral is land, because it is immovable. (vi) Land ought therefore to be negotiable. (vii) The way to make land transfers easier, and to keep track of them, is to issue titles of land ownership registered with the national government (Parker, 1992, Okoth-Ogendo, 1991, and Kenya Development Plan, 1974-78).

The evolution of land tenure systems and policy in Kenya can be divided into four phases:- (i) The pre-Colonial and early Colonial period, (ii) The Colonial period, (iii) The pre-Independence or late Colonial period and (iv) The post-Independence period (Odingo 1985).

(i) The pre-colonial and early colonial period was characterized by communal tenure where land was owned by the community though worked by individual farm families. Under this type of land tenure every member of the community regardless of gender had usufructuary rights to it (common property resource). Both men and women were free to till land. There were of course variations in tenure and use of land from one tribe, clan or kinship group to another, especially among those tribes that had kings or clan elders who would make land use decisions. Independent decisions were made by both males and females at household level as to what to cultivate, what to store for the family and what to exchange on the market. Some consultations took place between spouses as the husband was head of the household and priority in matters of food was given to the family first. The size of land cultivated by each individual depended on his/her ability to cultivate and on the availability of land. Communities moved from one area to another whenever the fertility of the soil deteriorated or when there was some slight increase in population. They also moved when they were attacked by other clans or tribes over land conflicts. This system was self evolving and it catered for the needs of all its members. The management of the environment was controlled because the communities were aware of what crops should be planted in what part of the farm. In most high-potential areas there were crops for the long rains and crops for the short rains (Davison 1988, Dankelman et al.1988, Pala 1978, and Hay 1976). However, the technology was very simple and it would be totally insufficient at present with large populations and small farm sizes.

(ii) The Colonial period was dominated by European settlement following the large-scale alienation of extensive areas of land that had formerly been occupied by African populations on a communal basis, or had been regarded as areas to which people could turn in times of ecological stress (e.g droughts, floods) as well as in peace time, when population growth required more land. This was the period when the 'African Reserve' and 'White Settlement' areas were created. The Europeans introduced new concepts of freehold and lease-hold tenure systems that were directly borrowed from the English system (Odingo 1973, 1985). The establishment of European settlers severely restricted African expansion which resulted in overcrowding in the 'African Reserves'. During much of this period, European land and agriculture was operated under freehold or leasehold land tenure systems whereas the great bulk of the African land areas continued to be governed by the customary land tenure systems with now restricted migration. This resulted in serious soil degradation and erosion.

(iii) The pre-Independence period could be taken to mark the ten years before the end of colonialism. During this period, the colonial power decided to introduce the European concepts of individual land tenure to the African population. Initially these concepts were forced on the Africans but later they came to be accepted and they were adopted as the basis of the official land tenure policy after independence. These concepts of individual land tenure were embodied in the Swynnerton Plan. This Plan was put forward as a land and African agricultural reform program in 1954. The Swynnerton Plan sought to stimulate growth and reduce land pressure in the African small-scale farming areas by such measures as irrigation, the introduction of new cash crops (previously prohibited), the provision of small-farm-credit, and the improvement of marketing facilities, especially through co-operatives. Furthermore, it sought to consolidate and enclose hitherto fragmented land, granting legal registration and individual land titles to Africans (Livingstone 1981, 1986 and Brack, 1977). The security of tenure established in this way was intended to provide (a) social security, (b) access to credit, (c) expansion of agricultural production, and (d) incentive for land reclamation or land improvement in the densely populated African Reserves.

(iv) The post-Independence period is characterized by the continuation of the land registration that began under the Swynnerton Plan, the break-up of European settler farms and the introduction of settlement schemes for Africans in the former White Highlands.

Kenya's land registration, launched under British rule in the late 1950s, was the first of its kind in Tropical Africa. Many authors have emphasized the link between agricultural production in Kenya and the benefits of the formalized land tenure systems that have been introduced into most of the areas of high agricultural potential in the last forty years (Sorensen 1992, Jungheim 1989, Odingo 1985, House and Killick 1983, Heyer 1981, Okoth-Ogendo 1976 and Moock, J. 1976). There has been a movement away from traditional systems of land tenure to an individualized tenure system. It was expected that with the titling of farmland as individual freehold property many goals would be achieved simultaneously. Apart from those mentioned above, investments would be stimulated, land would be allowed to pass into the hands of the most able farmers, land disputes would be quelled and prevented, husbandry would be improved through the consolidation of holdings, political stability would be promoted, and amongst other things a clear public record of ownership would be made available (Parker, 1992, Pala, 1984, Okoth-Ogendo, 1976). In terms of agricultural production, the justification for an individualized system of tenure was that it improved access to credit for the individual land holder, and positively influenced the farming decisions. The introduction of individualized land tenure in Kenya has ensured that landholders not simply secure utilization rights but also have freehold title. Thus they would be able to pledge their land for loans, and transfer their land not only by inheritance but also by sale.

One of the objectives of the colonial land reform was to ease population pressure in the 'African Reserves'. But the Swynnerton Plan did not include any division of the large farms and plantations owned by Europeans. Just before Independence the region known as the White Highlands comprised approximately three million hectares (7.41 million acres) of land of which 1.4 million hectares were taken up by plantations (Livingstone, 1986). To avert the seizure of European land and to minimize the chance of a possible of revolt against the new (African) government, a settlement scheme giving some land to Africans was initiated in 1961 (Migot-Adholla 1984 and Leys 1975). Britain, the Commonwealth Development Corporation, and the World Bank provided money to purchase land adjacent to the 'African Reserve' areas around the fringe of the White Highlands. The purchased land contained 180,000 acres of low density (high income) plots and 987,000 acres of high density (low income) plots (Bradshaw, 1990). This programme became known as the 'Million Acre Scheme' because of the amount of land allocated for high density plots. By 1971 when the scheme was completed, about 29,000 families had been settled on high density plots and 5,000 had been settled in the more lucrative low density areas. The cost of financing the Million Acre Scheme increased over the ten year period: between 1961 and 1969 the capital borrowed abroad to pay for the scheme accounted for one third of Kenya's foreign debt (Bradshaw 1990 and Leys 1975). The remaining the land in the former White Highlands has been shared between private individuals (the rich and the elite group, high ranking military personnel, businessmen and private estates), co-operatives, the government and a few settlers who had chosen to become Kenyan citizens (Hinderink and Sterkenburg, 1987).

Many of the technical objectives of the Swynnerton Plan had political consequences. Fewer people had effective ownership of land under the Plan than had previously been the case. Thus land consolidation actually accentuated rather than reduced the problem of land hunger. With land consolidation and registration, groups that had been squeezed into the 'African Reserves' suffered because they had very small parcels of land to register. As Brack stated (1970):

"The Swynnerton Plan implied a complete change in the basis of the economy and the disappearance of the idea that everyone must and can have some land. It implied a landed and a landless class... The success of the plan depended upon continued, rapid economic advance of the country in both agriculture and industry... Firm administration and maintenance of order were essential during the transitional period while the new situation was gaining acceptance. If order could be maintained through this period then the new situation would become the accepted mode of doing things and force would no longer be necessary." (Brack 1977:203)

After Independence the Kenya government's efforts to expand smallholder opportunities have had a considerable effect on the contribution made by smallholder production to total agricultural output. In 1958 smallholders accounted for only 19% of total agricultural output, and by Independence in 1963 the figure had barely increased to 22%. However, by 1968 the smallholder contribution had grown to 51%, and this remained until 1984 (Bradshaw 1990, Republic of Kenya 1985 and Heyer 1976). Undoubtedly, part of this increase was due to the fact that smallholders were allowed to cultivate export crops after the Swynnerton Plan had been implemented. By 1976 smallholders cultivated nearly two-thirds of all land used to produce tea and coffee, Kenya's leading export crops (Wyeth, 1981). Despite the gains made by smallholders in Kenya, the problem of transferring large-scale European farms to Africans has remained. Even after the Million Acre Scheme, between three-quarters to fourth-fifths of the White Highlands had not been distributed to smallholders (Bradshaw 1990, Migot-Adholla 1984). In fact, since Independence these large tracts of land have been purchased by a variety of African institutions and individuals with access to capital and credit. It is estimated that by 1977 34% of large farms were owned by a combination of individuals, 29% by partnerships, 25% by companies, and 12% by cooperatives (Bradshaw 1990, Livingstone 1986).

A number of criticisms have been levelled against the Kenyan land reform and tenure system. It is argued that the rapid growth of smallholder production after the inception of the Swynnerton Plan owed more to the removal of crop restrictions than to land tenure reform (Heyer 1976). It was pointed out by Livingstone (1986) that legal ownership may be unnecessary if an individual's right to use the land is unchallenged. As regards the reduction of fragmentation, Heyer observed that fragmentation can be useful in reducing risks by diversifying crop production in places where slope, altitude and type of soil differ widely within a small area. As far as facilitating access to credit is concerned, Livingstone maintains that land title deeds in small-scale farming have not had much impact because credit institutions find it politically and administratively difficult to give loans. Credit repayments have been poor. Most credit to smallholders is provided through crop authorities and cooperative societies and is not secured by land because the most effective way of obtaining repayment is by making deductions at the marketing stage. However, one had to have land and a title deed in order to be licensed to grow any cash crops and to be a member of a co-operative society. This was especially so in the case of coffee and tea cultivation.

Individualization of tenure does not, in fact, guarantee the full use of land, because large sections may be owned for their asset value or for speculative reasons, rather than for productive purpose as is the case in the former White Highlands. This is in contrast to the traditional customary system where the safest way to secure land was to use it and, so long as there was plenty of unused land, access was readily available. In most of rural Kenya today there is an increasing stratification as the well-to-do buy out poor farmers, and this leads to the increasing landlessness. Most of the landless have no other means of survival because most of them are unemployed. For most Kenyans farming is their source of employment and so long as one owns land then one is in employment (Hunt 1985).

2.3.2 Labour Input and Labour Constraints

Small farm operations in many parts of Kenya are not only land but also labour intensive. Apart from the availability of inputs, the judicious allocation of labour is essential for successful farm management and agricultural production. Assuming the availability of all necessary inputs, labour becomes the most important consideration. Labour can become more productive if time and effort is devoted to training and education. In small-scale agriculture however, what is more important is labour allocation and efficiency, the time devoted to different farm tasks and the precision with which the different tasks are performed.

Labour is an important cost item on farms, especially on small farms where in many cases it averages between 50 and 90% of total farming expenditures (Anaman,1988). Therefore, the proper allocation of available family labour is of great importance for efficient farm management. However, family labour supply is not always sufficient. In old times when there were labour shortages various forms of attracting labour from outside the household were called upon. Among the Luhya for example, labour was provided through the group method known as **obulala or lisango**. This group labour was organized at three levels, similar to the organization among the Kikuyu. The first level was **ngwatio** and usually consisted of three to ten women who would move and work as a group, from day to day, from plot to plot, until the tasks of all the participants had been completed. The second form was **wira**, in which a man or woman asked a number of participants (up to eighty) on a specific day to complete a specific task, such as digging up permanent grass and bushy land (men) and weeding (women). It was obligatory under **wira**, but not in **ngwatio**, for the host to provide either food and/or beer. The third form was **ndungata**, in which a man without a family, land or livestock would be taken into a household to herd his host's cattle, sheep and goats (Collier and Lal:1986:23). Women without family were also taken in by some families for domestic and agricultural work. Although traditional forms of labour still exist with modifications, today most of the labour drawn from outside the household is paid for.

Labour shortage is a constraint in many small-scale farming areas in Kenya. The paradox of large numbers of rural people living in poverty on the one hand, and shortages of labour on the other, is explained by various factors. Livingstone (1986) and Hunt (1985) relate labour shortages partly to seasonal variations in labour demand and point to the conflict between the estate and the smallholder sector. As to labour shortages in the estate sector Livingstone refers to the 'sponge effect', i.e the ability of the small-scale agricultural sector to absorb labour. Therefore, the observed scarcity may simply be because of the low wages offered by plantations or employers which are not attractive enough to draw away labour from the small-scale sector.

The apparent shortage of agricultural labour in spite of heavy rural population densities can also be explained in the context of the way men seek work away from the family holdings. Labour migration has an effect on agricultural management and production in the regions from which the migrant labourers originate. In 1979 there were about 175,000 rural households in Kenya without a male head. In another 560,000 the male was absent and if he was not working in the urban sector he was hoping to find work there soon (Livingstone 1986, Hunt 1985). It is suggested that one-third of rural households then were living without a male head. This has increased the number of female-headed and female-managed households in small-scale farming (Livingstone 1986 and Hunt 1985).

Despite the fact that the rural small-scale areas have helped to absorb population increase in Kenya, there is continuous migration from these regions. High agricultural growth in the small-scale sector became more difficult to achieve after the expansion of coffee, tea, dairying and the diffusion of high yielding hybrid maize varieties in the 1960s and 1970s. Thus, as agricultural incomes diminish and the purchasing power of the rural population is eroded, strategies for overcoming these constraints have to be devised. Migration especially to the urban areas is seen as the main alternative support for rural households by people (Livingstone 1986).

Remittances sent by migrants primarily serve to maintain the existing level of living of those left behind in the rural areas. Studies on migrant remittances and labour absence in Western Kenya revealed that more than half of the households surveyed spent most of the remittance money they receive on food, and the payment of school fees, before allocating any funds to paying farm labour and other necessities, such as the purchase of agricultural inputs. The obsession with school fees and a general investment in education has worked against female-managed households, because education is treated as a form of insurance against future uncertainties marginalizing agricultural investment. This situation confines the woman to the position of an overall supplier of labour in small-scale agriculture. One conclusion suggest that very little money from the migrant remittances is, in fact, used to improve farming in the small-scale sector (Pepels 1991, De Groot 1991). However, in a study of migration and agricultural innovation in the southern part of Kakamega District, it was found that those who had successfully expanded their farming enterprises were not those working in distant urban centres, but rather those in local, well-paid employment, such as teachers and government workers, whose year round presence enabled them to manage their farms effectively (Moock 1979).

Migration has been assumed to be a positive response to regional and sectoral differentiation in wages, employment, landownership and population density because it leads to an increase in productivity through the relocation of surplus labour and therefore benefits both the areas of origin and destination. As far as the area of origin is concerned the classic point of view stresses the positive effects of money transfers from migrants to their families, the direct relief of population pressure, and the modernization effects of frequent contacts with urban society. However, this may not always be the case. The transfer of money from urban areas to rural areas is not regular and the amounts remitted are usually insufficient to foster rural development (De Groot 1991). In fact, money transfers will depend on inter- and intra-household relationships between the migrant and those who have remained in the rural areas (see chapter 6). If a close relationship and understanding exists between husband and wife then money transfers will be maintained on a regular basis. Remittances also depend on the nature of the job and the remuneration a migrant receives. If the migrant has a well-paying job then he may remit money to his kin in the rural areas. The migration of a few family members does not ease the pressure of population on land in the rural areas as is often suggested. A common phenomenon in densely populated districts is the migration of men in search of employment. Since they are the ones who inherit land, land has to be sub-divided whether the male migrates or not. Moreover, the males who migrate to other areas have a wife with children taking care of the farm and therefore there is no major easing of population pressure. In fact, the rural families often have to feed the migrants as food is sent to them.

However, it has also been pointed out that labour migration has promoted long-term economic differentiation, not so much through its impact on agriculture but rather by funding investment in education through successive generations and by the purchase of capital assets wherever possible. Long-term patterns of differentiation are laid down primarily through the link between education and better paid employment which, over generations, promotes cumulative advantages and disadvantages. However, differences in access to labour markets have not been amplified through investment in farming (Francis et al.1993, Kitching, 1980) nor have imperfections in rural factor markets been offset by the use of remittances to fund investment in crops or livestock production (Francis et al.1993, Collier et al. 1986). As has been pointed out before, direct remittances to wives of men who migrate only tend to be enough to supplement food supply and buy basic household items, with larger amounts for school fees either paid directly by the man or earmarked beforehand. It has been asserted that, within Kenya, the process of growth has been shaped by the malfunctioning of rural markets, and the failure of labour, land and capital markets to equalize returns over farms of differing sizes and households (Collier et al 1986).

It can be concluded that the major constraints experienced by many small-scale farmers include labour shortage at the household level due to the migration of males; labour inefficiency and poor allocation; and the lack of capital and low remittances which make it difficult to attract labour from outside the household. These labour shortages and the resulting low labour productivity limit intensification and explain poor management levels and low yields.

2.3.3 Agro-Support Services in Small-Scale Farming

Many African governments have emphasized the important role played by the agricultural sector in their economic development. Most governments have developed agro-support services programmes to assist farmers maximize agricultural production. Since Independence, the Kenyan government has particularly emphasized the provision of two types of agro-support services to farmers: agricultural extension to enable farmers to acquire new scientific knowledge, skills and methods and to assist them to accept new innovations; and input distribution and credit for the purchase of farm inputs. The following two sub-sections discuss these agricultural support services in more detail. Given the limited resources available, such services are selective: choices are made between farmers - especially in the case of credit - and between regions. Geographically, the extension service is widely spread because it operates throughout the smallholding areas, whereas small farm credit programmes operate only in some of these areas, generally those that have a high potential. It has been observed that, within individual regions, the resources of both programmes have been concentrated on the better-off smallholders, mostly men - a pattern that is typical of peasant farm development programmes throughout Africa and in Kenya in particular (Hunt,1985).

Agricultural Extension

The widespread acceptance of the extension service in Kenya suggests that a more satisfactory method for communicating new farming ideas and skills to farmers has not been found (Pala 1984). It would appear that the extension service is by far the most widely used technique for disseminating new knowledge and skills to the farming communities.

In the traditional African setting, new technologies spread from one tribe to another or from one generation to the next through verbal communication during meetings necessitated by ceremonies or called by clan elders. African women interacted with other women during weddings, at water collecting points, when collecting firewood and at trading centres. Such occasions created opportunities for women to exchange new ideas and experiences. There was extensive traditional knowledge on crop management and animal production. Traditional pesticides were used and the concept of soil fertility was common among most of the agrarian communities (Hay 1976).

When Kenya was ruled as a colony by the British government, extension policy was biased towards European farms. These were large-scale enterprises specializing in plantation crops such as coffee and tea, or in pyrethrum and dairy farming. All extension messages were geared to these enterprises through publications, the media or during the course of farm visits. Extension, then, was important as the farmers were instructed on the most recent technologies and methods for combating pests and diseases. However, the traditional sector or 'African Reserves' where many Kenyans lived were completely forgotten because only the cultivation of subsistence crops was allowed there. When Kenya achieved its Independence, it inherited a structure of extension that was skewed towards large-scale farmers, cash crops and dairy farming. Agricultural research and written instructions were thus biased towards large-scale agriculture and a single crop approach (Heyer et al, 1976).

At Independence, the new Government of Kenya was under considerable pressure to maintain economic growth and therefore it continued to foster the policies that had been used by the colonial government. It can safely be said that not much has changed because most extension staff have been trained to discuss and emphasis cash crop farming, and in so doing they neglect other crops (Livingstone,1981/86). Yet, despite this continuing bias in favour of cash crops, extension can increase agricultural productivity and rural incomes by bridging the gap between technical knowledge and farmer's practices. Several studies have shown that extension is generally cost-effective, and has a significant and positive impact on farmers' knowledge and the adoption of new technologies and hence on farm productivity (Birkhaeuser, Evenson, and Feder 1991, Saito et al. 1994). Some studies have shown that extension in Kenya significantly and positively affects the overall gross value of farmers' output, when gender is not considered.

Extension services can increase agricultural productivity by: (i) stimulating farmers and researchers to collaborate in the development of new technologies (cultivation practices, varieties, chemicals, and tools) in response to rapidly changing circumstances; (ii) providing as many farmers as possible with these technologies and information in a timely and accurate manner, using a variety of communication and training methods; (iii) encouraging farmers to informally test, adapt, and adopt the technologies thus improving productivity; and (iv) taking into account information about farmers' concerns and problems with different technologies and conveying them to research and technology centres.

Though extension is vital for increased farm management and agricultural productivity, its diffusion to all regions and households in Kenya has been very slow and unpredictable since Independence. It has been noted that the extension machinery is not adequately geared to the rapid development of Kenya's small holders. Ashcroft (1976) observed that although Kenya had the best extension worker-farmer ratio in East Africa in 1976 (1: to about 500 farmers against 1:1500 in Tanzania and 1:1800 in Uganda), this vast machinery was not efficiently utilized for the rapid introduction of innovations.

Discussion of extension strategy usually centres on two alternatives: the individual 'Master Farmer' or 'Contact Farmer' approach focusing on the more progressive or receptive farmers in the neighbourhood, and the group approach based on addressing extension to average or less progressive farmers in groups or clusters. The extension programme in present-day Kenya has been criticized for its progressive farmer bias, its over-emphasis on individual farm extension methods, its inability to cope with large numbers requiring extension advice, and the poor quality of the extension advice (Heyer 1976). Furthermore, it has been pointed out by others that the junior assistant technicians or frontline extension workers, who have the most direct contacts with the farmers, have poor educational backgrounds themselves and in most cases have no formal training in extension work other than one or more week-long course at a Farmers' Training Centre. A large section of the present field staff are therefore not suited to bringing across sophisticated technical farm management advice. This is changing as more staff is trained but it is going to take a long time to train the large numbers needed to replace the old staff who were recruited after Independence.

A quick look at the methods used in extension might help us in evaluating its effectiveness as a development tool. At present, a variety of methods are being used. These include visits to contact farmers and individual farms through the Training and Visit method (referred to as T&V), 'Barazas' or chief's meetings, demonstrations, farm field days, agricultural shows, the use of the mass media and 4-K Clubs (formalized agricultural education in both primary and secondary schools so that the youths can grasp both practical and theoretical farming principals). The chief's meetings, demonstrations and farm field days are not very effective as they are quite rare, farmers are not informed in advance and, consequently, few farmers turn up to such meetings. In any case, those who attend are usually men and not the women who need the information.

Farmer's Training Centres (FTC) are another way of training farmers. It is observed that FTCs more often than not select more progressive farmers for training, though sometimes ordinary farmers from different localities are also recruited. The instructors at the Farmers' Training Centres are overworked. They have to run all the courses as well as managing the college farms and yet, their salaries are low and their chances of promotion very poor. This leads to a lack of commitment in the performance of their duties (Egsmose 1990, Heyer 1976, Pala 1974, 1976). It is further alleged that the Farmer's Training Centres which could be effectively utilized for group extension methods are not able to recruit enough farmers to fill their classes and usually work at less than 50% of their capacity because of poor advertisement. It was also noted that the same farmers are always invited to the Farmers' Training Centres, and there is no effort to involve 'new faces'. More often than not those invited to participate in the training are men and women play a very secondary role. Those women who are invited to Farmers' Training Centres are normally trained in home management or in women leadership and rarely in the running of farms. Actually, the majority of trainee courses in FTCs are not for farmers but for agricultural and livestock officers and for traders who deal in agricultural inputs (Egsmose 1990, Heyer 1976, Pala 1976 & 1984).

The conventional approach of concentrating on the progressive farmers who are wealthier, better educated, more receptive to innovation, and more willing to accept risks, widens inequalities because resources are transferred to the relatively well-off and favours them with a greater access to extension time. It also fails to take account of a communication and identification gap between progressive farmers and the mass of poor farmers. Extension staff claim that information dispersed in this way will spread to other farmers living and working in the neighborhood of the progressive farmers. However, it has been noted that diffusion may be limited if innovation requires resources not available to most farmers, or if the innovation involves risks which the poor or average farmer is not in a position to take or - even if neither of these is the case - if the ordinary farmer feels that the farming conditions of the progressive farmer are quite different from his/her own so that any innovations the latter may adopt are not relevant to his/her own situation (Bindlish et al. 1993).

The various opinions expressed on the agricultural extension service demonstrate that in terms of increasing per capita output and community involvement in agricultural production, substantial skepticism and dissatisfaction exist as far as the usefulness of the extension service as it functions now is concerned. Doubts have been expressed concerning the efficiency of communication between the researcher and the extension personnel, between the researcher and the farmer, and between the extension personnel and the farmer. A solution to the biased orientation of extension agents to progressive farmers who tend to have access to large tracts of land and who, needless to say tend to be men, has not been found.

Input Distribution and Agricultural Credit
Purchased inputs are still used relatively sparingly in smallholder agriculture in Kenya, but with the progressive modernization of agricultural production involving the use of purchased inputs, the need to develop an efficient input distribution has become apparent. The complementarity between new crop varieties and chemical inputs led to the idea of delivering an input 'package' to farmers in order to achieve rapid increases in agricultural output desired. This has been called the 'Green Revolution' package in some Third World countries. The package approach envisaged a major role for the state, delivering certified seeds together with the appropriate quantities of fertilizers and other farm chemicals, and advice concerning proper agronomic practices to farmers (Ellis 1992). There are two ways of assisting farmers in this respect: (i) the provision of credit in money form. This means that the government and other institutions involved should ensure that inputs are available at local market centres; (ii) the provision of input packages. A third option is the combination of (i) and (ii).

In Kenya the policy of input delivery is multifarious in that for each cash crop inputs are provided to farmers through the established co-operative channels who fulfill and satisfy the set conditions. The government has pledged to make funds available to the different co-operatives through its loaning system which allows co-operatives to purchase inputs at the right time. The government also provides inputs for marketing. Important here is the provision of rural access roads to secure the faster delivery of inputs and the marketing of products. The Agricultural Finance Corporation (AFC) is the main government parastatal body dealing specifically with credit arrangements, for small and large-scale farmers.

The question of farm credit is very much linked to input distribution by the government, the ability to purchase inputs by the farmers, and the improvement in agricultural production in both small- scale and large-scale farming. Credit is considered crucial for the Kenyan smallholder development programme, but the rate at which credit has been extended outside the settlement schemes has been slow and modest. It is suggested that on the settlement schemes, credit provision has been and continues to be generous. Loans are provided for the purchase of land, farm development and seasonal inputs, and the quality of credit is high. Yet, on small-scale farms outside the schemes credit use is limited because credit is unavailable. It has been pointed out that this is partly because the Kenya government had no established tradition of money lending in the small-scale sector until many years after Independence. Rural Kenyans did not have a tradition of borrowing money particularly not for agricultural development. Therefore, small-scale farmers have to rely on more formal sources for their credit requirements (Heyer 1976).

Most small-scale farmers are aware that they need credit to improve their farm management and production performance. The government has stressed the need for the intensification of small-scale agriculture to enhance agricultural production, improve food security and foreign exchange earnings. Credit has therefore been recommended for small-scale areas to help them increase their purchasing power so they can acquire inputs and intensify their agriculture (Mellor et al. 1987, Livingstone 1986, Heyer 1976, Von Pischke 1975). Therefore, a number of credit programmes directed at intensifying and improving agricultural production in the small-scale areas have been tried by the Kenya government. In the last thirty years, most of the agricultural credit to small-scale agriculture has been channelled through the Agricultural Finance Corporation (AFC). No attempt will be made to evaluate the success or failure of the AFC, but some of the problems associated with bringing credit to farmers deserve examination.

In the 1970s and early 1980s, the various Kenyan agricultural credit programmes could be divided into those that aimed at expanding commercial (cash) crop production and those that aimed at assisting poorer farmers. They can be listed as follows:-
1. The Cooperative Production Credit Services, established in 1972, was exclusively for coffee farmers and provided credit for seedlings, chemicals and other inputs.
2. The Smallholder Coffee Improvement Project (SCIP) was set up in 1979 for implementation in the period 1979-83. The objective of the SCIP was to improve coffee storage facilities and build more coffee processing factories.
3. The Commercial Farming Project also known as IDA III, financed largely by the International Development Agency (IDA) and the Kenya Government aimed at providing short and medium term credit for small and medium-scale mixed farming and was active in the period 1977/78 - 1979/80.
4. The Smallholder Production Service and Credit Project (SPSCP) began in 1975 and was a precursor of the Integrated Agricultural Development Programme (IADP)
5. The Integrated Agricultural Development Programme (IADP) was seen as combining the provision of credit and technical advice to smallholders with rural infrastructure development (Livingstone 1981, 1986).

The IADP is the most relevant initiative in the context of this study because it focused on relatively poor farmers, particularly those outside the commercialized cash economy. In this programme, land was not required as collateral for loans. Instead a 'security-crop' was identified and formed part of a crop package for which inputs were provided. To facilitate repayment, the security crop was generally a cash crop marketed along one channel only i.e. a co-operative society. Extension and credit were based on comprehensive farm planning that embraced all farm activities. This facilitated assistance being given to subsistence food crops as well as cash crops. However, in most areas this credit scheme was not very successful. As Livingstone points out, the main problem was the weakness of the cooperative marketing institutions and their lack of storage, transport and other facilities for the efficient marketing of crops produced. This negatively affected the disbursement of loans at the right time. Throughout the programme and in all areas where the programme was implemented, the loan money was distributed late leading to the late delivery of farm inputs. This meant that all farming operations were delayed and led inevitably to poor productivity and low repayments. Many farmers were recruited and trained but very few farmers were given loans. This discouraged farmers from enrolling for training for loans and goes a long way to explaining the low turnout for second loan application. The supply of inputs was delayed because the Co-operative Bank of Kenya was late in releasing funds. Poor organization and planning at the headquarters and the unclear policies to the benefiaries also contributed to the failure of credit programmes. In general, agricultural credit in Kenya has certainly favoured progressive and richer farmers (Livingstone 1986, Heyer 1976). The unfavourable conditions that characterized the IADP, and other credit programmes in the past as well as the present credit system makes it easy to understand why very few small-scale farmers in Kenya have been able to benefit from credit.

Another method of providing credit to farmers is through co-operative societies. Co-operatives are encouraged because of the poverty of the mass of the rural population. In the rural areas most farmers do not have enough money to purchase the tools and inputs they need, and they also lack collateral to apply for individual credit. Diallo (1986) holds that in such a setting co-operatives enable communities to get credit, acquire machinery, obtain needed inputs and sell their produce. However, experience with co-operatives has raised some serious questions about how well they serve the interests of their various members. There is evidence to suggest, for example, that in some situations co-operatives actually disadvantage the poorest farmers. Heyer (1976) points out that although the new co-operative credit programme is relatively centralized, it is only available to the limited number of small-scale farmers who farm in areas where the co-operatives are strong. These are the areas where small-scale farmers have taken cash crop farming and dairy farming seriously. Even then, the co-operative credit programme provides only a very partial solution to smallholder credit provision problems.

The current credit programme not only favours privileged areas and selected farmers but it also favours inappropriate resource combinations in small-scale farming. Unlike in the large-scale sector, credit in the small-scale farming areas is generally provided in material form. It is difficult for small-scale farmers to obtain credit in cash and to use it for the employment of labour. This shows the disregard for labour as an input and the bias towards inputs other than labour in short-term credit. The problems that keep occurring in both input distribution and agricultural credit use are inappropriate input packages, insufficient information accompanying the new inputs, late and inadequate supplies, and rigid regulations for the use of credit. According to one author, credit is more easily available if one wants to purchase a four-wheel drive tractor than if one wants to buy an oxen, ox equipment, or even simple tools like hoes (Heyer 1976).

2.4 Gender and Farm Management

Studies in African countries have shown that women do most of the preparation, planting, weeding, harvesting, transportation, drying, storage and processing, and the marketing of the commodities produced there. In Tanzania, it was found that 80% of the above activities were performed by women with the help of their young children (Bahemuka 1987, Feldman 1983, Yacink 1978 and Mbilinyi 1974). They also participate in sowing, weeding, harvesting and transporting cash crops such as coffee, cocoa, tobacco, cotton, rice and tea, which are the main stay of many African economies. Over the past decades there has been an increasing realization on the part of both national governments and international agencies that women are a central element in Africa's agricultural systems and that their limited access to agro-support services and their neglect in agricultural policies may be a major constraint to raising agricultural production (Cloud 1988, Davison 1988, Boserup 1981). Boserup challenged the prevailing notion that economic development and modernization would automatically improve women's status by replacing traditional values and economic backwardness with new opportunities and an egalitarian ethos. Instead, technical innovations have often replaced women's traditional economic activities with 'more efficient' forms of production controlled by men. In such a situation women often find themselves with more work to do and less authority.

Many considered the main constraint in African agriculture to be the negative gender division of labour (Christensen 1981, Norman 1979, and Cleave 1974). In many parts of Africa men have migrated to the urban areas, and because of this labour bottlenecks manifest themselves during peak farming periods when several operations (planting, ridging, thinning, weeding and harvesting) have to be carried out. Consequently, the labour available to meet these peak requirements limit the amount of land a family can farm, the level of intensification and the ability of the farm household to adopt labour increasing technologies. If the migrant husband does not send remittances to cater for such labour bottlenecks, the wife finds herself burdened with work and has to make painful choices and difficult compromises in managing the farm.

2.4.1 Gender Relations and Farming in Tribal Society

A number of authors have attempted to reconstruct the nature of production and the division of labour in African tribal societies in the period immediately before their incorporation into the colonial system (Boserup 1965, 1981, Dankelman and Davidson,1987, Davison,1988, Collier,1974, Nelson,1974, Clark, 1980, Pala, 1974, 1978, 1984, Parpart and Staudt 1989). Women were found to be predominantly responsible for food crop production which in most tribal societies with subsistence economies was carried out with relatively simple technology. Women's farm tasks which included planting, weeding and the harvesting of different crops and the storing and caring for the food family supply to were spread throughout the year. Land clearing and preparation which were men's main tasks were seasonal. Women had a thorough knowledge of crop growing and men a complementary knowledge of animal husbandry, fishing and hunting. However, both groups participated in the agricultural production system. There is evidence that in some tribal societies women have been the primary decision makers in matters of crop production, although their position in the structure of decision making in the family and in the community at large is not clearly indicated.

In most parts of Kenya prior to colonization land was plentiful and was viewed as an economic resource that people had an obligation to use wisely for the good of the community on behalf of the ancestors. To some extent therefore the users of the land were accountable to the head of the community as to how they managed their farms (Davison, 1988). Among the Luhya, Luo and Kikuyu land was allocated to male heads of households according to need and availability. Male heads of households were obliged to provide each wife with sufficient land on which she could grow food crops. Consequently, all women had guaranteed rights to arable land, and once the land was allocated to them they had overall control on it. Both men and women contributed to the food security of the family. However, each of them could dispose of their products as they deemed fit so long as it did not compromise the welfare of the family. Each individual was therefore independently making decisions about how to use his/her farm. The diversity of soil conditions was taken into consideration when allocating land. If land was becoming degraded then farm households moved to other places so that the land concerned could regain its fertility. Land was cultivated in rotation. Moreover land was set aside for various other uses including the cutting of firewood, ceremonial rituals, grazing, salt licks, and public meeting places. Land was also reserved for future expansion and as a buffer between different ethnic groups (Khasiani 1992, Hulsebosch (1990), Jungheim (1989), Davison (1988), Nasimiyu (1985), and Hay (1976),

In much of Kenya, the family consisting of a man, his wife or wives, and their children, was the basic land using-unit. In certain areas the larger extended family or kinship group functioned as the basic unit, and where in this case part of the land was tilled in common by members of the group. The head of this group had the power to dispose of the output. Such a clan leader had an important voice not only in determining the way land was apportioned and utilized, but also on the subject of adopting innovations. As pointed out earlier, women's primary responsibility and preoccupation tended to be the cultivation of the food crops required to feed the family. Husbands could not arbitrarily take the produce from their wives' granaries (Clark 1980) and women could also engage in activities such as trade that enabled them to control some cash of their own. Thus, in the traditional system the woman had her own crops.

In some tribal societies apart from their important role in food production, women contributed to the size of the family herds by trading food crops for livestock. Women travelled long distances to trade with people of other tribes like, for instance, the Kikuyu women who traded with the Maasai. The livestock they brought home after such a trading expedition was kept in the woman's dwelling and could not be disposed off without her consent. Women were also entitled to inherit resources. If a man died without a close male patrilineal kin and made a verbal will that his widow might inherit a share of his herd and land, this will was adhered to. This shows that in traditional tribal societies women could make decisions on the use of resources and could have property of their own.

2.4.2 The Impact of Colonialism on Gender Relations

Many authors have pointed out that the impact of colonialism and the integration of Africa into the world economy have weakened women's access to land, labour and capital and therefore marginalized women's social position. With the colonization of Kenya, the traditional roles and rights of women were eroded. The introduction of colonial administration consolidated and precipitated the rapid entry and integration of men into the colonial economy. Men acquired certain skills, such as the use of the plough and became orientated to the colonial market, commercial farming and the growing demand for labour. As a result food production sector where women exercised their skills and decision-making power was neglected and stagnated. Thus women continued to use backward cultivation techniques and these have become less and less productive in the context of the increasing pressure of population on land and the increasing competition between commercial agriculture and food crop production (Pala,1974, 1978, 1984).

> "The commercialization of agriculture through the introduction of cash crops altered the customary gender division of labour in ways most disadvantageous to women. Men were taught to grow new crops such as coffee and tea for export, while women continued to grow food crops for the family and local consumption. Men were forced into the wage economy to work on plantations or in towns; mostly women remained in the rural areas, often assuming the responsibilities their absent menfolk could no longer perform. Schooling and the teaching of new skills were made available primarily to males. All in all, although both men and women were exploited within the colonial economy, men gained some access and control to important resources such as money, skills, land, and education less available to women." (Gordon and Gordon, 1992:205)

In addition, land became a scarce resource, because of land alienation and the subsequent concentration of Africans in the 'African Reserves' (Khasiani 1992, Hinderink & Sterkenburg 1987, Leys 1975, Zwanenberg & King,1975, Lamb 1974, Ghai & McAuslain 1970, Sorrenson 1968). With the increasing alienation of good farm land to settlers in the 1920s and 1930s, women in the 'African Reserves' not only found themselves cultivating poorer quality land but they also had smaller plots on which to grow food. At the same time, they also had more farm tasks to perform because their men had to work for the settlers in order to raise money. Until the early 1930s this was primarily for taxes and later other needs were added. In the ensuing commoditization of land and commercialization of agriculture, women found themselves increasingly marginalized in their agricultural and domestic activities. In many ways they found it difficult to maintain any meaningful control over their access to and utilization of available resources. Where the introduction of cash crops in the 'African Reserves' was permitted, women's labour hours increased because they had to combine subsistence and cash crop production with a multitude of domestic chores. To make matters worse, with the growing population density of the African Reserves, land there quickly became depleted from overuse, causing soil erosion, and the degradation of soil fertility.

Davison (1988) comments that although colonial land policies had an equally detrimental effect on both sexes, the Swynnerton Plan, in particular, undermined women's relative autonomous economic position in rural areas. There were three reasons for this. First, it marginalized the usufructuary rights of women formerly guaranteed under traditional forms of tenure. Second, it created disadvantages in women's abilities to secure credit for agricultural improvement because land as collateral is required for credit. Third, it fostered the capitalization of agriculture by encouraging, for the first time, massive export crop production by Africans, marginalizing the labour of women in food production.

The deteriorating economic position of women under colonialism is also emphasized by Parpart et al (1989:12). They state that both for economic and ideological reasons colonial officials incorporated African men into the colonial system. This incorporation was not brought about in solidarity with men per se but to foster increased male productivity and thereby to spur capitalist accumulation. Colonial policies were geared to pressure men into entering wage labour force, to train men for a commercial economy and the civil service, to register land in men's names, and to support cash crop farming by men through credit and extension. This implies that women were excluded from the system from the very earliest days.

This early incorporation of men was the starting point of male out-migration from the rural areas. There are now increasing numbers of households headed by women as a result of disintegrating traditional patterns of family and kinship. Yet, households headed by women are often excluded from programmes and projects simply because of current laws and statutes which only recognize males as heads of households. Legal, cultural, religious, institutional and economic constraints restrict women's access to have any title to land. Until now, the commoditization and commercialization of agriculture has in many cases, created more work for the women but has not resulted in increased access to the proceeds of their labour neither has it improved their farm management practices (Khasiani, 1992).

2.4.3 Gender and Agricultural Extension and Credit

Women's role in agricultural technology and labour is an important factor when considering the improvement of farming systems. Women's labour is essential during planting, weeding, harvesting and in the post-harvest handling of produce (small-scale processing, storing, drying e.t.c.) (Fresco 1982 and 1985). Women are faced with time constraints because their roles as domestic reproducers, agricultural producers, caretakers of livestock and, in some cases, off-farm employees conflict as far as time is concerned. Therefore, close attention should be given to the relevance of improved technology to women in as far as this helps to reduce their labour constraints and improve both agricultural management and production.

It has often been assumed that when men leave agriculture to their wives, agricultural production suffers because women are constrained by serious labour shortages and the lack of agricultural knowledge and skills. But the effect of the 'feminization' of smallholder agriculture on agricultural production depends on the extent to which agricultural policies, institutions and programmes are geared to the needs of women farmers (Safilios-Rothschild 1990:101, Saito et al. 1990). When women farm managers and female heads of households have access to agricultural extension, credit and markets for their products, their productivity and incomes are high, in fact can be higher than those of male-managed farms of the same size. This is clearly shown in Safilios' study of two villages in Kenya (Safilios-Rothschild 1990). Safilios found that where women had full control of their farms, produce was higher and farms were better managed than those farms where females worked under the control of their spouses. She also found that women invested as much as they could in their cash crop farms (mainly coffee in this case) and in their dairy farming. They bought fertilizers, new seeds, and animal feed for their cattle. They also sprayed, pruned, and hired extra labour because the cash income from the farms belonged to them. Therefore, the prevailing stereotypical belief that men should cultivate cash crops because they are better farmers and women should cultivate food crops no longer holds true. However, Safilios points out that in the village with highest number of men staying on the farms and who participated in farming because they lacked wage employment, women had no autonomy and their production was very poor.

However, the production of the men was also poor too (Safilios-Rothschild 1990).

In the past, the extension staff have concentrated their efforts only on male farmers hoping that the men would pass on the information they received to the women. This however, was not the case. For the effective transmission of information both men and women have to be approached, because a household is not a homogeneous unit of production and because women in Kenya contribute more than 70% of the labour in the small-scale agricultural sector (Egsmose, 1990, World Bank 1990). Therefore, if the government aims to increase the productivity of agriculture and the effectiveness of extension performance, more attention should be paid to reaching women farm managers and food producers as targets for agricultural extension.

It is observed that while financial means facilitaties access to land, labour, and such recurrent capital inputs as seed and fertilizer, a proper management of the production process is required to make such an investment pay off (Moock 1971). Under conditions where men are involved in labour migration, the maintenance of a close relationship between the manager and the migrant head of household is vital for productive farming. In this case, the manager is the wife left behind to maintain the rural home. However, Moock suggests that sometimes women are confronted by conflicting roles as they face opposition from their relatives in-law who appoint themselves managers of the rural homestead. This affects farm management. Moock suggests that one of the most important requirements for good farm management is access to adequate information concerning efficient use of capital inputs and farming techniques that are relevant to the farmers's own land, labour and marketing situation. This requirement will not be met until male agricultural staff stop avoiding female-headed farms and the female-managed farms of absent labour migrants.

Apart from adequate extension information, credit is an important factor for improved management. But access to credit for rural women has proved to be an almost insurmountable difficulty in the Kenyan context mainly because of lack of collateral and restrictive loan conditions. The conditions for commercial credit from banks and financial institutions (for example Agricultural Finance Corporation) are particularly difficult for women. The requirements for operating a bank account are closely related to wage employment and/or having some other source of income, being able to provide security in the form of land or a house, and being able to pay high interest rates. High interest rates can only be met by a select group of borrowers and indeed there are very few women amongst them. Therefore, it is very difficult for women in rural areas to improve the management of different crops and livestock since they have almost no way of getting loans.

2.4.4 The Household as a Unit of Production and Consumption? Four Debatable Assumptions

When it came to land consolidation, registration and allocation, the colonial administration in Kenya assumed farm households to be single production units (Safilios 1985:113). The males who were seen as the providers and heads of households were granted land title deeds. Women were seen as an appendage to men and anything provided for them was provided through the head of the household. This view of the farm household as a single production unit has been taken over by the Kenya government as its policies are implicitly based on the following assumptions:
(a) The household is considered to be one unit with supposedly strong intra-household relations

(pooling) and with supposedly balanced inter-household relations (exchange);
(b) This unit is seen as a unit of production and consumption without taking into account variations such as production and consumption patterns along kinship lines outside the household;

(c) Such a unit is considered to have one male head to represent it, assuming that his interests, problems, and decision- making patterns supposedly reflect those of other household members;

(d) Household duties are taken care of by the wife whose work is considered as 'given' and 'natural' and does not need to be investigated.

These assumptions underlie a very influential theoretical model of household organization, the 'New Home Economics Model', which conceives of farm households as single production and consumption units headed by individual males who command the resources of individual household members (Fresco, 1985 in Safilios-Rothschild not dated).

However, it has become increasingly clear that 'the household as a homogeneous decision-making unit within which members pool resources and have utility functions' (Becker 1981), often does not conform to reality: the pooling of resources by household members is not the rule; it is not uncommon to have two or more production units within a household; and the head of the household does not necessarily the command resources of the other members (Guyer and Peters, 1987:208,210). Yet, it would appear that the policies are formulated with such a model in mind, and thus greatly influence farm management and agricultural development. Many authors, including Safilios, Backer and Barnes believe that to understand farm management practices and the impact of agricultural policies, or any other institutional policy on women producers in the smallholder households, it is necessary to make and the following distinctions.
(1) A typology of farm households on the basis of gender, by dividing households into three groups: male-headed, female-headed, and female-managed households. This helps us to understand the various agricultural situations in the rural areas, especially as far as the multifariously gendered inter-household differences in farm management are concerned.
(2) A compartmentalization of the farm household, exposing the often contradictory interests, relations and decision-making patterns between the individuals and the production units within these households. This helps us to understand the intra-household differences in farm management.
An alternative household model based on bargaining, negotiation or contracting between household members has been proposed by some economists as one that supposedly fits reality (Fapohunda, 1987; Foibre, 1984; Jones, 1983 -in Safilios not dated pp, 3 - 5). This model was clearly depicted in Safilios study of intra-household relations and interests in two Kenyan villages. Such an alternative household model helps us explain a wide range of observed behaviours and dynamics within the farm-households. In the words of Safilios:

"The bargaining power of the different members in the household varies according to their socio-economic characteristics, and the importance they attach to the maintenance of family stability, their organizational membership and access to economic and agricultural resources, the availability of profitable occupational alternatives, and the degree of legitimation of their rights to benefits provided by prevailing cultural norms and influential institutions. Their access, on the other hand, to economic resources and profitable occupational alternatives as well as to agricultural services and resources depend upon a number of social-structural factors such as the sex ratio of migration, the existence of differential producer price policies for male and female controlled crops, the type of national and project policies regarding women's access to co-operatives, extension, credit and agricultural training and the degree of adaptation of recommended technical innovations to the needs and interests of men and women farmers."(Safilios-Rothschild, not dated, pp. 3 - 5).

This implies that policies should be formulated with such a model in mind. Governments must be flexible in their policy formulation and implementation in order to cater for the interests of all the farmers and farm-households regardless of gender. This calls for the greater sensitivity of governments to the requirements of women as small-scale farmers for support services.

2.4.5 Women Groups and Agricultural Production

Since the mid-1970s scholars concerned with women and development in the Third World have shown an increasing interest in women's organizations and their role in the development process as a strategy of rural development. Different approaches to rural development, (for example the top-down District focus for rural development, policy in Kenya) have been experimented with but have not brought much improvement to the lives of women in the rural areas, nor have they been able to incorporate women as active participants in the social, economic and political programmes of the country. Generally, these approaches were not gender-specific and therefore are unable to incorporate women into the main stream development and policy issues. It is important to introduce a strategy of rural development that incorporates women more directly and improves their social and economic well-being. As the case is clear that most governments in Third World countries, especially in Africa South of the Sahara, suffer from a scarcity of finance and therefore are not able to reach most farmers in the rural areas, the group approach is recommended. The government and its agents can reach many women farmers through women groups. Different forms of women activities have always existed in many parts of Africa. It is therefore possible that through womens' networks and organizations much can be achieved in the rural areas, especially in the field of agriculture and in the general social and economic well-being of women. In Kenya, women in both rural and urban areas are nowadays encouraged to form Women Groups. Through these organizations women can gain access to some of the resources and services the country has to offer and which were out of their reach in the past. Creating access to such resources as land, income and agro-support services is seen as a way of empowering women and making them decision makers in their own right (Were, 1985, Sorensen 1992). An effort must be made to incorporate all rural women because many are very poor and cannot afford the money and time these groups require.

However, it is observed that women's participation in women's organizations is primarily related to their practical needs and, as yet has not led to any broad political and economic awareness or consciousness among women. This was one of the findings in a study of women groups in two divisions of Kenya and may well be true for many other areas in rural Kenya. In the words of Thomas:

"Women do not question existing structures of stratification and authority. Nowhere in Weithaga and Mbiri did anyone suggest that an association should join with others to achieve common objectives. The issues about which these women are concerned are not the big ones such as rights to landownership, inheritance laws or even female education in its broadest sense. Instead they pursue immediate and pragmatic concerns affecting each and every household: food, shelter, school fees, water, health care etc. These concerns stem from the needs and interests of the family." (Thomas, 1988:418)

The success of rural women groups in Kenya may therefore depend on their ability to transform existing structures of stratification and authority which hamper women's access to agricultural resources and services. The question, therefore, is whether women groups in the rural areas can be used to improve farm management and agricultural production.

2.5 Conclusions and Research Hypotheses

This chapter has attempted to build a framework of the major components of small-scale farming and the role of women in farm management in Kenya. In the past, land was abundant and all adults regardless of gender had access to land to some degree. Individuals made independent decisions on land use and there was concern for maintaining soil fertility. The farmers planted crops according to the suitability of the soil and the seasonality of the rains.

This situation changed very rapidly with the colonization of Kenya. Then land was alienated for European settlement and many Africans were enclosed in what was known as 'African Reserves'. African men had to work on European farms to raise money and women had to work on the family farm. This brought about changes in the roles they played within the family. Women lost any grip they may have had on land ownership with the introduction and implementation of the Swynnerton Plan in 1954. This Plan encouraged the consolidation and registration of land in the names of the male head of household. Since then, agricultural policies, extension and credit have been geared to men as heads of households and land title holders. In many areas cash crops were introduced and registered as men's crops and the technology associated with cash crops passed on to the male as the assumed managers of the farms.

However, in many parts of Kenya, men migrate from the rural areas in search of wage labour. This out-migration makes the woman the de facto heads of households and the decision makers on the farms. At the same time, male labour migration increases labour requirements on the farms. To a large extent this appears to have put constraints on women as far as farm management is concerned.

In Kenya the issue of land ownership, especially in small-scale farming, needs to be addressed again. Since land ownership has been privatized and purchase is now the way to acquire more land a solution to low agricultural productivity may lie in improving management practices within small-scale farming. Women who are managers on many farms should be given more access to agro-support services and agricultural inputs. Government policies and other institutional programmes should become more gender-sensitive and labour saving techniques should be introduced to help rural women manage their time properly and efficiently.

Research Hypotheses

The subject of this study are the problems of small-scale farming in a densely populated rural area in Kenya and in particular, the problems women face as farmers and managers in various types of households. Four hypotheses are proposed to guide our research.

(1) Female-headed and female-managed households have less control over land and less access to agro-support services than male-headed households, which negatively influence their farm-management practices.

(2) Female-headed and female-managed households experience more labour constraints than male-headed households which leads to poor farm management practices.

(3) Female-headed and female-managed households benefit less from non-farm sources of income than male-headed households and this has negative effects on farm management.

(4) Female-headed and female-managed farms compare unfavourably with male-headed ones in investments, crop yields and farm incomes.

Notes

1 In many countries of the Sub-Saharan Africa, apart from population density, natural resources (the physical environment), location in relation to markets, roads and railways, institutions relating to the land and land tenure, and technical knowledge and capital resources greatly influence the system of farming (Upton 1987).

2 One of the features which placed Sub-Saharan Africa in a special category was the slowness of its governments to respond to the deteriorating economic results produced by these weaknesses. In many cases, deficient policies were and are sustained for many years. The position of many modern African rulers and their governments is based on family and ethnic loyalties with followers rewarded with preferential access to loans, import licenses, contracts and jobs. Rulers are less interested in the long-term economic growth of the country than maintaining their power base. Moreover, to meet part of government financial obligations, it is generally the politically marginalized rural sector, characterized mostly by women, that have to bear the cost, not only in terms of depressed market prices for their produce, but also through the taxation of cash crops (ODI 1993, London).

3 In the words of Zwaneneberg et al "Given the general situation of a land surplus in the pre-colonial era every individual could expect to have some right to occupy and use land. This ability arose either from his/her position within a kinship group or from his/her own entrepreneurial abilities in opening up new tracts of land. Every large agricultural group had rules governing, firstly, the forms of land ownership, which were usually based on kinship with the founder who claimed the area for cultivation, secondly the distribution of lands to wives, sons, and daughters, and thirdly, the distribution of lands to friends and outsiders for the temporary cultivation of a plot. The general principle of land tenure seems to have been that, within clans land was generally equally divided." (Zwanenberg and King:29)

A Framework 37

4 Before Independence large-scale farming (the White Highlands) comprised 3 million hectares. Of this 1.2 million hectares were redistributed in the course of the 'Million Acre Scheme'. If the rest (1.8 million hectares) were redistributed and each small farmer were to get 4 ha (10 acres), this would create greater room for 450,000 farm households. Hunt (1985) is of the opinion that land on farms in excess of 10 hectares should, in principle, be considered eligible for redistribution. It is further suggested that the holding size of newly created farms should not exceed that which can readily be farmed by family labour without the hire of agricultural labourers (Hunt, 1985: 258).

5 Yet, the men who migrate to urban areas find it very difficult to send regular remittances to their rural families because their wages are low. The ILO report (1972) showed that at the time the minimum wage of 200 Kenya shillings per month in Nairobi could hardly be expected to support an individual let alone allow him to save a small amount to send to his family back home in the rural areas. The same applies to the present situation when the minimum wage is 800 Kenya shillings. Despite the low wages in the urban sector people still migrate because agricultural labour and pay is not attractive and lacks incentives.

6 If any investments are done for productive purposes, these are usually confined to either the purchase of land, the building of a house or the opening-up of a small business. Studies also indicate that most migrants to Kenyan cities seek wage employment not because they lack land or any prospects of acquiring it, but because the size of the land available was insufficient to support them and their families. Better paid workers often spend money improving their rural holdings (Stichter 1982).

7 To improve the flow of information to farmers, the government's extension policy, implemented through the Ministry of Agriculture, Livestock Development and Marketing emphasizes that by the expansion of programmes for staff training, the dissemination of research stations findings to extension officers and the more effective demonstrations of approved management systems for both crops and livestock, more farmers should be visited enhancing agricultural production. The policy further notes that greater attention should be given to more effective and widespread group extension techniques involving demonstration farms and field days. Emphasis is given to on the spot training of farmers in the basic principles of crop husbandry, use of fertilizers and other inputs, crop rotation, on-farm storage for subsistence crops, record keeping and financial management (GOK,1990, 1983, 1989).

8 The group approach seems more equitable in its effects. Its effects on output are likely to be greater because of more effective diffusion. It appears to be more cost-effective, reaching many more farmers in a given area while economizing on transport and fuel. It is also applicable to different types of groups including soil and water conservation groups, self-help groups and women groups. The proportion of farmers from female-headed and female-managed households receiving extension advice through groups is increasing, yet the frequency at which the group approach is used is very low compared with the contact farmers' approach (Bindlish et al. 1993).

9 For instance Leonard (1977) states that in one province in Kenya, extension agents spent 57% of their visits with progressive farmers (the most advanced 10% of all farmers) and only 6% with traditional farmers.

10 For farmers to make effective use of the improved supply of agricultural and livestock inputs, Government policy is to provide adequate financial resources. These resources will be provided by

expanding seasonal and long-term credit programmes. Particular emphasis will be placed on the timely disbursement of seasonal credit for land preparation and for the purchase of seed, fertilizer, and other inputs. Where possible seasonal credit will be paid in kind rather than in cash. In the longer term, policies will be developed to strengthen the institutional framework with the aim of mobilizing rural savings and providing increased agricultural finance. The policy of the government is to move towards a decentralized agricultural finance system (GOK 1990, 1994).

11 In a World Bank study to evaluate the effectiveness of the T&V method in Kenya in the sampled small-scale farming areas, it was found that of those farmers applying for credit, 64% failed to obtain any credit at all, and only 11% received what they had applied for (Bindlish et al. 1993).

12 In a comparative study in Kenya, it was found that only 13% of the total female population involved in agricultural production was reached by agricultural extension (Saito, et al 1994).

13 An example of divergent interests were the expectations of men that their wives would give them some money from their coffee and milk earnings.

3

Vihiga District: Background Information

3.1 Introduction

This chapter describes the physical resource base, the human resource base and the socio-economic situation of Vihiga District. It provides the basis for understanding the research area, the North Maragoli location, one of the administrative units of Vihiga District. The Kenyan administrative map changed drastically between 1990 and 1994 because of the creation of more districts, divisions and locations. Vihiga District is a new administrative entity which was carved out of Kakamega District and commissioned as a new district in February 1992. It is one of the most densely populated agricultural districts, and is located in one of Kenya's high potential zones. The creation of new districts and the redrawing of boundaries has been politically motivated in many countries, although sometimes high population density may necessitate the creation of new administrative units which can serve the populations more efficiently.

Vihiga District together with Kakamega, Bungoma and Busia Districts form the southern part of Western Province. Vihiga District is bordered by Kakamega to the north, Nandi to the east, Kisumu to the south and Siaya to the west (see Figure 3.1). The district is dissected by the equator and is divided into five administrative divisions namely Vihiga, Sabatia, Tiriki, Emuhaya and Luanda (see Figures 3.2 and 3.3). At the time of this study, the district had 24 locations and 110 sub-locations but it was anticipated that the district will be further sub-divided to create more divisions, locations and sub-locations for administrative purposes. The district is inhabited by the Abaluhya ethnic group which, in this region, is sub-divided in three sub-groups, viz. the Abalogoli (the Maragoli), the Abanyole (the Banyore) and the Abatiriki (the Tiriki). The Maragoli inhabit Sabatia and Vihiga Divisions, the Abanyole occupy Emuhaya and Luanda while the Tiriki populate Tiriki Division. The three sub-groups are culturally related and speak the Luhya language with slight variations in intonation.

Vihiga has an area of approximately 521 square kilometres (52,100 ha) of which about 409 sq.km (80%) is arable, supporting approximately 84,995 farm families (households). The average size per holding in the district is 0.6 ha per family of between 6 to 10 persons. About 90% of the total area is under crop cultivation or is used for livestock production. The rest is mainly government-reserved forest area such as Maragoli and Kaimosi, and very rocky and stony areas unsuitable for agricultural productivity (Kenya 1994).

Figure 3.1
Location of Vihiga District in Kenya (source: Kenya (1994), Vihiga District Development Plan (1994-1996).

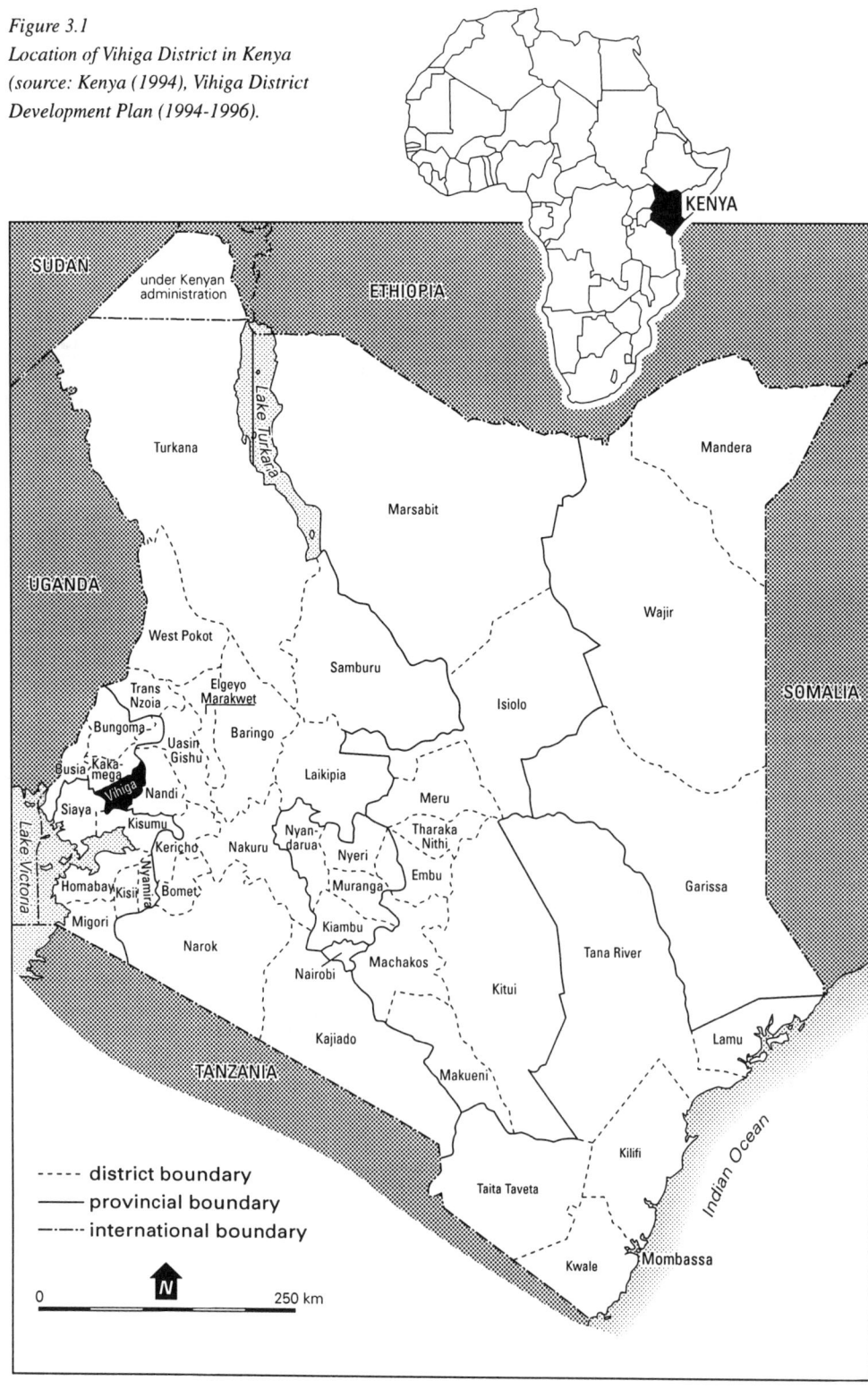

Vihiga District: Background Information

Figure 3.2
Administrative Boundaries of Vihiga District
(source: Kenya (1994), Vihiga District Development Plan 1994-1996).

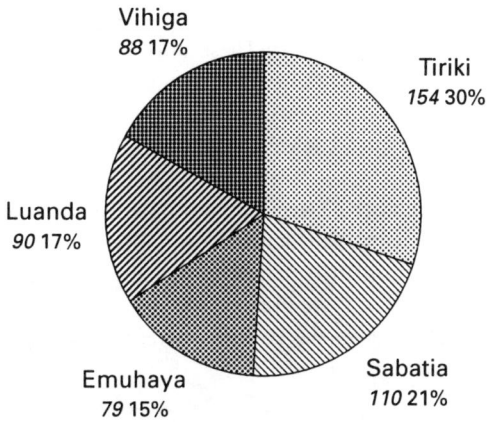

Area distribution in Sq. km

Figure 3.3
Administrative Divisions of Vihiga District
(source: Kenya (1994), Vihiga District Development Plan 1994-1996).

3.2 The Physical Resource Base

3.2.1 Topography and Geology

Vihiga District is situated to the north of Lake Victoria. It is part of the extensive Western Kenya plateau that descends towards the Lake Victoria basin (Ojany and Ogendo 1973). The height of this plateau varies from 1250 m to 1800 m above sea level.
Rocks of the Basement and Pre-Cambrian systems invaded by igneous or granitic intrusions underlie the whole district. The Pre-Cambrian rocks in this area have been subjected to weathering and erosion and this has produced isolated rock outcrops and ridges. These residual rock outcrops have been castellated into tors, finger-like and doom-shaped inselbergs. These inselbergs are more prominent to the western part of the district, especially in the Luanda area, and in the areas bordering the Nandi escarpment to the east. The Maragoli hills are composed of resistant granites and giant white quartz veins giving rise to pronounced ridges which are the main feature of the Pre-Cambrian system still visible.

Overlying the Pre-Cambrian system are strata of the Nyanzian and Kavirondian series (Morgan 1973 and Ojany & Ogendo 1973). These series occur in close juxtaposition and have a number of similarities. Rocks of the Nyanzian series consist of great thicknesses of various types of ancient volcanic materials in which basalts, tufts, and trachytes are very common. The volcanics have been interbedded with a number of coarse-grained sediments such as conglomerates, quartzites and banded ironstones (Ojany and Ogendo 1973).

The Kavirondian series consists of younger rocks, sedimentary derivatives of the Nyanzian System (Ojany and Ogendo 1973). Lithologically, they consist of alternating bands of sandstones, grits and mudstones with water-lain conglomerates. The Pre-Cambrian system, the Nyanzian and Kavirondian series have greatly influenced the topography and soils, and the economic base of the area. In general, the land rises towards the south east eventually forming three hills; Buyonga, Namesa and Dabwango hills (see Figure 3.4). These hills are steep and stony and make both communication and cultivation difficult in this part of the district. Structurally this area is part of the Lake Victoria Basin, the genesis of which is associated with tectonic faulting disturbances that caused the formation of the Rift Valley (Lavrijsen, 1984).

Relief has also had a considerable influence on the drainage pattern. Streams flow from east to west and the area is well watered. Numerous natural streams flow through deep valleys cut into the plateau. The main streams are the Wamondo, Edzawa, Ezava and Mazi-Ma-Mwamu which join the Yala river system and finally empty into Lake Victoria (see Figure 3.4). The streams have water throughout the year. However, the volume of water in the streams decreases markedly between December and February during the dry season.

3.2.2 Soil Characteristics and Agro-Ecological Zones

The above-mentioned rocks have been weathered and eroded and this has produced generally good, well-drained fertile soils. There are three dominant soil groups derived from the parent rock material: granitic, volcanic, and igneous.

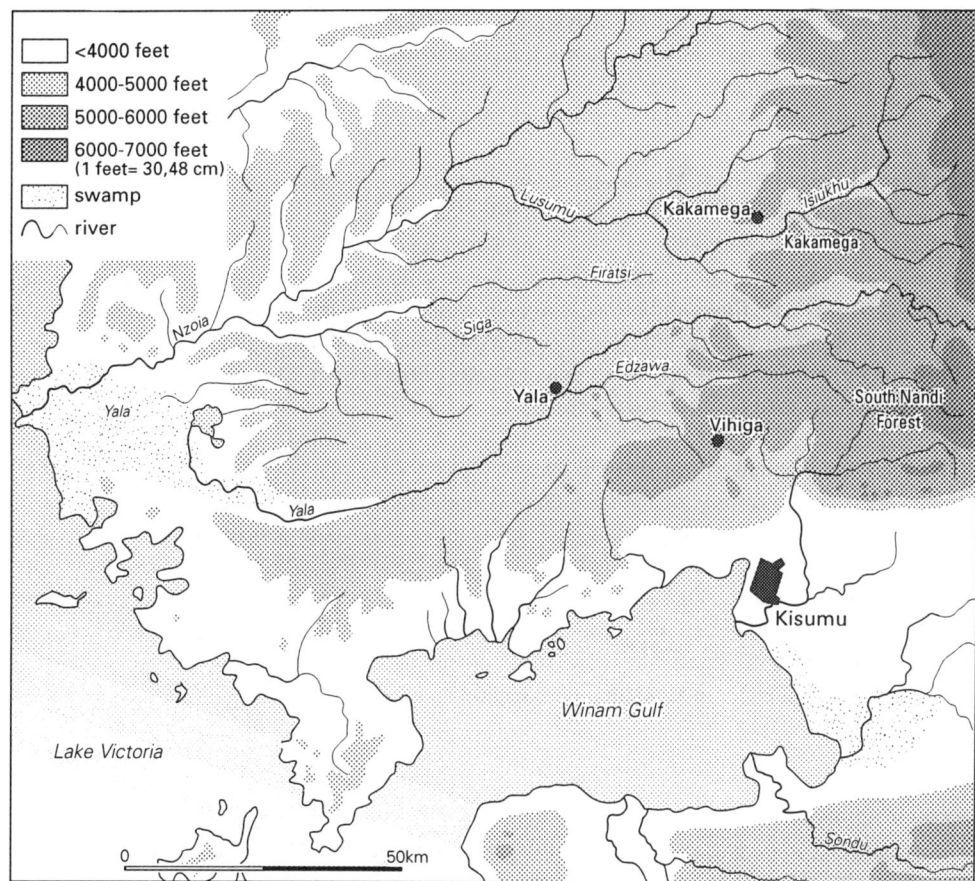

Figure 3.4
Drainage and Relief of Vihiga District (based on: Survey of Kenya, National Atlas of Kenya, 1970 page 9).

(i) Soils developed on granites (classified as 131U and 132U) are well-drained, very deep, dark red to yellowish red, friable to firm, sandy clay to clay, with acid humic top soil and therefore referred to as **humic ACRISOLS**. (ii) Soils developed on basic igneous rocks (basalts, classified as 143U are well-drained, extremely deep, dark reddish brown, friable clay and called **dystric NITOSOLS** (Jaetzold and Schmidt (1982). (iii) Soils developed on acid igneous rocks classified as 177U are well-drained, moderately deep to deep, yellowish red to strong brown, friable clay, over petrophlinthnite or rock, in places shallow over petrophlinthnite (**orthic FERRALSOLS**, partly petroferic phase, with murram cuirass soils (Jaetzold and Schmidt 1982). Both (i) and (ii) are very good soils, able to support sustained agriculture, provided rainfall remains high and reliable and the soils are well managed. Type (iii) has a low soil fertility characteristic of the marginal sugar cane zone but still found within the LM1.

The district can be divided into two main Agro-Ecological zones on the basis of the above soil types. These are the Upper Midland zone (UM1) (131U and 132U-**humic ACRISOLS**) and Lower

Midland zone (LM1) (143U and 177U-**dystric NITOSOLS**). The Upper Midland Zone has fertile well-drained dark red soils which support tea, coffee, maize, finger-millet and cassava cultivation. The zone covers the western slopes of the Nandi escarpment, Sabatia, Vihiga and Tiriki Divisions. The Upper Midland Zone (UM1) covers about 82% of the district (Jaetzold & Schmidt 1982).

The Lower Midland Zone (LM1) covers the western parts of the district in Emuhaya and Luanda Divisions. The Lower Midland Zone is less fertile (less productive) when compared to the Upper Midland Zone and covers 18% of the district (see Figure 3.5). The soils are red loamy sands derived from sediment and basement rocks. These soils support the growing of sugar cane, maize, coffee, beans, finger-millet and sorghum. The soils in the district have generally been subjected to considerable leaching due to high rainfall and continuous cultivation over the years. Gully and hill erosion is also common in many parts of the district. The soils need fertilizers if a high productivity is to be sustained.

Tables 3.1 and 3.2 make clear that the greater part of Vihiga District falls in the fertile UM1 Zone. However, the fertility of the soils there is under threat. Despite the fact that the soil is a vital component of agricultural production and requires good farm management, some farmers still do not fully appreciated its importance and have tended to take it for granted. This may partly explain why there has been over-utilization and why preservation methods have been inadequate. High rainfall and lush vegetation seem to support the erroneous popular conception that the whole district is one of inexhaustible potential. The farmers show greater interest in rainfall as a critical factor in agricultural production than in sustained soil fertility.

Table 3.1
Agro-Ecological Zones in Vihiga District (source: Kenya (1994), Vihiga District Development Plan 1994-1996).

Zone	Area in sq.km	Location	Classification	Crops grown
UM1* 131U and 132U	427	Vihiga, Sabatia, Tiriki/Hamisi divisions	Humid	Tea, coffee, maize, beans, bananas, millet, sorghum, fruits etc.
LM1* 143U and 177U	94	Western parts of the district, Emuhaya and Luanda divisions	Humid	Coffee, sugar-cane, maize, beans, bananas etc

UM1* and LM1* stand for UM_1 and LM_1

Table 3.2 Agro-Ecological Zones by Division (in sq.km) (source: Kenya (1994), Vihiga District Development Plan 1994-9196).

Division	UM1*	LM1*
Vihiga	80	8
Sabatia	104	6
Tiriki/Hamisi	138	16
Emuhaya & Luanda	105	64
Total	427	94

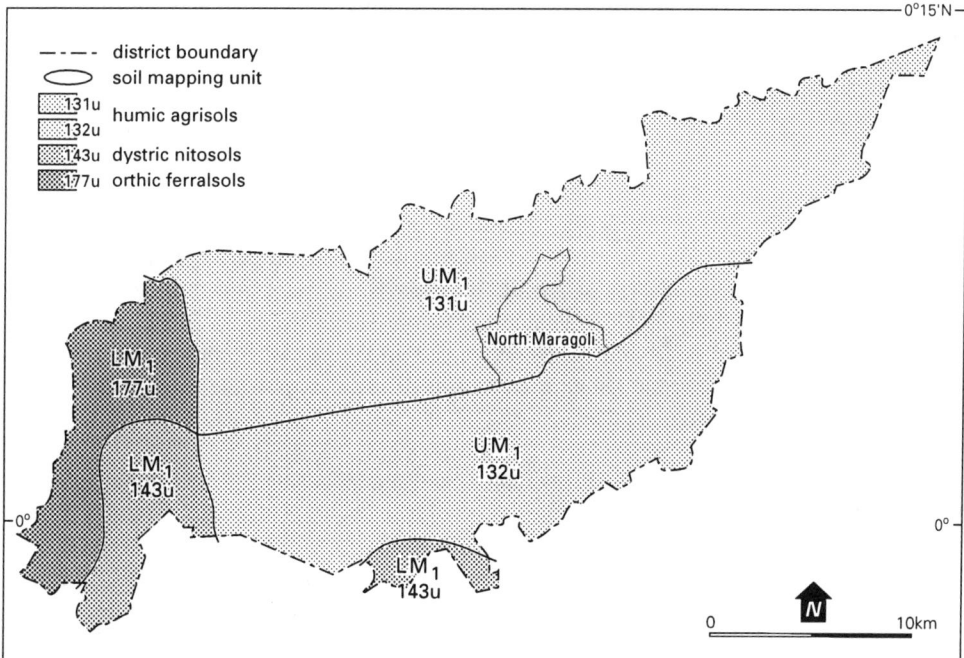

Figure 3.5
Soil Map of Vihiga District (source: Exploratory Soil Map of Kenya (1980)).

3.2.3 Climatic Characteristics and Agro-Ecological Zones

Monthly rainfall, monthly potential evaporation and average temperatures place the region's climate clearly in equatorial climate zone (Lavrijsen 1984). The climate of the area is characterized by heavy, reliable, and well-distributed rainfall throughout most of the year. The pattern of rainfall is characterized by two peak seasons which occur soon after the equinoxes (Lavrijsen 1984, Jaetzold & Schmidt 1982). The long rains begin in February through July and the short rains between August and November with a short dry spell between December and February. The fact that the region receives continuous rainfall from February through to November and sometimes into December supports the idea of a monomodal rainfall pattern. The rainfall distribution over the area is conditioned by a number of factors: monsoonal winds that carry moist air-masses from Lake Victoria and the Congo Basin, and the orographic lifting of monsoonal winds which give rise to heavy afternoon thunderstorms (Groenenboom & Drift 1991).

Average rainfall varies from between 1600 mm to 2000 mm per year. The district is situated in a belt of high rainfall with pockets of even higher rainfall (over 2000 mm) and receives more than 100 mm rainfall per month in 10 months of the year. The pattern and amount of rainfall is almost similar for all the divisions in the district but it decreases, though not significantly towards the extreme south, close to Kisumu District (see Table 3.3 and Figures 3.6, 3.7, 3.8 and 3.9)

Minimum average temperatures vary from 14 0C to 18 0C, and mean maximum temperatures vary from 26 0C to 28 0C. The mean annual temperature is approximately 22 0C. These fairly high and uniformly distributed temperatures produce warm conditions throughout the year (Groenenboom & Drift 1991, Jaetzold and Schmidt, 1982). Temperatures do not vary much during the year because the district is close to the equator (Table 3.3 and Figures 3.6, 3.7 3.8 and 3.9).
What is the relevance of the observed rainfall pattern to agricultural production? In the tropics, agricultural seasons and yields depend, among other things, on the length of time when water is available. For unimpeded growth, crops require an optimum transpiration during the whole period of growth. This can only be achieved if water is continually available to the roots. Crop-specific water requirements are also a function of the stage of plant growth. Potential evapo-transpiration (Et) indicate the water requirements of crops (Lavrijsen 1984). These must be met from rainfall in a farming system that is wholly rainfed. In Vihiga, this is not a problem because rainfall is higher than evapo-transpiration rates. Thus, there it is possible to cultivate a wide range of crops including perennial ones.

Given the prevailing soil and climatic conditions, the two agro-ecological zones for agricultural purposes are:-
(i) Tea - Coffee Zone or Upper Midland Zone (UM1) at an altitude of between 1500 and 1800 m, annual mean temperature of 18 0 to 20 0C and annual rainfall between 1600 to 2000 mm (Jaetzold & Schmidt 1982). This zone is considered the most productive in the district. It has permanent cropping possibilities, divided into two to three variable cropping seasons and with very good yield potential. Climatic conditions are good for coffee, tea, maize and many other crops. More than two-thirds of the district is found within this agro-ecological zone.
(ii) Lower Midland - Coffee-Sugar cane zone (LM1) at an altitude of 1500 m with mean temperature of between 20.8 0 to 22 0C. Annual average rainfall is 1600 to 1800 mm (Jaetzold & Schmidt 1982). Climatic conditions are good for coffee, sugar cane, maize, beans and many other crops. It is warm and humid throughout the year with permanent cropping possibilities, - divided in two variable cropping seasons - and has good yield potential. This zone is located to the west of the district (Jaetzold & Schmidt 1982).

As a whole, Vihiga is humid with a moisture availability of over 80 % (rEO, evapo-transpiration rate). The evaporation in the area is not high, the annual total varies between 1800 and 1900 mm (Groenenboom & Drift 1991, see Table 3.3). The district receives both orographic and

Table 3.3
Climatic Data of Vihiga District (source: Groenenboom & Drift (1991:12) Soil Survey in West Kenya).

Temperature	Jan	Feb	Mar	Apr	May	Jun	Jul	Aug	Sep	Oct	Nov	Dec	Total
Daily mean	21.2	21.7	22.0	21.1	20.6	19.9	19.5	19.9	20.1	20.7	20.7	20.8	20.7
Mean min.	13.9	14.3	15.0	15.4	15.0	14.1	13.5	13.6	13.5	14.4	14.6	14.1	14.3
Mean max.	28.5	29.2	29.0	26.9	26.2	25.7	25.5	26.2	26.7	27.1	26.7	27.6	27.1
Rainfall (mm) Mean	81.6	105	162	263	214	182	163	189	161	130	117	91.3	1859
Surpassed 6 out 10 years	26.3	37.8	87.3	169	187	119	107	125	103	75.4	53.1	32.9	1123
Potential evapo-transpiration mean (mm)	165	172	183	145	153	132	137	147	149	158	145	164	1850

Kaimosi Mission (20 yrs), Eregi Teachers T.C. (24 yrs)
Western Agricultural Research Station - Kakamega (15 yrs)

convectional rainfall. With this type of rainfall, there is no limit to the range of crops that can be grown. In fact, the heavy rainfall received in this district has influenced farmers there to accept any new crop for example hybrid maize, French beans, coffee and tea that is introduced. The high rainfall combined with rather good soils have helped farmers to diversify the crops they grow.

Because of its high rainfall received the district has a high ground water potential. The water table is less than 6 m in many parts of the district. The ground water resources are not fully exploited and the district agricultural staff believed that this water is now polluted by farm inputs although there has been no study in this area to analyze pollutants in the ground water.

3.2.4 Forestry Resources

Vihiga District was originally in a transitional area between the lowland forests of the Congo and the Afro-montane forests of the Western Rift Valley. Due to human interference most of the original climax vegetation has disappeared and nowadays exotic trees predominate. Original tree species that remain include the Elgon teak, Celtic and Cordia. Eucalyptus now forms about 70% of the trees grown on farms and it is the major source of wood fuel and timber in the district (Kenya 1994). Kaimosi and Kabiri (small extension of Kakamega Forest) Forests comprise of both exotic soft woods for timber production and indigenous forests of high timber value. The Maragoli Forest comprises of exotic soft woods planted during the 1950s for the protection of

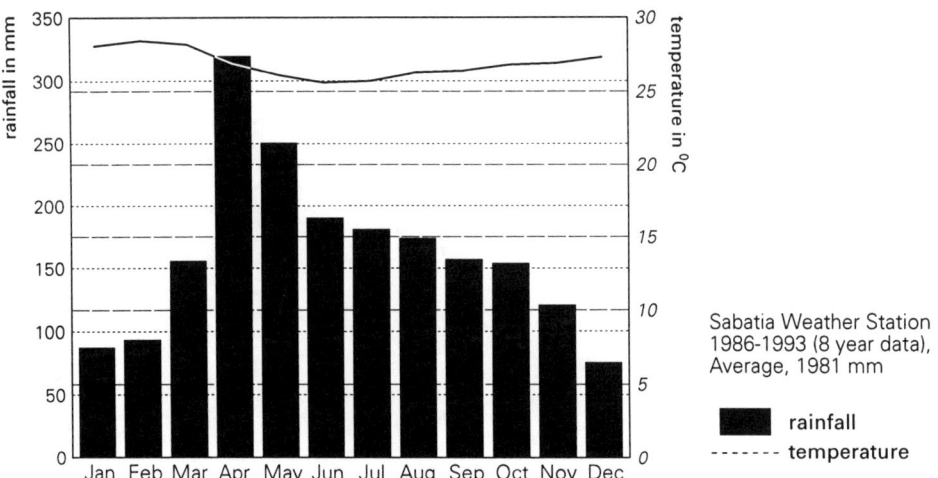

Figure 3.6
Vihiga District Annual Average Rainfall and Mean Temperatures (source: Sabatia Weather Station, Annual Rainfall Records, 1994).

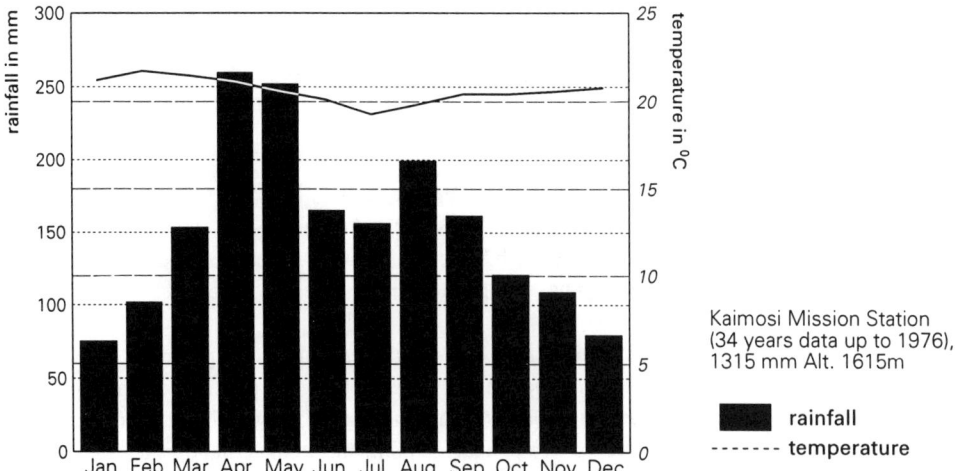

Figure 3.7
Vihiga District Annual Rainfall and Temperatures (source: Jaetzold & Schmidt, 1982, Farm Management Handbook of Kenya).

the hills and therefore it is exploited selectively. The growing demand for building poles, fencing poles and fuelwood because of the rapid growth of population has caused an extensive clearing of most of the indigenous forests and trees.

In the mid-seventies the energy crisis and environmental degradation resulted in the establishment of the International Centre for Research in Agroforestry (ICRAF) with the proviso that it should

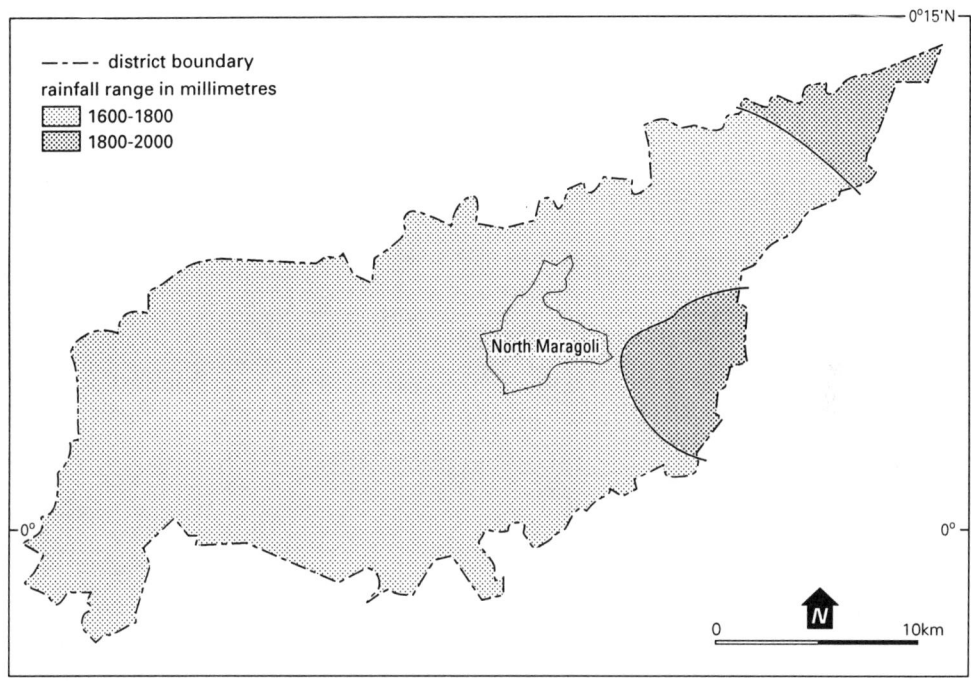

Figure 3.8
Mean Annual Rainfall Map of Vihiga District (source: Agro-climatic Zone Map of Kenya (1980)).

co-ordinate national, regional and international research on agroforestry in the interest of meeting local fuel energy demands in rural areas and at the same time minimize environmental degradation (Dietz et al 1994). In Kenya, the fuel wood cycle study jointly carried out by the Ministry of Energy (MOE) and the Beijer Institute of Sweden (1980-82) was a direct consequence of the national desire to search for alternative sources of energy. The objective of the fuel wood cycle study was to establish the role of biomass in the total energy environment of Kenya.

The Kenya Woodfuel Development Programme (KWDP) was established by the Ministry of Energy (MOE) between 1983-1988 as a research and development project on the energy situation at household level in two pilot districts, Kakamega (of which Vihiga was then part of) and Kisii Districts (Chavangi 1984). Parts of present-day Vihiga District, specifically Bunyore (Ebusikhale) and South Maragoli (Kegoye), were selected as target areas because of their very high population density, the small land parcels there (average 0.6 ha), agro-climatological considerations, the relative importance of commercial and 'subsistence' agriculture and indices of an agricultural economy undergoing a rapid transformation (Bradley 1984, Gelder et al 1984 & Chavangi 1984). These circumstances led to the decimation of the woody biomass at a vastly accelerated rate and to an increasing demand for fuelwood and timber at the household level.

The KWDP in conjunction with extension staff from the Ministries of Energy, Agriculture and Environment and Natural Resources sensitized many households to the importance of

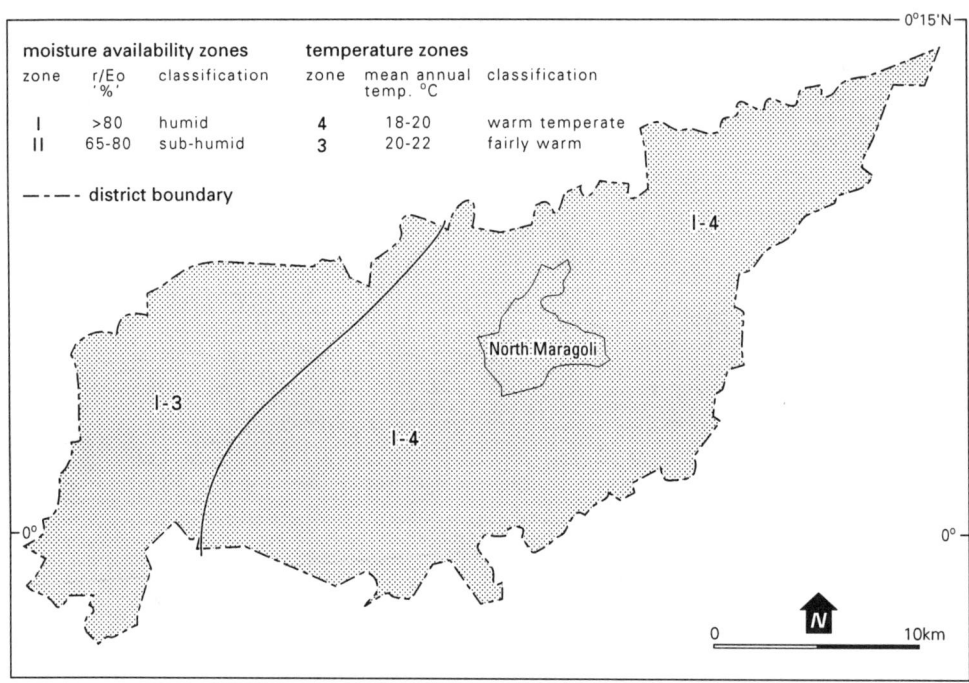

Figure 3.9
Agro-Climatic Zones Map of Vihiga District (source: Agro-climatic Zone Map of Kenya (1980)).

establishing trees and controlling environmental degradation. Farmers were provided with seedlings and were taught methods of establishing tree nurseries. This may partly explain why large parts of Vihiga District are now under eucalyptus trees. The promotion of agro-forestry has also had an influence on many farm households who subsequently began to plant fruit trees and tree species intended for fodder as well as firewood.

The Kenya Wood fuel and Agro-forestry Programme (KWAP1) continued with KWDP projects until 1993. Though farmers in Vihiga had already established their own woodlots on their farms prior to these projects, KWDP and KWAP helped to encourage farmers to increase the area under trees. However, these programmes met with several barriers for example; (i) the land tenure system, which prevented some young men being recruited to plant trees because they lacked land title; (ii) gender conflicts because women could not be easily incorporated because tree planting and ownership is culturally dominated by men. In fact, the shortage of wood fuel continues to be critical in many households as men harvest trees mainly for sale and not for domestic use. Thus the objective of alleviating wood fuel shortage has not been met. The KWDP and KWAP programmes were not prolonged because no government Ministry could be found to take responsibility for them.

3.3 The Human Resource Base

3.3.1 Population: Settlement and Distribution

Settlement

The people of Vihiga District belong to the Bantu people. The Bantu of Western Kenya consist of the Abaluhya (the Luhya), Abagusii (the Kisii) and Abakuria (the Kuria). The Bantu presence in Western Kenya stretches back in time to the Early Iron Age (Ochieng' 1990). The earliest Iron Age sites in East Africa have been discovered in the area around Lake Victoria and date from the third and fourth centuries A.D in the Winam Gulf in Nyanza. It has been pointed out that, by the tenth century A.D, most of present-day Nyanza had become a predominantly Bantu domain, after the assimilation of most of the earlier societies living in the area (the Stone Age hunter-gatherers and the Southern Cushites). Probably by the fifteenth century Bantu communities could be found from the Mara district in north-eastern Tanzania to the region around the Winam Gulf, extending northwards to the foothills of Mount Elgon (Ochieng' 1990). This is the region of present day Nyanza, Western and parts of the Rift Valley Provinces.

The period 1500 and 1850 saw the immigration of many Bantu clans and families from Eastern Uganda into Western Kenya and the emergence of the present-day Abaluhya, Abagusii and Abakuria ethnic communities (Ochieng' 1990, Were 1967). The formation of the Abaluhya entailed the incorporation of the pre-1500 inhabitants of Western Province with Bantu groups that arrived in the region in the sixteenth, seventeenth and eighteenth centuries. The Abaluhya who are the dominant group among the Western Bantu consist of about seventeen sub-tribes which include the Abalogoli (the Maragoli), Abanyole (Banyore) and Abatiriki (the Tiriki) who inhabit the Vihiga District today (Were 1967, Osogo 1966).

Other writings indicated that the earliest Bantu inhabitants arrived in Vihiga District from 1250 onwards (Mutoro 1985, Were 1967, Osogo 1966 and Barker, 1950). There is speculation that they originally came from the southern part of Eastern Uganda, having originally settled there after the long migrations of the Bantu group. This period of arrival is based on estimated genealogical length (Were, 1967). The Abalogoli, Abanyole and Abagusi descended from three brothers named Mulogoli, Munyole (Anyole) and Mugusi (Barker, 1950). The families are said to have come across Lake Victoria from Uganda in boats and were separated by a storm. Mugusi and Mulogoli landed somewhere near present-day Kendu Bay and later moved inland. Munyole (Anyole) is said to have landed near the site of present day Kisumu and to have moved and settled in the Maseno area (the western part of present-day Vihiga District) where the Abanyole people are found today. Mugusi moved on from Kendu Bay to the present-day Kisii Highlands where the Kisii people are found. On the other hand, Mulogoli and his people re-crossed the lake (present-day Winam Gulf), to central Nyanza and finally to the Vihiga area (present-day Vihiga and Sabatia Divisions) (Osogo 1966 and Barker 1950).

Other sources have it that the Maragoli and the Abanyole came from southern Africa, along the eastern shores of Lake Victoria in Tanzania (Shirati area) towards the north across the Winam Gulf to Vihiga District (Wagner, 1949). It is said that these people were both cattle herders and crop cultivators and so their movements were dictated by the search for fertile lands and plenty of rainfall. Prior to the settlement of the Maragoli and Abanyole in this area it is said that much of the land was under thick tropical (rain) forest, remnants of which can still be seen in the Kaimosi and Kakamega Forests (Osogo, 1966, Wagner 1947). As the human population has increased rapidly from 1900 onwards forest clearance has been relentless and is now virtually complete. As a result, much of the land to-day is fully settled, apart from a few restricted areas covered by forest plantations established by the government in the 1950s when environmental deterioration was identified as taking an alarming trend.

Between 1580 and 1733 there was another large-scale immigration of Bantu clans from Eastern Uganda into Western Province. Among them were the Tiriki (Ochieng' 1990). The Tiriki settled in the Kaimosi area around the same time as the Maragoli and the Abanyole. Some narrators have suggested that there was a man called Mudiriki from whom the name Tiriki comes. However, it appears that the name Tiriki was derived from the Nandi name Terik. Since Tiriki lies on the border of the Abaluhya and the Nango'ori, some of its inhabitants have borrowed much from the Nandi or are of Nandi origin (Osogo 1966, Wagner 1947, 1955).

Many writers claim that the migration and settlement of the Abaluhya was the result of a combination of factors, the most important of which being hostilities from neigbouring tribes and a search for fertile land. It is further claimed that the Abaluhya were not been able to expand any further because of hostile neighbours, and later because of colonial and post-independence land policies (Lavrijsen 1984, Were 1967, Swynnerton 1955, Wagner 1940 and 1970). During the pre-colonial period the population in this area is believed to have risen very slowly, where conditions favoured a natural increase. Present day increase is the result of an accelerated process of demographic growth and a decreasing death rate set in motion during the colonial period.

Population Size, Density and Growth
Overall population density figures in Kenya in 1989 were 37 persons per sq.km (Kenya 1994, National population census 1989). This conceals a rather uneven regional distribution. Approximately nine-tenths of the population lives in less than a quarter of the country. Densities range from a mere two persons per sq.km in the climatically hostile North-Eastern Province to a dense 307 per sq.km in the Western Province. The unevenness of population distribution and density is even more pronounced at the district, division and location level.

Vihiga District has a total population of 457,647 living on an area of 521 sq.km (Kenya, 1994). The district has the highest population density in Kenya with 878 people per sq.km. This contrasts sharply with other districts in Western Province such as Bungoma with 221, Busia 243 and Kakamega 339 with persons per sq.km. There are other districts in Kenya that have a high population density. These include Kisii with a density of 517 persons per sq.km; Kiambu 353; Muranga 340 and Kisumu 320 (Kenya 1994, National Population Census 1989). The current annual population growth rate in Vihiga is 3%. In the recent past, the district experienced very high annual population growth rates that ranged from 3.9% and 4.1% (Kenya 1979, 1989, 1994).

Nearly every part of the district is inhabited except for the rocky hills in the southern region and the Maragoli and Kaimosi Forests. The population figures provide a general indication of the scarcity of land in Vihiga District. There is some internal spatial variation with respect to population size and density within the district and, consequently, the amount of agricultural land available varies considerably (see Table 3.4).

Table 3.4
Population Size and Density per Division, Vihiga District 1989 (source: Kenya (1994), Kenya Population Census 1989).

Division	Males	Females	Total	No of hhs	Area sqkm	Den/ sqkm	Ha per capita
Sabatia	50,105	58,752	108,857	19,264	110	990	0.10
Hamisi*	57,593	64,653	122,246	21,374	154	794	0.13
Vihiga	34,960	40,748	75,708	14,094	88	860	0.12
Emuhaya/ Luanda	69,476	81,360	150,836	30,263	169	893	0.11
Total	212,134	245,5134	57,647	84,995	521	876	0.11

* Hamisi is currently called Tiriki

The age structure of the district is similar to that of the population of Kenya as a whole (see Figures 3.10 and 3.11). About 53% of the population is below 15 years of age while those who are 60 years old and older make up 4% of the population. Therefore, 57% of the population is dependent and the dependency ratio is 1:1.3. Sex ratios of specific age categories however, reveal significant difference between the pattern for Kenya as a whole and the Vihiga District. In 1979, in the age category 15-49 years there was a sex-ratio of 105 females to every 100 males, whereas Vihiga's sex-ratio for the same age group in 1979 was 134:100. In 1989, the population pyramids by sex and age of Kenya and Vihiga District show sex ratios in the category 15 to 59 years as 104:100 and 126:100 respectively. Vihiga District has one of the highest sex-ratios in the country. The major reason for this unbalanced ratio is that more men than women migrate from the district.

The population in Vihiga District has grown steadily since 1962 when data first became available (see Figure 3.12). Population growth rates of 3.3% 1.7% and 2.7% were recorded between 1962 and 1969; 1969 and 1979; and 1979 and 1989 respectively.

Whilst the population of Vihiga District grew each year, the amount of land available per capita declined. For Vihiga District as a whole it shrank from 0.23 ha/cap in 1962, to 0.20 in 1969 and in 1979 it was 0.15 ha/cap. In 1989 it stood at its current level of 0.11 ha/cap. This low average figure of 0.11 ha/capita is also found in our study area, North Maragoli, as shown in Table 3.5. The figures showing land per capita give a better overview of the extent to which the agricultural population puts pressure on the land than the statistics indicating a simple density per sq.km do. When population figures are compared to farm size, a gloomy picture emerges as far as agricultural production and household food security are concerned, especially if account is taken of present farm management and agricultural production levels in the district. (see analysis in the following chapters).

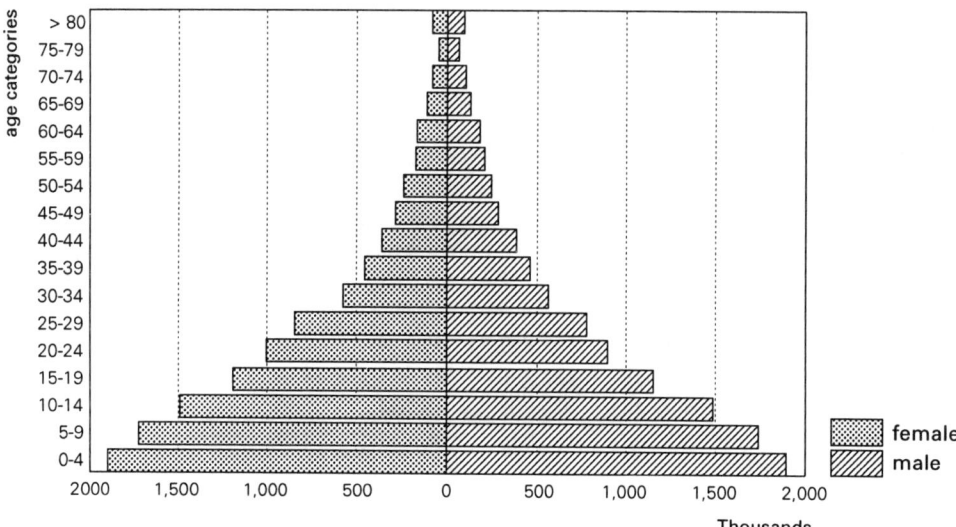

Figure 3.10
Kenya Population Pyramid by Sex and Age (source: Kenya (1994) Kenya Population Census, 1989).

Population structure

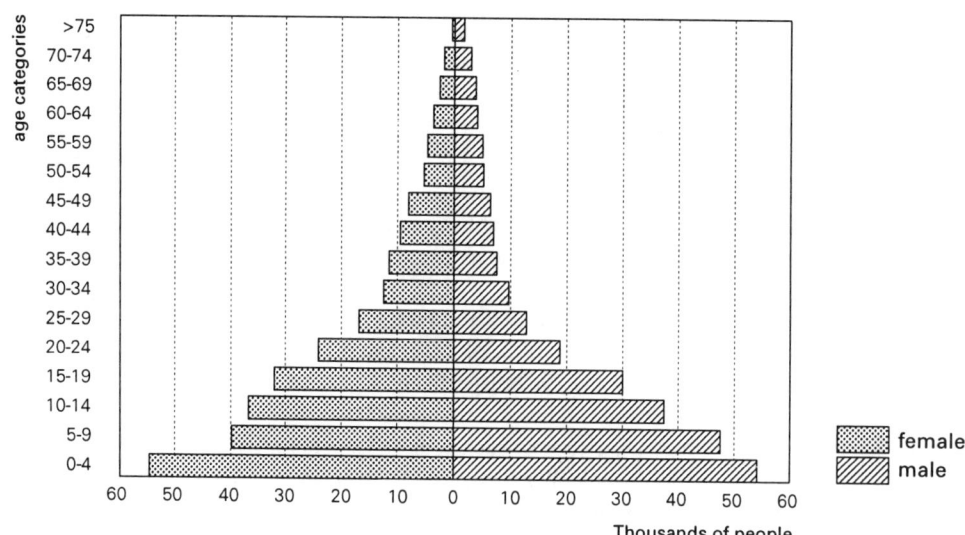

Figure 3.11
Vihiga Population Pyramid by Sex and Age (source: Kenya (1994) Vihiga District Development Plan 1994-1996).

Most of the district's population is rural and only a comparatively few people live in urban and market centres such as Majengo, Luanda, Vihiga, Mbale (the district headquarters), and Chavakali (see Table 3.6). The urban population represents a mere 2% of the total population of

Vihiga District: Background Information

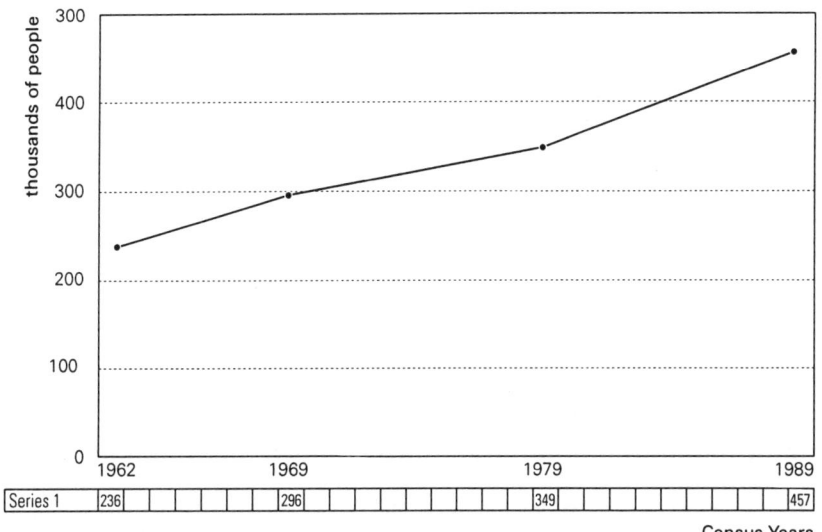

Figure 3.12
Vihiga District Population growth 1962-1989 (source: Population Censuses 1962, 1969, 1979 and 1989).

Table 3.5
The Population of North Maragoli in 1989 (source: Kenya (1994), Kenya Population Census 1989).

Sub-location	Male	Female	Total	Number of hhs	Area in sq.km	Density sq.km	Hectare capita
Mudete	2666	3200	5866	1036	6	978	0.10
Kigama	2686	3125	5811	972	7	830	0.12
Mambai	2026	2521	4547	819	5	909	0.12
Gaigedi	1352	1568	2920	492	4	730	0.14
Gavudia	1593	1847	3440	602	4	860	0.12
Vokoli	1497	1804	3301	565	4	825	0.12
Kivagala	2058	2397	4455	771	6	743	0.13
Losengeli	1809	2087	3896	665	3	1299	0.07
Total	15,687	18,549	34,236	5,922	39	878	0.11

the district. A striking characteristic of the Vihiga urban population is the fact that, contrary to the Kenyan situation, the female population is larger than the male population.

The people of Vihiga are hemmed in by tribal borders: the Nandi to the east and the Luo to the west and south, and other Abaluhya to the north. The people of Vihiga District form a nucleus of a heavily populated region, with no contiguous area into which they can expand unopposed. Any significant resettlement that has occurred in recent times, alleviating the pressure of people on land

Table 3.6
The Population of Urban Centres in Vihiga District (source: Kenya (1994), Kenya Population Census 1989).

Urban centre	Males	Females	Total	Sex-ratio F/M	No hhs 1989
Majengo	2,000	2,335	4,335	1.2	855
Luanda	1,613	1,748	3,361	1.1	617
Mbale Mkt	1,320	1,551	2,871	1.2	565
Chavakali Mkt	186	234	420	1.3	125
Total	5,119	5,868	10,987	1.1	2,162

has been to far away places, such as South Nyanza and, under government auspices, to the settlement schemes eighty kilometres or more to the north of Vihiga (Moock 1973). More recently, populations have been moved from Mbale, Mbihi, Mudete and Vokoli to parts of Nandi,Tiriki and the settlement scheme areas to create room for the construction of district offices, an airstrip, the proposed tea factory and the expansion of Moi Vokoli secondary school and the Vokoli mission.
Labour migration is perceived by most of Vihiga's population as the solution to land shortage. However, severing ties with the land of one's fathers is a decision that is not taken lightly by any Luhya. In this way even those who have purchased farms in the settlement scheme areas often retain their tiny holdings in Vihiga so that, when they die, they can be buried at home.

3.3.2 Out-migration and Wage Employment Opportunities

Vihiga District was already experiencing population pressure in the late nineteenth century. Thomas remarked on the intensity of land cultivation while travelling through the district in 1883 (Thomas, 1885).

> "What surprised me most was the surprising number of villages and the generally contented and well-to-do air of the inhabitants. Almost every foot of ground was under cultivation... the people seem to have some idea of the rotation of crops, for they allow land to lie fallow occasionally, such parts being used as pasture ground for the cattle and flocks " (Thomas 1885)

Vihiga District, like many other regions in Western Province has long served as one of Kenya's major labour reservoirs, economic necessity driving adult males to seek employment opportunities outside their over-populated homeland (Moock, 1976). "The history of wage employment dates back to the beginning of the twentieth century when external forces (political) combined with internal pressures drove large numbers of males into the colonial labour market. This was the time when European farmers and businessmen were rapidly settling in Kenya. As their farming estates and new companies grew, they required greater inputs of local labour. Thus when a joint hut and poll tax was introduced in 1910, the same year in which a rinderpest epidemic killed 50% of the cattle in North Kavirondo District (as Western Province was then called), the stage was set for massive waves of labour migration among the people of Vihiga as a means of easing their precarious economic situation" (Moock, 1976:46-60). Also the restriction of Africans to the "Reserve areas" led to growing population density, the degradation of the soil, poverty and a considerable reduction in the size of farms. These factors combined to push males from this area, forcing them to migrate to ease the pressure on land and to find a way of supporting their families with remittances. Another contributory factor stimulating migration was the occurrence of several acute famines in 1906, 1907 and 1910 (Sangree 1966).

The most wage labour opportunities at the beginning of the twentieth century lay in the categories of unskilled and semi-skilled labour and there was a demand for those willing to work as agricultural labourers and domestic servants. Christianity came early to Vihiga. The Friends African Mission was established at Kaimosi in 1902 and about the same time the Church Missionary Society arrived at Maseno (Kima). With the churches came education (Moock 1973 and Moock 1976). The people of Vihiga, especially the Maragoli, responded quickly to literacy and the chance to learn trade skills that the missionaries offered alongside theology. Many quickly qualified for more skilled work and, as they gradually began to fill positions such as office clerks, estate headmen, and factory supervisors, accelerated the migration of adult males. With a head-start both in relation to other Kenyans and the Luyia themselves, as far as the unskilled and intermediate job categories were concerned, the Vihiga people have been able to maintain their advantage up to the present time.

Throughout the 1920s and 1930s, the European settler demand for labour to work their plantations and their industries increased at a far greater rate than the supply. Tax payments in money soon replaced those paid in kind as the labour market expanded. It seems that from its inception, the Vihiga people have looked upon labour migration as more of a necessity than a free choice. Although population pressure made this people more prepared to migrate and earn wages, they too originally entered paid employment as 'target income workers' to finance limited needs (Moock, 1976). However, it was often less a question of a limited desire for goods, than a low wage level that failed to attract large numbers of labourers or make long periods of work away from home worthwhile for many people.

"After almost a century of exporting labour across the country, people from Vihiga have been completely incorporated into the cash economy. Entry into the labour market on the basis of limited cash needs has ballooned into dependency upon monetary income for actual survival for both the migrant and his rural family. When labour migration first began, the major cash needs of the Maragoli were taxes, clothing along with assorted manufactured items, and cattle and cash for bridewealth. But now, the payment of bridewealth is only one of the numerous costly basics faced by these migrants. Of importance in many homes is the purchase of food for the rural households as many farms have become too small to support rural families. For many years the people of Vihiga have not been able to sustain their growing population on agriculture alone. Given small land holdings, even profits from cash crops cannot provide sufficient cash to stretch much beyond family consumption needs let alone cover school fees, maintain the upkeep of the home, pay fares, and other extended obligations and investments. Therefore, dire economic need has long acted as both a necessary and sufficient propellant for the massive movement across the gap between small farm income and non-agricultural wage employment." (Moock 1976:58-60).

After Independence, investment in education became even more attractive than it had been in the past (Moock 1976). Many parents in Vihiga were and are willing to make many sacrifices to enable their children to finish school. It is a common saying among the Vihiga people that a good and successful education is the best inheritance parents can give their children because education may help them secure well-paying jobs enabling them to purchase more land for themselves. In this way they can augment the tiny pieces they have inherited from their parents and give them a way of supporting their rural parents financially.

It has been argued, rightly, that one of the effects of rural - urban migration is the return of money and resources by the migrants to their respective home areas. Such remittances are assumed to be a significant way of removing the supply constraints that impede improvements to agricultural productivity. It is noticeable in Vihiga that some smallholders tend to look upon agricultural activity as merely a way of supplementing alternative opportunities for employment. The community's economic life-blood seems to consist, in fact, of a flow of labour to town centres across Kenya. Agriculture is stigmatized and labelled as an activity for the illiterate and particularly women and on some farms it is performed to fulfill a social function inherited from the forefathers. This explains why most of the remittances are channelled into investing in education, and why agriculture is the last to benefit from this source of funds. Even though remittances have been an important source of income for many rural households in this area, the current economic hardships, recession and inflation brought about by the inter alia structural adjustment policies have made it difficult for many migrants to send remittances home regularly and the amounts sent have become increasingly inadequate.

"Therefore, the mere existence of substantial urban to rural financial flows cannot serve as evidence that remittances are flowing into investment in agriculture. Farmers are diverting funds which would normally have been used for farm improvements into investments in formal education for family members" (Moock, 1973:306-7; see also Guyer 1972). The proportion of urban income remitted varies directly with the strength of the social and economic ties to the rural area and inversely with how well migrants are established in urban areas, the latter being measured in terms of the amount of urban income earned, the length of urban stay and whether the migrant is accompanied by members of his immediate family or not (Rempel et al. 1978:333).

3.4 The Economic Situation of Vihiga District

The people of Vihiga District derive their local livelihood from primary production activities which include small-scale farming, and to some extent forestry-related activities and tertiary activities such as commerce and trade. There is hardly any industry and there are no minerals to boast of. The district lacks local economic prospects and small-scale farming is expected to provide the people with food, income, shelter and other basic needs.

3.4.1 The Agricultural Resource Base

Agriculture is the main economic activity and supports about 90% of the district's population.

Table 3.7 Production Figures for Major Crops in Vihiga District 1991/92 (source: Kenya (1994), Vihiga District Development Plan 1994-9196).

Crops	Area (ha)	% Arable land	Production (tons)	Yield in kg/ha
Tea	1,582	4	5,430	3,432 (green)
Coffee	23,090	57	14,243	617
Maize & beans	14,833	37	25,291	1,705
Beans	43	0.11	5,700	384
Bananas	38	0.09	70	1,842
Avocados	38	0.09	702	18,473
Paw Paws	30	0.07	468	15,600
Tomatoes	75	0.19	413	5,507
French Beans	316	0.78	543	1,718
Kales	430	1.06	1,389	3,230

The remaining 10% of the inhabitants derive their main livelihood from other activities such as commerce, trade, fishing, quarrying, woodwork and industry.

As we have mentioned above, the district has a high rainfall and soils that are suitable for the growing of cash and food crops. Table 3.7 displays a summary of the main crops recorded by area and production.

However, most of the above statistics presented in the Vihiga District Development Plan (1994-96) are very misleading and often quite ridiculous. For example, the population of tea growers is given as being approximately 7900 farmers, yet, field observation shows a larger number of farms under tea and more farm households involved in tea production. Tiriki, Sabatia, Vihiga and some parts of Emuhaya Divisions are mostly covered by tea, and this makes up some two-thirds of the district. Our field findings and measurements presented in Chapter Four and Five also indicate that farmers have more land under tea than coffee. Second, there was very little coffee on most of the farms as many farmers have neglected the crop. Yet according to the statistics the area under coffee (23,090 ha) is much larger than the area under tea (1,582 ha) and under maize (14,790 ha) which is the staple crop. In fact, according to the district's statistics, coffee is grown on 57% of the arable land available in the district. Third, in the District Development Plan, the area under tea is shown to be 1,582 ha and production was shown to be 5,430,427 tons a figure which would mean 3,432,634 kg/ha in Vihiga District, a return not physically possible anywhere in the world. In my text I have changed this figure to 5,430 tons to make it appear more reasonable. Four, adjustments have been made to most figures to correct obvious arithmetical mistakes.

Food Crops: Approximately 40% of the arable land in Vihiga District is under food crop production. The two most widely cultivated food crops in the district are maize and beans. On average 0.2 ha of land per household is devoted to maize production and this is sometimes intercropped with beans. All the maize produced is consumed within the district.

Vihiga is one of the districts (Machakos, Kitui, Isiolo, Kilifi, Kwale, Taita Taveta etc) that is classified as food deficient-areas, despite its location in one of the country's high agricultural potential zones. In fact, the maize crop produced in the district is only sufficient for about four months of the year at most. For the rest of the year maize is brought in from areas where maize is grown on a large scale such as parts of Kakamega, Bungoma, Nandi, Trans Nzoia and Uasin Gishu districts.

Generally, the average yields for maize are about 2,000 kg/ha against a potential of 4,500 to 6,000 kg/ha. The question is bound to arise "Why can't this district produce enough to feed itself?" After all it is located in one of the zones of high agricultural potential. Moock's study (1973) in this district showed (i) that maize production is correlated to and influenced by population pressure on agricultural land and labour migration, resulting in the absence at any given time of many heads of farm household and leaving much of the decision-making responsibility to a secondary member of the farm family, usually the wife; (ii) that the exceptional demand of parents for educational services dictates that income use is for investment in education rather than agriculture. This is a demand which shifts ever upwards in response to the rising standards of educational qualification required for off-farm wage positions.

Moock's analysis exposed a positive relationship between yield level and female management. Female-managed farms had higher maize yields than male-managed maize farms because women tend not to migrate and are more experienced farmers than men. Also farm managers with four or more years of schooling obtain generally higher yields than do managers with less, despite the fact that education in Kenya is intended, by both government and parents, principally for non-agricultural purposes (see Chapter 5.5 and 6). Finally, it is claimed that there are apparent inefficiencies in the use of certain inputs which may indicate that increased output is possible at no increase in cost, simply by an improved allocation of production factors (Moock, 1973). Management, education, extension information and availability of technology are identified as being some of the important factors that influence maize production.

Other subsistence crops are bananas, millet, sorghum, cassava, sweet potatoes, and horticultural crops. Horticultural crops such as tomatoes, French beans, avocados and bananas are sold both inside and outside the district. An explanation for the poor performance of the food crop farming sub-system will be given later.

Cash Crops

Between 50 and 60% of arable land in Vihiga is under cash crops, mainly tea and coffee. Coffee was the first cash crop introduced into this district in the late 1950s and early 1960s and many farmers converted part of their land for coffee cultivation (see Table 3.7). According to official figures, a total of 23,090 ha of land is under coffee cultivation with a production of 617 kg per ha against the minimum expected of 1,000 kg per ha (Vihiga District Development Plan 1994). Over the years, the benefits to farmers in Vihiga District derived from coffee have dwindled and been disappointing. The coffee societies are mismanaged and riddled with corruption, and as a result farmers are paid badly in comparison to other coffee-growing areas. The coffee factories which were built expensively through big loans serviced by farmers, are running at less than half their capacity. Others are not operational for technical or mechanical reasons or because they do not receive coffee beans from the farmers. On the farms, coffee is poorly managed because of low and irregular (or even absence of) payment and a lack of incentives. This means there is little investment in the crop. Many coffee plots in the district are intercropped with other crops, others are totally neglected and some are being uprooted to show farmers defiance of the government policy which states that once coffee is established it should not be uprooted.

According to the Agricultural Information Centre and the Ministry of Agriculture, tea after coffee, is Kenya's second most valuable export crop. Tea production in Kenya is the responsibility of the Kenya Tea Development Authority (K.T.D.A) and a few private companies. Tea in Kenya is grown on large estates and by small-scale farmers. Tea is grown on large estates at Kericho, Sotik, Limuru, Nandi Hills, Endebess, Subukia, Makuyu and Meru. The small-scale tea-growing areas are in Kisii, Kericho, Nandi, Kakamega, Vihiga, Kiambu, Murang'a, Nyeri, Kirinyaga, Embu and Meru districts (see Figure 3.1). All the small-scale growers are under the management of K.T.D.A. There are more than 75,000 hectares of tea in Kenya of which 40,000 hectares (53%) are cultivated by more than 250,000 smallholders with an average holdings of 0.3 hectares (Daily Nation 1994, KTDA Annual Report 1993/94). There has been a major increase from the modest 4,471 hectares and 19,775 smallholders in 1964, to 54,000 hectares of tea and 138,000 smallholders in 1980/81 (KTDA annual report, 1994; Daily Nation 1994 and Lamb & Muller 1982). The amount of green leaf produced annually by smallholders has increased from 624,583 kg in 1964, to 146,000,000 kg in the crop year 1980/81 to over 483,000,000 kgs in the year ending June 1993 (KTDA 1994, Daily Nation, and Lamb & Muler 1982). The number of smallholder tea processing factories has increased from the original three to 43, while processing capacity has risen from 3 million kgs per annum to 450 million kgs. Tea earnings have increased over the years from 1.69 shillings per kg of green leaf offered in 1964 to 15.79 shillings per kg paid in some factories in 1993. The total net payments to smallholders per annum has risen from 29 million shillings to 76 billion shillings. Smallholder tea exports were virtually non-existent before 1960 but have accounted for about one-third of Kenya annual tea exports since the mid-1970s, the remainder coming from the long-established commercially-owned tea estates. The smallholder tea programme can therefore be categorized as being extremely successful and excellent quality teas are produced.

Tea in Vihiga District was first planted in Shamakhokho Location of Hamisi (Tiriki) Division in 1958 (see Figure 3.2). From here, it spread slowly to parts of what is now Sabatia and Vihiga Divisions so that by the early 1960s the greater part of these divisions were officially gazetted tea-growing zones, with the Kakamega-Kisumu road marking the western boundary beyond which tea was not to be planted. However, tea spread to the entire district from the mid 1970s when restrictions were lifted and it is still expanding stimulated at the moment by the price of tea on the world market. The cultivation of tea is found in practically all the divisions of Vihiga District as the good soils and climatic conditions there are particularly suitable for tea production. The average yields of this cash crop range from between 2,500 to 4,500 kg/ha, well below the potential production capacity of between 6,600 to 10,000 kg/ha for this area (see Table 3.9). In fact, tea production in 1991/92 (3,432 kg/ha) was much lower (see Table 3.7). Vihiga and Kakamega Districts are reportedly still the lowest producers of tea per hectare in the whole country (Kenya 1993).

In 1990 tea farmers in Kakamega/Vihiga Districts earned a total gross income of Ksh. 46 million. However, according to the KTDA field staff in Kakamega, the tea earnings would have

Table 3.8 Land Suitable and Under Tea in Vihiga District (source: (KTDA 1992), Kakamega District Tea Annual Report, 1992).

Divisions in Vihiga District	No. of farm families	Total land suitable for tea farming -Ha	Total land under tea 1991 (ha)	25% of available land suitable for tea farming (ha)
Sabatia	14080	8700	865	2175
Tiriki	17485	14452	488	3613
Vihiga	13500	6600	193	1650
Emuhaya/Luanda no data	4400	37	1100	
Total	45065	43152	1583	8538

Table 3.9
Green Leaf Production in Vihiga District (in kgs) (source: (KTDA 1992), Kakamega District Tea Annual Report 1991).

| Divisions in Vihiga District | Hectarage 1991-ha | Growers 1991 | Green Leaf Production of Tea in Kg | | | Ave. per ha 1991 |
			1988/89	1989/90	1990/91	
Sabatia	865	4098	2,640,700	3,764,071	3,904,210	4,514
Tiriki	488	2449	1,457,599	1,939,199	2,063,400	4,228
Vihiga	230	1429	373,089	539,376	658,220	2,862
Total	1583	7976	4,471,388	6,242,646	6,625,830	4,186

been three times higher from the same hectarage if farmers had to intensify their production and improved their management. Massive untapped tea growing potential is available. Only about 4% of the area suitable for tea farming in Vihiga District is actually under tea (see Table 3.8). The land that is suitable for tea is either being cultivated for coffee or for food crops. If 25% of the area suitable for tea farming were brought into production, then the 8,538 ha in Vihiga District will comfortably sustain about four factories at the current low production unit level (personal communication with resource persons).

Not only average yields of tea but also yields of green leaf per hectare are poor in both Kakamega and Vihiga Districts. Low unit productivity generally implies the less the intensive use of land, low return per unit resource, low total production and low income. Table 3.9 depicts the production in the last few years and yields per hectare for the various tea-growing divisions in the district in 1991.

Table 3.10
Comparison of Green Leaf Tea Production of Vihiga and Other Tea-Growing Districts in Kenya (source: (KTDA 1992), Kakamega District Tea Annual Report, 1992).

District	Production Kg Green Leaf per Ha	Would be Total Production for Vihiga District
Embu	7000	11,074,347
Meru	4794	7,585,095
Kirinvaga	10496	16,605,666
Kiambu	6700	10,552,557
Murang'a	7682	12,215,353
Kericho	6390	10,110,032

Table 3.11
Zebu And Grade Cattle Population 1993 (source: (Vihiga 1994), Vihiga District Livestock Annual Report, 1994).

	Zebu			Grade			
Division	1989	1993	% Change	1989	1993	% Change	Tot. % Change
Vihiga/ Sabatia	30,000	42,950	43	2,500	2,970	19	41.2
Hamisi/ Tiriki	30,000	22,017	-26.6	1,500	1,540	2.6	-25
Emuhaya/ Luanda	28,000	27,700	-1	400	930	132	0.8
Total	88,000	92,667	5.3	4,400	5,440	23.6	6.2

National average production in 1990/91 was approximately 6,000 kg green leaf per hectare. If the hectarage under production in Vihiga were to approximate the national average then about 9,498,000 kg of green leaf would be produced there, i.e one-third more than the actual production. A further look at the performances of other small-scale tea-growing districts further demonstrates the poor production score of this crop in Vihiga as displayed in Table 3.10. The third column compares what would be the total green leaf production if Vihiga District had the same green leaf production per hectare as the other smallholder tea-growing districts in Kenya.

Table 3.12
Livestock Production in Vihiga District (source: (Kenya 1994), Vihiga District Development Plan 1994-96).

Division	Livestock reared	% hh with HV cattle	HHs/sq.km	Number of hhs
Vihiga	Cattle, pigs poultry	5	160	14,094
Sabatia	Cattle, pigs poultry	8.3	175	19,264
Emuhaya/ Luanda	Cattle, pigs poultry	1.9	179	30,263
Tiriki	Cattle, pigs poultry	8.4	39	21,374
Total	Cattle, pigs	5.7	163	84,995

HV = High Variety Cattle
HHs stands for households per sq.km

Tea, coffee and French beans (horticultural crops) appear to be the main crops upon which many farmers depend for their cash incomes. Horticultural farming is growing fast, stimulated by the existence of the large markets of Kisumu, Kakamega, Kapsabet and Eldoret which are close by and the integration of women in horticultural projects.

Livestock

Livestock farming is a tradition among most people of Kenya and also among the Luhya people of Western Province. It plays an important role in cultural and socio-economic life in many communities. Livestock is sold when cash money is required for family obligations and it is very important for dowry and other traditional functions/ceremonies such as funerals and weddings. However, due to population pressure and shrinking farm size many farmers have had to reduce their herds. On average, each farm household in Vihiga keeps two local zebu cows for milk production. The national policy on livestock states that farmers should be encouraged to abandon the traditional zebu cattle and adopt the improved cross-breeds or pure dairy grade cattle if they can afford to do so because improved cattle require less surface area and produce more milk than the local zebu breeds.

Table 3.12, shows that only one-twentieth of the farmers keep high value cattle (grade or cross-breed cattle). The district is deficient in livestock for slaughter and milk production. Cattle for slaughter and milk are brought in from neighbouring Nandi District. At the time of writing no accurate and reliable data was available on milk production and on other livestock populations as no census had been carried out at district level.

3.4.2 Other Primary Production Activities

Apart from agriculture, the mining of sand, clay, stones for building and rocks for ballast are other primary production activities. Pyrite, an important source of sulfuric acid, is found in large deposits in the Kaimosi area. Quartz crystals occur to the north-east of Kaimosi. Granite is found in abundance to the south of Tiriki and possesses a high potential for quarry development. At present ballast and building stones are obtained but on small-scale. All these provide a livelihood for a small proportion of the population. Brick-making is also an important activity in the district as a whole. The Lake Basin Development Authority has established a brick-making centre at Solongo where local people are taught how to make cheap and good quality bricks which they can sell or use their own construction work.

Fishing is carried out on a small-scale. There are 300 fish farmers with 500 fish ponds. The Lake Basin Development Authority promotes fish farming by selling fingerlings to farmers at a subsidized rate.

3.4.3 Commerce, Trade, Industry and Services

Between 6 to 10% of the adult population are engaged in non-primary production activities. Many of these activities are located in the market centres and include the manufacture of food (maize milling and butchery), beverages (beer-brewing), fibre products, wood and forestry products (carpentry, saw-milling), pottery and metal products; construction; wholesale and retail trading; repair and various commercial services (bars, tea houses, garages) (see Table 3.13). The only important industry is the Papaya Wine Industry located in Gambogi area (South Maragoli) which has helped to enhance pawpaw growing in the district.

The labour force in the district was estimated to be 209,000 in 1993 and was expected to rise to 228,000 in 1996. This implies a growth rate of 2.9% during the plan period. Since indus-

Table 3.13
Distribution of Business Enterprises by Division, 1993 (source: (Kenya 1994), Vihiga District Development Plan 1994-96).

Division	Whole sale	Catering	Garages	Petty Trade	Jua Kali	Total
Emuhaya/ Luanda	18	23	1	281	40	365
Sabatia	20	14	-	248	48	430
Vihiga	9	25	3	295	43	375
Tiriki	14	45	2	416	56	533
	61	107	6	1,340	187	1,703

Jua Kali[11]

trialization is a almost non-existent, agriculture has to absorb most of the labour force. Table 3.14 shows the number of people employed in the various economic sectors. Agricultural employment involves the small-scale farm households who work on their farms. The rural self-employed are those who engage in either agricultural and non-agricultural related activities including timber-sawing, brick-making and charcoal burning, and these activities may not necessarily be located on the individual's farm.

The public sector provides work for government employees such as teachers, government administers, health workers, civil servants, and those working for parastatal bodies. It is projected that the public sector will expand four times the present number of employees because of the growth taking place in government institutions, services and facilities in this new district. The private sector consists of the salaried, formal private sector employees such as banks, NGOs and factory workers. The urban self-employed are those who work within the informal sector mainly in the Jua Kali sector. This sector is characterized by low wages, high job risks, instability of jobs, use of crude or cheap technology and tools (for example transport, carpentry, car repairs, petrol attendants, hawkers, vendors, tin smiths and those working in saloons).

The few urban, rural and market centres that exist in the district offer some public and private sector wage employment and work to the urban self employed. These centres have the potential to employ more (the growth rate of urban employment is estimated at 5.6% at the moment). Agriculture offers seasonal employment mostly during the seasons of long and short rains. During the dry season unemployment is common. Unemployment is persistent among the school leavers due to the unpopularity of agricultural work. Permanent non-agricultural employment in the district is very limited because of the small size of the public and private sectors.

Table 3.14
Employment Profile Vihiga District in 1994 (source: (Kenya 1994), Vihiga District Development Plan 1994-96).

Sector	1993	1994	1996- Projection
Labour on small farms	200,327	205,946	213,345
Rural self-employed	1,400	1,441	1,525
Salaried employment			
Public sector	1,639	1,736	6,692
Private sector	1,194	1,228	1,301
Urban self-employed	4,751	5,030	5,188

3.5 Physical and Social Infrastructure

The growth and development of urban and market centres and agricultural production in the rural areas largely depends on the number as well as the quality of the public services and whether they stimulate greater economic activity. The major public services are transport and communication, water supply and sanitation, energy and power, health facilities, education, banking and finance, cattle dips and the cooperatives.

Transport and Communication
Transport is a major factor in agricultural and industrial production, in the exploitation of natural resources, in the marketing of products and in facilitating socio-economic mobility. Therefore, transport influences the efficiency of services and enterprises considerably and it facilitates the implementation of policies in the health, education and cultural sectors. As the Development Plan (1994) points out, an adequate supply of efficient, safe and affordable transport is critical to increased productivity in all sectors of the economy and for sustainable development. Thus, for improvements in the agricultural, industrial and to ensure tertiary sectors and the accessibility of all parts of the district, all weather roads are necessary.

The district is fairly well opened-up. It is situated close to the port and railhead of Kisumu and it enjoys good external road connections to the major urban centres of the western region of Kenya, i.e Kisumu, Kakamega, Kapsabet, Eldoret, Bungoma and Busia. Vihiga District has a better road network as compared to other districts in Western Province. The road network serves all parts of the district but the total length of classified roads reflects an unbalanced and inadequate coverage (see Figure 3.13). There is approximately 550 kms of classified and tarmac roads (Vihiga 186 kms, Sabatia 32 kms, Tiriki 215 kms and Emuhaya/Luanda 117 kms), and about 200 kms of unclassified roads (Vihiga District Development Plan 1994 and Kakamega District Development Plan 1989).

The most important road is the Class A national road which runs through the centre of the district, connecting Vihiga with the towns of Kisumu and Kakamega. Other tarmac roads are the Chavakali to Kapsabet and Eldoret road in the north-eastern part of the district, and the road from Majengo to Luanda, Luanda to Kisumu and Luanda to Busia in the western side of the district. From these tarmac roads branch many murram roads (secondary rural access roads that are gravelled and have improved bridges and culverts, but are not tarmaced) (see Figure 3.13). In addition to tarmac roads, there is a good network of earth roads (feeder roads), connecting market centres in various parts of the district. However, during the rainy season these earth roads deteriorate rapidly and often vehicles have great difficulty in using them. In this way the collection of agricultural produce, tea in particular and the delivery of farm inputs is impeded. Generally, bridges are in good condition and no area becomes completely inaccessible (Utrecht 1989, North Maragoli Locational Profile). All the roads are heavily utilized throughout the year. The Kakamega-Kisumu road was re-carpeted in 1993-94 and therefore allows for smooth driving. However, a heavy traffic flow of buses, lorries, trucks/trailers and 'matatus' (taxis) compete for the narrow road. The roads between Majengo-Vihiga and Luanda, and from Luanda to Busia are in very poor condition. The tarmac has been washed away in many sections and there are big potholes in others.

The western part of Vihiga District is served by a rail connection that runs from Kisumu to Luanda and all the way to Butere in Kakamega District. The rail line provides both passenger and goods services to the populations in the part of this district.

The government has moved a section of the population from South Maragoli (Mbihi sub-location) to create space for the construction of an airstrip. The airstrip will be mainly for small passenger planes. Feasibility studies have been done but construction depends on funds becoming available.

Telephone services are modernized to Standard Truck Call Dial (STD), and one can make international calls directly without the assistance of post office personnel. Most of the urban centres and market centres are served by both telephone and postal facilities. However, the district is still connected to the Kakamega District Code number for telephone purposes. The district has about 10 sub-post offices and one main post office at Mbale. The quality of the postal services is but modest.

Water Supply and Sanitation
Water has a crucial role domestically, and in agriculture and industry. Water supply influences the pattern of human settlement and the distribution of economic activities. Since this district receives adequate rainfall well distributed throughout the year, one would be tempted to assume that water supply presents no problem. The district is served by six water supply projects, namely Mbale, Sosian, Hamisi, Kaimosi, Vihiga and Maseno. The water supply projects mainly serve the urban and market centres while most rural households still have to rely on river and stream water. The water schemes themselves are overutilized and are unable to supply enough water to meet the needs the growing population. Water flow is interrupted by the frequent pump break-downs, the lack of diesel, and the disorganization and mismanagement of the water supply projects. Many residents complained that they are often supplied with untreated water and many of the urban and market centres in Vihiga District experience dry taps from time to time.

The poor operation of the water supply system has resulted in a lack or shortage of water in many of the educational and health institutions in the district. Some institutions in fact are not even connected to the water supply lines. Consequently, institutions (health and education; pupils, students) have to fetch and use untreated river and stream water thus exposing a large population to the danger of waterborne diseases including intestinal diseases, skin diseases, cholera, typhoid, diarrhoea and amoebic dysentery.
In order to alleviate the existing water supply problems, the Kenya - Finland Corporation (KEFINCO) has been drilling boreholes near homes, schools, and health centres. An adequate water supply would allow the irrigation of horticultural crops during the dry season and a faster development of the various town and market centres. It is common practice for eating places to close down when the water supply situation becomes very bad, and investors are detracted from businesses that require a plentiful supply of water. The government's objective is to develop and distribute sufficient safe and clean water to all households within a reasonable distance. A clear implementable policy direction is therefore required if the population, especially institutions in the rural areas are to enjoy treated piped water by the year 2040 as government objectives imply in the 1994 National Development Plan. This will require serious organizational and financial management.

The sanitation situation is not yet well organized. Many of the small urban centres are not connected to a proper sewerage system. In fact, pit latrines are used in all urban centres. The rural population are conscious of the benefits of good sanitation and more than 70% of the homesteads have a pit latrine and a shed for bathing. Sewage is disposed of in pit latrines both in rural and urban areas. These do not conform to any construction standards and therefore there is nothing that protects against sewage seepage into ground water. The few coffee processing factories in the area pump their waste directly into streams and rivers, further increasing water pollution and contributing to the poor quality of the water available to the population in Vihiga District. However, this situation is expected to improve when a large sewerage system is constructed in Vihiga Municipality.

Energy and Power

Energy is a vital input for economic development. Hydro-electric power is one of the primary sources of energy apart from wood fuel and petroleum. Electrical power generation in Kenya has been developed with an extensive national grid network covering the major population centres in the country. In addition, the Rural Electrification Programme, established in 1974, has also been growing steadily and supplies electricity to consumers at a government subsidized price. Many rural households and institutions situated inland are expected to obtain electricity through the Rural Electrification Programme.

Vihiga District is covered by these two supply networks. The district and its urban and market centres are fairly well served by the electricity grid. The electricity grid follows the road network connecting Chavakali, Mbale and Majengo on the Kisumu-Kakamega road. Another grid supplies Kaimosi, Shamakhokho and Sabatia on the Kapsabet-Eldoret road. Shamokhokho is connected to Serem, Hamisi and other centres in Tiriki Division and the Majengo line runs through Vihiga to Luanda. Surprisingly, many of the residents and even some commercial premises in these centres are not connected to the electricity lines because the houses are not wired. Households living along the main road and who can afford it have been able to get electricity connected for domestic use. Most educational and health institutions are not supplied with electricity, in fact, the percentage of the population using electricity is very low (approximately 5% of the total population). Those that cannot afford electricity use generators, lanterns (especially the case in institutions), oil stoves, lamps or improvised tin lamps (in the villages). The majority of the population (approximately 90%) in Vihiga District still relies on wood fuel - mainly firewood and charcoal- and paraffin for both cooking and lighting since electricity is very expensive. The consequence of this is the relentless deforestation in the district and it is spreading to Nandi District where firewood and charcoal are being processed on a commercial basis and transported to market centres in Vihiga District. Though this may be happening on a small-scale at the moment, if it is left unchecked it may cause considerable environmental degradation in Nandi in the future.

Within the agricultural sector, the coffee processing factories are all connected to power lines (electricity). Many of the tea buying centres are not supplied with electricity, especially those in North Maragoli which means that these have to close down by 5.00 p.m. The supply of electricity has enabled the installation of electric grinding mills all over the district and this eases congestion.

Health Facilities
Since Independence health services in Kenya have largely been provided by the government through the Ministry of Health and the local authorities, with considerable contributions from the private sector and voluntary agencies. These agencies include church missions, industrial health units, private institutions, individuals (medical doctors, clinical officers and nurses in private practice) and NGOs.

The health service is organized by the government and includes mobile clinics (specially organized once in a while by district health staff and dealing mainly with immunization, child care, family planning, nutrition education and primary health care); dispensaries; health centres; and hospitals. Whilst other districts have a government district hospital, Vihiga does not have. However, there are two mission hospitals in the district; the Kaimosi Mission Hospital and Kima Mission Hospital. Kaimosi and Kima hospitals have a 150 and 120 bed capacity respectively giving a ratio of 1 bed to 1,695 people (Vihiga Development Plan, 1994-1996). Kaimosi hospital also trains enrolled (community) nurses (lowest category of nurses). These two mission hospitals are private and patients have to pay for consultation and medical attention. The government is currently building a district hospital at Mbale while a private nursing home is being build at Majengo.

The district has twelve health centres and three dispensaries operated by government. The distribution of health facilities in the district is uneven with a large concentration of services in Tiriki Division (see Figure 3.14). There are only a few private clinics and these operate in the urban and market centres. It is anticipated that the number of these clinics will increase now that Vihiga has been made a district.

An evaluation of the quantity and quality of health facilities shows that much has still to be done. The district has about five doctors to serve 457,647 inhabitants (1 doctor to 91,529 persons) and about one health centre to 55,000 people. Each health centre is staffed by one clinical officer and a few nurses. There is a great shortage of qualified medical staff in the district which results in congestion, long queues and long waiting times for patients.

The mission hospitals and the health facilities are overcrowded. Many of the health centres and dispensaries lack electricity and this poses a problem for the treatment of patients and the general running of these centres. Water is a day-to-day problem. Even when centres have piped water, they are frequently faced with dry taps; this is especially the case in the health centres at Sabatia, Mbale, Vihiga, Hamisi and Luanda. The lack of electricity and water explains why these centres are not much used.

At Independence it was a proclaimed government policy that medical care would be provided free because more than three-quarters of the Kenyan population could not afford to pay for health care. But with the increase in the country's population the demand for health facilities grew rapidly and the government found it increasingly difficult to run and maintain these facilities. Therefore, the services offered by the government health institutions have deteriorated over the years and the perpetual shortage of drugs, materials, equipment and staff are well-known and nagging obstacles to medical care in Kenya. Frequently, patients in Vihiga wait in desperation at the health centres where there may not even be paper on which to prescribe treatment. The situation worsens when the only clinical officer is absent and all the work is left to the nurses who may not be as qualified. Recently, the situation in government health institutions has deteriorated further because the government has cut down on the budget allocated to medical care among others, which was one of the conditions of the structural adjustment programmes (National Development Plan 1994). This will encourage private practice, generating local business and employment. However, for rural people who lack the money to consult private practitioners it means they will have to rely on traditional medicines, consult witch doctors and traditional medicine men/women or simply suffer without treatment.

The infant mortality rate has been improving over the years as a large part of the population receives family education, medical care and better nutrition. The child mortality rate is 98:1000 which is above the national average of 87:1000 (Vihiga District Development Plan 1994-1996). It is expected to decrease further because of increasing literacy among mothers and the provision of improved health facilities. The major causes of child mortality in the district have been identified as intestinal diseases, malaria, respiratory diseases and water-born diseases (diarrhoea, typhoid, dysentery) (Vihiga District Development Plan 1994-1996). Among children malnutrition has been identified in many parts of the district and is especially prevalent. This is caused by poor and unbalanced diets which cause **kwashiorkor** and **marasmus**.

The most prevalent diseases in the district are malaria, respiratory track infections, skin diseases, pneumonia, ear/eye infections and the diseases of old age such as high blood pressure and arthritis. In 1992, 74,236 cases of malaria and 52,313 cases of respiratory track infection were reported (Vihiga District Development Plan 1994-1996). However, it is believed that the real figures for the number of people afflicted are larger because many people do not report for treatment when infected. Diarrhoea which is a common killer among children, is attributed to drinking polluted and untreated water. The high rainfall in the district and the infestation of open water with malaria-carrying mosquitos contribute to the high incidence of malaria, while overcrowding in the houses and at funeral gatherings spreads respiratory diseases (Vihiga District Development Plan, 1994). Many agricultural production hours are lost due to the high incidence of these diseases.

Though about 80% of the district's population is aware of family planning (use of contraceptives such as pills, condoms, Intra-Uterine Device (IUD), injectables) , the acceptance rate is only 25% (Vihiga District Development Plan, 1994). The low acceptance rate is attributed to the negative attitudes of both men and women. Men discourage their wives from using contraceptives and women fear the side effects of contraceptive methods. In the villages many people believe that using contraceptives result in impotence, and increases unfaithfulness and prostitution. The district has 15 family planning service delivery points offering modern contraceptive methods. Because of the high birth rate and rapid population growth the government has established these family planning units to educate the population and encourages reductions in family size. However, as observed earlier, the majority of people in the rural areas of Vihiga has not accepted these methods. Maybe the communication (transmission) techniques used to deliver these services have not been effective and have not appealed to the local people.

Education

The government's aim is to promote literacy among as many Kenyans as possible. Employment in Kenya is pegged to one's standard of education. As already mentioned, making sure the youth of Vihiga District is educated is seen as a necessary investment by every household since education paves the way to wage employment. As Moock rightly points out, education no longer guarantees a salaried job but the possible 'payoffs' are so great that the investment remains attractive despite the high risk of unemployment (Moock 1976:24). A large proportion of the household income is therefore invested in education.

The district has 417 pre-primary schools, 332 primary schools, 72 secondary schools, 12 youth polytechnics, and one teacher training college (see Table 3.15). Many of the pre-primary and secondary schools are private and receive only partial assistance from the government. Ten of the youth polytechnics are government-assisted, and two are fully private. The only big government training institution is Kaimosi Teachers College where primary school teachers are trained. In North Maragoli, our research area, there are 30 primary schools and 7 secondary schools.

Although it would appear that there is a concentration of schools in Vihiga District, they are inadequate. This can be judged from the ratio between students and schools, and teachers and students. The schools and classes are overcrowded with a student population of more than 400 children per school and a minimum class enrolment of 50. The teacher - student ratios differ sharply between schools, and larger schools have a high ratio. The percentage of inadequately trained teachers is quite high and accounts for about a third of all teachers.

Table 3.15
Distribution of Educational Institutions in Vihiga District (source: Modified from Vihiga District development Plan 1994-1996).

Division	No. of pre-primary schools	No. of primary schools	No. of secondary schools	Youth Polytechnics	Teachers training college
Emuhaya	113	89	18	4	-
Sabatia	118	90	21	1	-
Tiriki	117	92	19	5	1
Vihiga	69	61	14	2	-
Total	417	332	72	12	1
Ratios=totals school/pop teacher/pupil	417:109,058 1:262 1:130	332:162,384 1:489 1:50	72:62,599 1:869 1:60		

Both primary and secondary schools are evenly spread over the district. High population density has meant that the average distance to primary school is around one kilometre and secondary school children have to travel less than two kilometres. The problems experienced in many schools include (a) insufficient classrooms and workshops, (b) uncemented floors, lack of shutters and not enough desks, (c) lack of teaching materials, books and equipment, (d) shortage of teachers, (e) scarcity of land for further expansion, and (f) lack of recurrent finance. These problems have resulted in poor performance at school. Schools in the Vihiga District were among the ones that performed the worst in the primary school examinations of 1994.

Banking and Finance
The district has two banks: the Kenya Commercial Bank and Barclays Bank of Kenya. The former operates branches at Mbale and Luanda, while the latter has one branch at Mbale. The Kenya Commercial Bank has mobile banking services which serve Majengo, Cheptulu and Serem. Sabatia and Tiriki Divisions are underserved and there is a need both for more branches and to encourage farmers to open bank accounts.
The district does not have institutions which provide business loans and hire-purchase facilities. These services are offered from branches in Kakamega and Kisumu. The Kenya Industrial Estates and the District Joint Loans Board that are responsible for making small-scale loans to businesses and for industrial development are not yet operational in Vihiga though their services are urgently required to spur development (Vihiga District Development Plan, 1994).

District Urban Services
Vihiga Municipal Council located at Vihiga has electricity and a fairly good road transport system. Telephone services are modernized to Standard Truck Call Dial (STD). Postal services are modest. The town's commercial and industrial activities are limited to small retail businesses, welding and blacksmith workshops. There is a lack of adequate water and land in the urban area. Luanda Town Council is an old commercial centre of the western part of the district. Owing to its strategic location along the Kisumu-Busia road, commercial activities have been expanding although rather slowly. At the moment Luanda has electricity, a modest postal service and modern telephone system. Its physical infrastructure is not particularly well developed. Given the agricultural potential of the area, there are possibilities for agro-processing industries and a further increase in collection and marketing services. Towns along the Kakamega-Kisumu road such as Chavakali, Mbale and Majengo are expanding quite fast as can be seen by the many business ventures opening up and there are many new hotels and eating places, commercial buildings and residential houses. Many rich people from this district who work in other areas, are investing in these towns.

In order to strengthen the role of these urban and market centres as a foci of rural trade, employment and production, postal and banking facilities must be improved because such facilities can mobilize local capital and stimulate investment in the district. At present these facilities are very weak and in many centres they do not exist at all.

Cattle Dips
There are 40 cattle dips in the district (see Figure 3.15). Thirty-one of these are operational but their utilization level is low because of the belief that traditional breeds are immune to diseases,

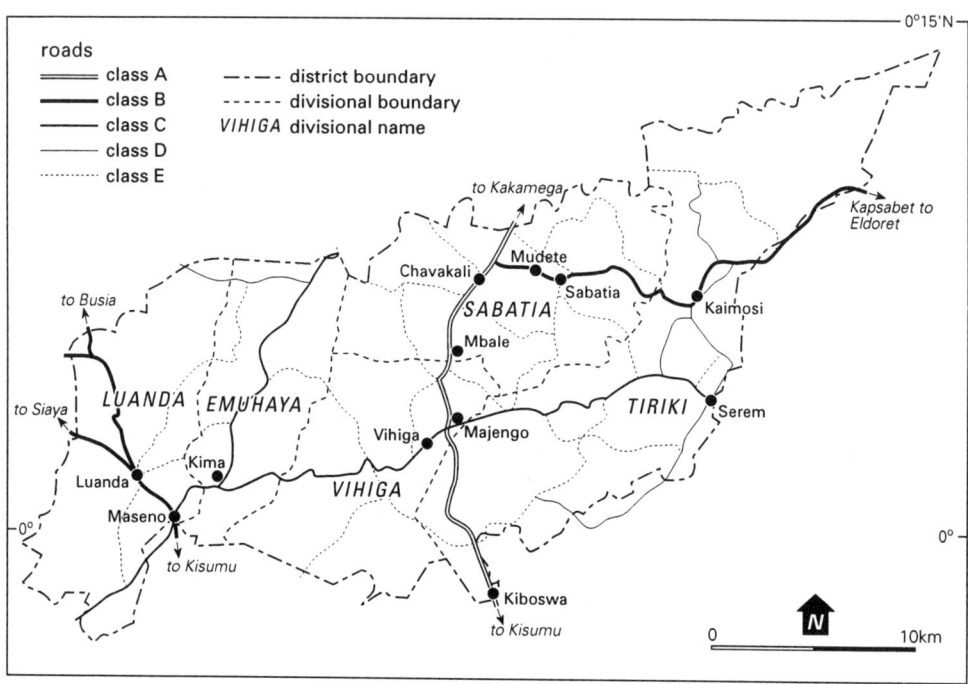

Figure 3.13
Communication Network of Vihiga District (source: Kenya (1993) Vihiga District Development Plan).

and because the dips are mismanaged and poorly operated. Tiriki, Emuhaya and Luanda Divisions have the largest number of operational dips. These dips have been handed over by the Livestock Department to management communities for their operation and maintenance under a system of cost-sharing strategy.

The computed average number of district cattle dippings is very low reflecting the inadequacy of this service. Tick-borne diseases have increased because of the poor management of dips and the fact that little revenue is generated because dips are not used very much. Under a new policy farmers are supposed to manage their own dips and this has given rise to the following problems: sharp rises in the cost of acaricide because it is no longer subsidized by the government with the result that dipping water is not made strong enough; low rate of dipping due to high prices and the farmers poor attitude towards dipping traditional cattle; increase of tick immunity because of low concentrations of acaricide in dips.

3.6 Recapitulation of the Main Problems in Vihiga District

As pointed out above, Vihiga District is located in one of Kenya's high potential agricultural zones of Kenya. The district enjoys a favourable amount of rainfall (between 1250-2100 mm per annum), which is reliable and evenly distributed. There is no distinct dry season at such. The district has over-worked and often poorly managed soils. Soil fertility has decreased and requires

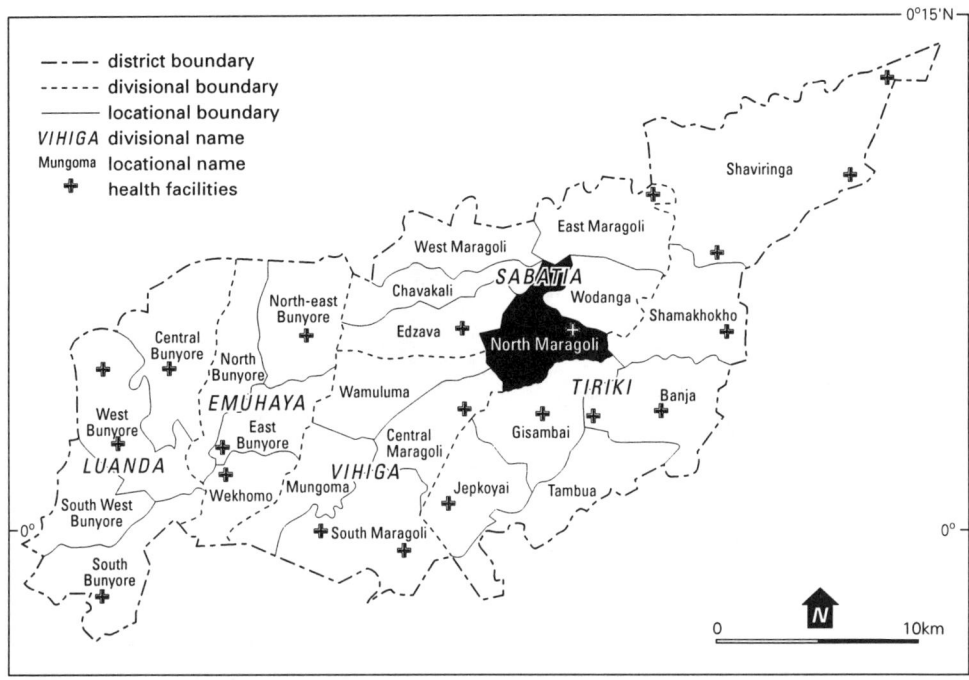

Figure 3.14
Health Facilities in Vihiga District (source: Kenya (1994) Vihiga District Development Plan).

proper fertilizer use, conservation and good farm management. The district has the potential to grow most of the crops cultivated in Kenya and to produce high yields provided that management is efficient and appropriate.

Being one of the most densely populated rural districts in Kenya and probably in Africa as a whole (densities of 800 persons per sq.km and above in most divisions and locations), there has been an increasing amount of holding fragmentation. The average holding per family is now 0.6 hectares and with an ever growing population farm sizes are bound to keep on shrinking. For many farm households there is no longer enough land to provide food and income for the large families of between 6 and 10 people per household. The land question is bound to become more serious in future. The high rate of population growth has created a large, dependent population. Productive resources are directed to consumption and sustenance rather than investment. This trend has made Vihiga District a labour reservoir. Most men migrate to other areas in search of employment to supplement the produce of the small farm units. Migration is a safety valve for many farm households. However, migration also increases labour constraints which lead to poor farm management and low production in small-scale farming. Due to the extensive migration of men from this rural area, many households are managed by women who face the task of providing all the labour the family farm needs. Though farm sizes are small and the district is under population pressure, crop production in many cases is below the potential yield for the prevailing conditions. This implies that the explanation for poor agricultural production may have other causes. Therefore, we should not put the blame on climatic hazards, population pressure, and small farm size alone.

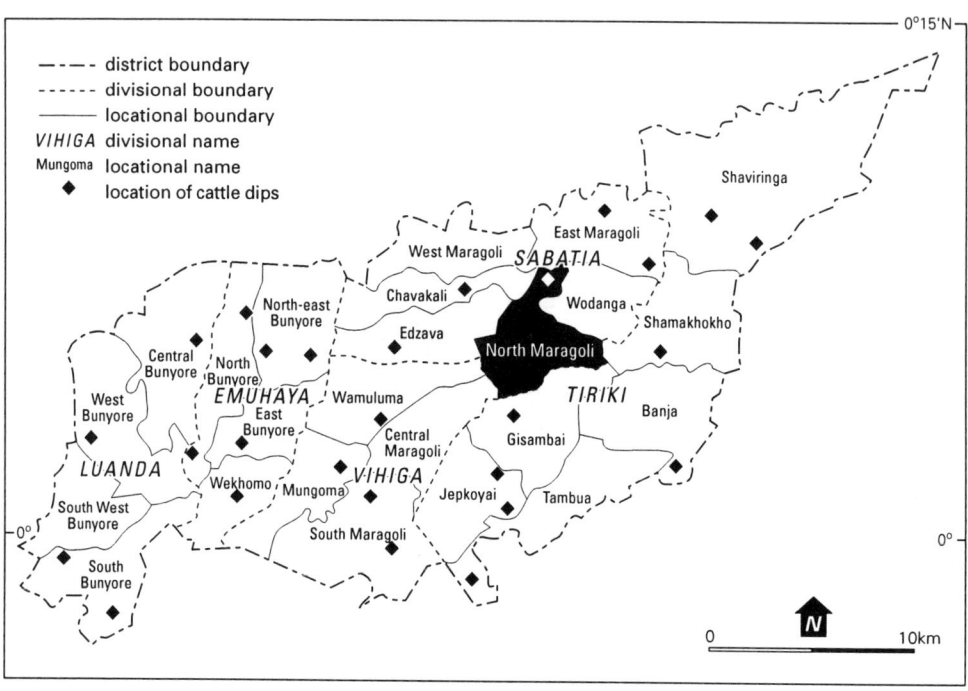

Figure 3.15
Location of Cattle Dips in Vihiga District (Kenya (1994) Vihiga District Development Plan).

Despite the problems mentioned above, the district relies heavily on agriculture for the growth of its economy and employment because commerce, trade and industry are poorly developed. The district has a weak physical and social infrastructure, and the quality of this infrastructure is poor. Public services are mainly found in market and urban centres. To a large extent, these services are still inadequate and are poorly managed. If these services were expanded and better managed it might help the social and economic development of the district. In the following chapter the general situation of small-scale farming in our research area, North Maragoli, is described in more detail.

Notes

1 The reader may experience some statistical gaps and discrepancies while reading this chapter. In 1962, Vihiga District was an administrative division in North Nyanza District and Nyanza Province. From the late 1960s to 1992, it was part of Western Province. At the time of conceptualizing this study in 1991 Vihiga was still part of Kakamega District. Vihiga became a new district in February 1992 when the research was already in progress. However, the area covered by the district has remained almost the same, but the different administrative units within Vihiga District have been sub-divided and renamed over the years. For instance, North Maragoli of 1962 and 1979 is the present Sabatia Division. North Maragoli is a Location in the present Sabatia Division and therefore smaller in area with fewer households. Much needs to be done to create an independent database for Vihiga as a district. This chapter will serve as the beginning of creating such a database.

Vihiga District: Background Information 77

2 Tiriki was formerly called Hamisi, and in some documents the name Hamisi is still used. Emuhaya division was divided into two to create Luanda division after 1992. Sabatia was formerly part of Vihiga division. Due to these changes some tables miss data for Luanda, Sabatia and Tiriki.

3 Interchangeably spelled as Luhya and Luyia in many documents.

4 The name Luanda is the Luhya word for an extensive rocky area and Vihiga means 'rock outcrops'. There are a number of names related to rocky surface indicative of the nature of the topography of Vihiga District.

5 The high price of land in Vihiga lends credence to an argument that people attach a social, cultural and emotional value to their land which is over and above the agricultural value (Moock 1973). An acre of land was selling for between Kshs.45,000 to Kshs. 75,000 at the time of this study. If it is good agricultural land, level and free of boulders or close to a growing market centre, the price will be well above this.
6 Quoted from Osogo 1966:19

7 "Internal migration is likely to improve the distribution of incomes in rural areas.... and accelerate capital formation and technical change on small peasant farms. Migration in effect, enables the peasantry to overcome the imperfections of the rural credit by creating opportunities to amass finance in the cities for subsequent investment in agriculture" (Griffin, 1976:359).

8 A total land area of 14,790 ha was under maize in the period 1991/92. This realized 25,291,000 kgs for an estimated population of 457,647. Per capita of this is 55 kgs. Statistically, each person in Vihiga District could only receive 55 kg of maize against the possible cereal consumption of 200 to 250 kg per annum. This implies that at least 150 kgs of maize equivalents per person per annum must be imported from outside the district.

9 No data for Emuhaya/Luanda

10 No data available on milk production.

11 In Kenya, the Jua Kali sector is an informal sector that operates from open space or with very little shelter and therefore has to adapt to the climate. It was officially termed 'Jua Kali' (hot sun) in May 1988. The Ministry of Technical Training and Applied Technology which is responsible for this sector was set up in March the same year (King & Abuodha 1991).

Female Farmer Weeding Maize

4

Small-scale Farming in North-Maragoli

4.1 Introduction

This chapter attempts to clarify and analyze the reasons for low agricultural production and income in a small-scale farming area in Kenya. It also tries to explain how the farm households go about solving farm management constraints. The various farming sub-systems found in this area are critically examined, namely the food crop, the commercial crop, livestock and tree production sub-systems. This chapter provides the basis for Chapter Five in which a comparison is made of farm management practices, agricultural production and incomes at the household level.

4.2 Characteristics of North Maragoli

Selection
North Maragoli was selected because it had all the characteristics found in small-scale farming areas in the zones of high agricultural potential. These characteristics include:
1 Good soils and a suitable climate with high reliable rainfall,
2 High population density of more than 500 persons per sq.km,
3 An adult population dominated by women and a predominance of households headed and managed by women,
4 Low levels of education
5 Small farms
6 Individualization of land tenure.

In Kenya, North Maragoli falls within the zone of high agricultural potential (Tea-Coffee Agro-Ecological Zone, UM1 p or 2/3). Because of its climate and soils, it is an area suitable for both arable farming and livestock production (Jaetzold & Schmidt 1982)(see Sections 3.2.2 & 3.2.3).

Population
North Maragoli, with a surface area of 39 sq.km, has a population of 34,236 and a population density of approximately 900 persons per sq.km. It is one of the areas with the highest population density in rural Kenya. The total number of households is 5,922 and these households have an average size of between 6 and 10 persons (Kenya 1994).

Demographic Structure
Fifty-four percent of the total population in North Maragoli is female and 46% is male (Kenya 1994). One reason for this under-representation of males appears to be the out-migration of men over the age of 20 years. They leave their home area to look for employment elsewhere. Table 4.1 and Fig 4.1 show the distribution of the sample population drawn for our study by sex and age.

Table 4.1
Population Distribution by Sex and Age (source: North Maragoli Sample Survey, 1992).

Gender	0-14	15-19	20-29	30-39	40-49	50-59	60+	Total
Females	157	109	192	183	126	58	30	855
%	18.4	12.7	22.5	21.4	14.7	6.8	3.5	100
Males	190	106	178	61	43	50	11	639
%	29.7	16.6	27.9	9.6	6.7	7.8	1.7	100
Total	347	215	370	244	169	108	41	1494
%	23.2	14.4	24.8	16.3	11.3	7.2	2.7	100
F/M	0.83	1.03	1.08	3.00	2.93	1.16	2.73	1.34

The total population in the sampled households was 1494 with 639 males (43%) and 855 females (57%), slightly more females than according to the locational statistics. The distribution of the sample population according to age and sex displays a higher percentage of males in the age category of 0 to 14 years than females; the same pattern is exhibited in the national population census in both 1979 and 1989 (Kenya 1981 & 1994). Forty-eight percent of the sample population consisted of dependents as compared to a national figure of 52% in 1979 and 51% in the 1989 population census. The high percentage of dependents appears to be a common feature in many of the small-scale farming areas in Kenya. The data also reveals a large out-migration of males in the age category 20 - 49 years. Such a large out-migration creates problems when it comes to the availability of farm labour (see Sections 4.8 and 5.9).

North Maragoli

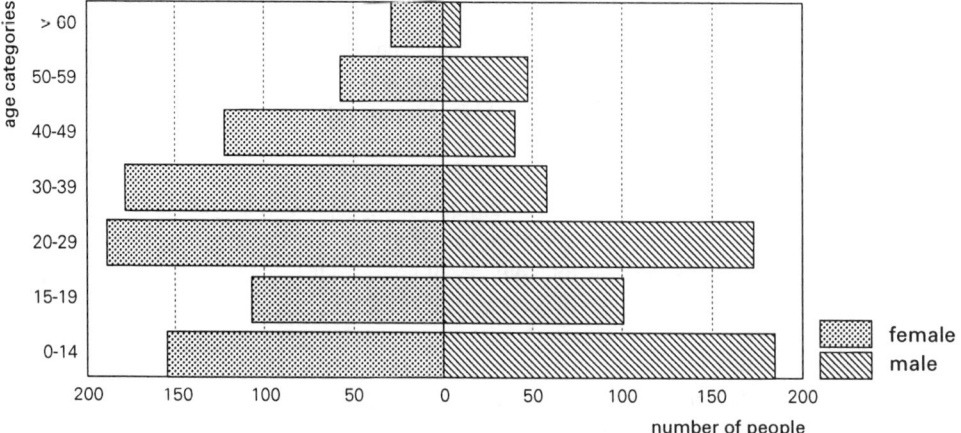

Figure 4.1
North Maragoli Population Pyramid by Sex and Age (source: North Maragoli Sample Survey, 1992).

One consequence of this skewed demographic structure is that almost half the rural households are headed and managed by women (Khasiani 1992 & Moock 1986). In our household random sampling listing we found that approximately 55% of the households are headed or managed by women.

Level of Education

Many older rural Kenyans have not had the opportunity to benefit from formal education and so the level of education of many small-scale farmers is quite low. This is the case with many of the heads of the sample households (see Table 4.2).

More than one tenth of the heads of the sample households had no education, and about one third had only lower primary education. Therefore, almost half of the household heads were to some extent were illiterate. Another 40% of household heads had no more than upper primary education.

Farm Size

Farm holdings in North Maragoli location are small. The Kenya Government considers all farms below 20 hectares as belonging to the small-scale sector. Table 4.3 presents the frequency distribution of farm sizes over the sample population.

The sample had a total farm area of 200.5 hectares distributed over 180 farm households resulting in an average farm size of 1.1 hectares. Forty-four percent of the farms were found to be smaller than 0.8 hectares. Many households (71%) in the study area have farm sizes in the range 0.41 to 1.2 hectares.

Land Tenure

According to the Luhya tradition, properties and resources belong to men. Women acquire user rights to resources through their husbands or male relatives. Therefore, land belongs to men and they are the holders of land title deeds. Land is inherited from parents in accordance with the

Table 4.2
Levels of Education of Heads of Households (source: North Maragoli Sample Survey (n=180)).

Standard of Education	Frequency	Percentage	Cumulative
No education	24	13.3	13.3
Lower primary (class 1-3)	61	34.0	47.2
Upper primary (class 4-8)	73	40.6	87.9
Secondary (form 1-4)	17	9.4	97.2
High school (form 5-6)	1	0.6	97.9
University	3	1.8	99.5
Other	1	0.6	100
	n = 180	100 %	100 %

Table 4.3
Frequency Distribution of Farm Sizes (source: North Maragoli Sample Survey 1992).

Farm Size - ha	Frequency	Percentage	Cumulative %
0.4 ha & less	4	2.2	2.2
0.41 - 0.80 ha	76	42.2	44.4
0.81 - 1.20 ha	52	28.9	73.3
1.21 - 1.60 ha	28	15.6	88.9
1.61 - 2.00 ha	15	8.3	97.2
>2.01 ha	5	2.8	100
	n = 180	100	100

customs of the area. A dual system exists as far as acquiring and owning land in this area is concerned. There is the legalized government system where land is registered on an individual basis and the societal organization where land is inherited from parents. Therefore, both legally and traditionally men are the recognized owners of land in every household. This provides men with the overall control and mandate in decision-making in the farm-household system.

Virtually all the farm holdings visited were formally registered as private individual holdings in the name of a man. There was one exception, a widow. Each parcel of land had a plot number. Only 38% (68 farmers, all men except one woman) had collected their land title deeds despite the fact that the possession of land certificates (title deeds) is essential if collateral is required for a loan.

4.3 The North Maragoli Farming System

Many researchers have worked with the concept of the farming system as an approach, inter alia to improve existing farming (for example FAO 1989, 1990, Moock 1988, Upton 1988, Fresco 1988, Fresco and Poats 1986, Moock at al. 1986, Gilbert 1980). A farming system embraces all human and technical factors which affect the life of a farm family. It consists of farm households, cropping and livestock systems that produce crop and animal products for consumption and sale. Farming systems may be zoned according to agro-ecological divisions (Fresco et al. 1992, FAO 1990). Using FAO's Farming Systems Development Concept, any farming system in a country is visualized in a system's hierarchy which comprises a number of components (see Figure 4.2)

The Farming System Development Concept, according to the FAO, focuses on the farm-household system as the basic unit for analysis and development. The farm-household is the decision-making unit which ultimately controls the transformation of inputs into basic agricultural outputs (production). Agricultural production, especially in small-scale farming areas like North Maragoli, takes place in a range of farm-household systems that belong to a

Small-scale Farming in North-Maragoli

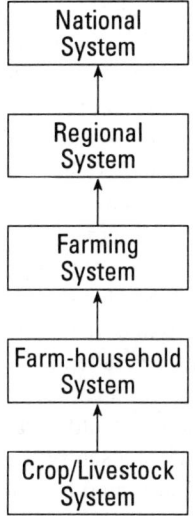

Figure 4.2 Farming System Hierarchy (source: Adapted from FAO 1990, Farming System Development: Guidelines for the Conduct of a Training Course in Farming System Development).

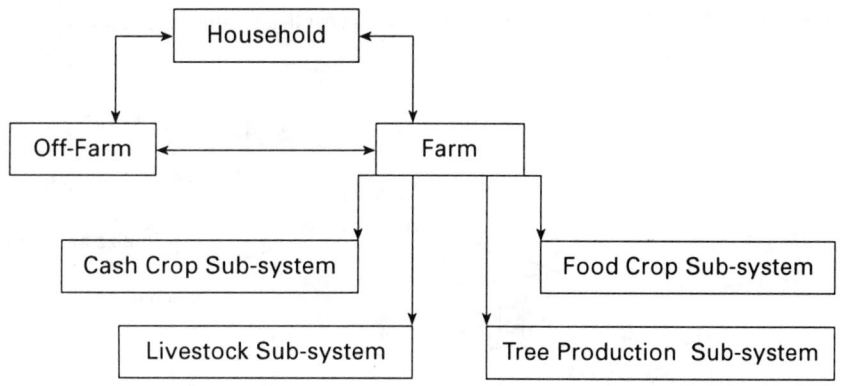

Figure 4.3
The Farm Household System (source: Modified from FAO 1990, Farming System Development).

particular farming system. According to FAO (1990), the farm-household system consists of three components which are closely interlinked and interactive: (i) the **household** which is the decision-making unit; establishing goals for the system; controlling the system; providing labour; demanding food and cash in fulfillment of set objectives; (ii) the **farm** and its crop and livestock activities; providing employment, food and cash for the farm family; (iii) the **off-farm component**, competing and/or complementing/supplementing with farm activities for labour; providing employment and income-generating activities; becoming increasingly more important in supplementing the well-being of farm families (FAO 1989, 1990:15, see also Figure 4.3).

A farm-household system is influenced to a large extent by internal household management. This may incorporate headship of the household (male or female), membership of the household and the relationships and interactions of the members within the household. It is evident from our fieldwork that there is an overlap between farm management tasks, individual decision-making and control over resources within the farm-household. This makes the conventional notion of a simple coordinated household group in which welfare gains are distributed among all members somewhat misleading. The farm-household interacts with the larger community (socio-cultural environment); different institutions such as agricultural extension, government administrators, research institutes, seed producing companies, groups/programmes concerned with extension of credit, and policy implementors in rural development (the policy/institutional environment); as well as the physical environment. These inter- and intra-household relationships are translated into overall farm management situations, agricultural production and incomes.

In North Maragoli the farms are made up of different farming sub-systems, namely the food crop, cash crop, livestock and tree production sub-systems (see Figures 4.3 and 4.4). Practically all the farm-households include these four farm sub-systems in their production. Many farms are no more than narrow pieces of land that extend from a road or a divide (the summit) down to a river or stream. Tea, coffee and French beans are the income-generating crops and the main food crops are maize, beans and bananas; grazing land is for livestock. Trees are found in the lower lying areas mainly along the banks of streams (see Figure 4.4). In this chapter we analyze these farming sub-systems in order to gain a picture of the general agricultural situation (farming system) in our research area.

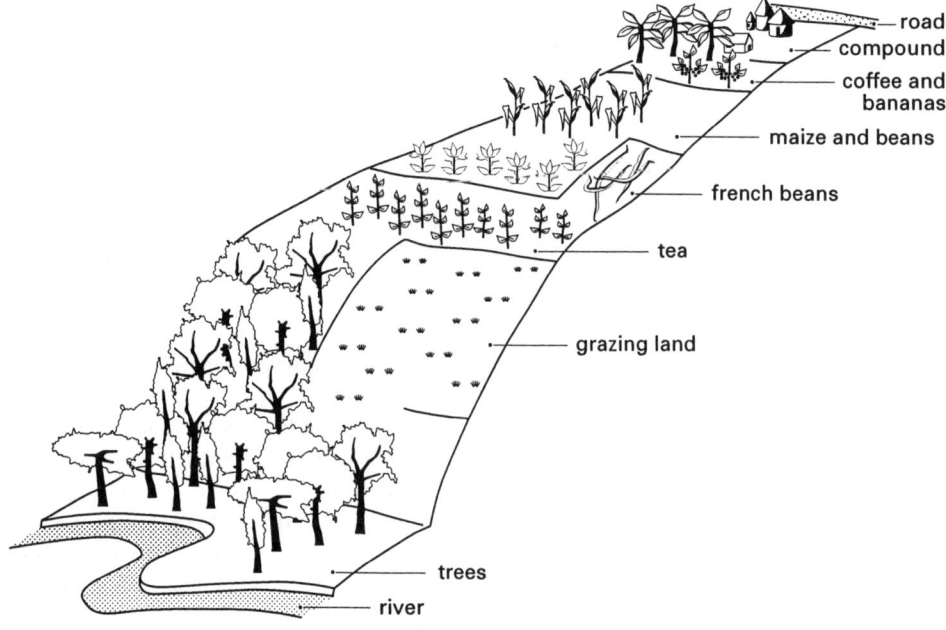

Figure 4.4
Simplified North Maragoli Farming System (source: Study Results).

Land Allocation to the Different Sub-systems

Tables 4.4 shows the allocation of land to various farming sub-systems and to various crops in the North Maragoli sample survey. Both commercial and food crops are considered important and this is shown by the land allocated to both sub-systems (see Table 4.4). Many households place considerable emphasis on these two sub-systems with food and cash crop sub-systems occupying 50% and 39% of the total farm-household land respectively.

Though farms are small, many farmers in North Maragoli have all the sub-systems mentioned. But these farming sub-systems are farmed permanently on the same plot and there has been no rotation or fallow over the years. Maize and tea appear to be the most important crops as they occupy 33% and 30% of the total farm- household land respectively (see Table 4.4).

Table 4.4 Cropping Patterns for Food, Cash Crops and Livestock Production (source: North Maragoli Sample Survey 1992 (n=180)).

Sub-system	Components	Hectares	% total land	Tot & % sub-ss
Food crop ss	Maize & beans	65.5	32.7	
	Bananas	19.3	9.6	
	Vegetables	15.3	7.6	100.1 ha 49.93%
Cash crop ss	Tea	59.3	29.6	
	Coffee	14.7	7.3	
	French beans	4.8	2.4	78.8 ha 39.30%
Livestock ss	Napier grass	16.3	8.1	16.3 ha 8.13%
Other ss	Tree production	5.3	2.6	5.3 ha 2.64%
Total		200.5	100%	200.5 100%

In the following sections we will take a closer look at the various farming sub-systems in North Maragoli.

4.4 The Food Crop Sub-system

North Maragoli is suitable for a wide range of food crops. The main staple crop is maize followed by beans and bananas. The other food crops are sorghum, millet, cassava, sweet potatoes and an assortment of different types of vegetables and fruits. More than 80% of the farmers cited the unreliability and unpredictability of the market as far as food availability and pricing are concerned as being the main reason of the insistence on food production. It is also a long standing tradition amongst the Luhya that each household provides for its own food needs before any surplus is traded or before land is allocated to other uses. Many farm-households, therefore, aim at food self-sufficiency as the main objective in the food crop sub-system. However, on more than 90% of the farms in the survey, this self-sufficiency was never achieved.

4.4.1 Maize

Maize cultivation was enforced in Kenya in the 1920s by the British colonial administration. The millet crop failure due to locusts in 1918 contributed to increasing acceptance of maize and it soon surpassed millet as the most feasible and widely used, if not the most highly cherished crop (Barker 1958 & Wagner 1949 & 1956). In the 1940s and 1950s, the Maragoli (Vihiga District population) produced enough maize to feed themselves and there was a surplus for them to trade (Wagner 1956).

The hybrid maize introduced in 1964 had a higher yield potential and more surplus became available for trade. Land preparation, weed control and harvesting all required little labour in comparison to some of the indigenous cereals. Maize was also found to be more palatable and this led to the abandonment of a number of other food crops (such as millets and sorghum for example), especially in those areas where there was land pressure such as North Maragoli and Vihiga District (Gerhart, 1974, 1975). One consequence of this was a decline in crop diversity and a shift in the eating habits of the population. For a number of years maize monoculture was emphasized by agricultural officers as the best approach to the management of the crop. This was the approach adopted in the large-scale farming areas (Edalia 1980, Gerhart 1974, 1975 & Acland 1971). However, since the mid-1970s maize production has declined. At the present time maize production is low on many farms if we compare it to the potential of which is capable.

There are three levels at which maize yields and production can be measured in North Maragoli and in other high agricultural potential small-scale farming areas:- (i) the **potential** production which is the maximum production determined by climatic and soil conditions (agro-ecological zone). This is between 10,000 and 14,000 kg/ha/year; (ii) the **expected** or anticipated production from maize farms which is calculated at between 60 to 80% of the potential because of farmer resource disposition and other intervening variables, hence 6,000 to 8,400 kg/ha/annum; (iii) the **actual** production of the farmers (Luning 1988 & 1989, Jaetzold & Schmidt 1982, Acland 1971 & Kakamega Agricultural Research Station).

According to calculations and findings by various authorities, it is estimated that maize yields of around 6,000 kg/ha/year (two seasons) can be achieved if the right farm management practices are followed in North Maragoli and Vihiga District in general (provincial and district agricultural personnel). Therefore, in our case with average farm sizes of 0.4 hectares under maize cultivation, it is expected that farmers should be able to harvest average of about 2,400 kg (Acland 1971, Rundquist 1984). To gauge the standard of farm management in this area it is important to compare maize yields from the North Maragoli sample survey with potential and expected yields (see Table 4.5).

Table 4.5
Maize Production 1991 (for both long and short rains) (source: North Maragoli Sample Survey, 1992).

No of Bags (90 kg bags)/hh	Frequency	Percent	Cumulative percent
0 - 3 bags	99	55.0	55.0
4 - 7 bags	52	28.9	83.9
8 - 11 bags	20	11.1	95.0
≥ 12 bags	9	5.0	100 %
	n = 180	100 %	100 %

The long rains harvest is considered to be the major harvest and represents three-quarters of the total annual harvest. An average of 1,260 kgs/ha was produced from the total of 65.5 hectares of maize cropped land of our sample farmers which is far less than the expected averages of 6,000 to 8,400 kgs/ha. A meagre 6 bags (540 kgs) was harvested from the 0.4 hectares which was the average production per household instead of the expected production of 27 bags (2,400 kgs). Slightly more than half of the farmers (55%) harvested three bags (270 kg) or less in the two seasons and more than 80% produced less than 7 bags (630 kg) in 1991. Our analysis reveals that on average each farm household produces 5 bags (450 kgs) of maize in a year. This production can feed an average family for less than 4 months given that one adult requires 200-250 kgs of cereals in a year (World Development Report 1992). These figures explain why North Maragoli is importing maize.

4.4.2 Reasons for Low Production and Low Yields of Maize

The reasons for the low production and yields of maize in our study area appear to be related to:-
(a) Farm size, (b) Certified hybrid maize seeds, (c) Crop husbandry practices, (d) Fertilizer use, (e) Lack of credit and general poverty and (f) Green plucking.

(a) Farm size
Farms have been shrinking in size from the time when the African populations were restricted to 'African Reserve Areas' under British Colonial rule. Population growth rates of between 3.4 and 4.1% per annum were recorded in this area from the 1970s to the early 1990s (see Section 3.3.1 & Kenya 1981 & 1994). A high population growth rate and density have led to increased farm fragmentation and sub-division into small units and this has caused a reduction in the size of plots under maize. A correlation of 0.4266 exists between maize production and farm size. The results of a multiple regression indicate that we can explain 18% of the variation in maize production by farm size (Multiple Regression .42657 and R Square .181196). This analysis shows that farm size is not a major factor in explaining low maize production because the present maize plots do not produce half of what might be expected.

(b) Certified hybrid maize seeds

Another factor which may help to explain the low level of maize production is the wrong use of hybrid maize seed in spite of the fact that farmers in this area were among the first to accept the use of certified hybrid maize because of its high yielding effect (Gerhart 1974 & Allen 1969). It is suggested that the problem of low maize production has to do with using hybrid maize seeds incorrectly (Rundquist 1984, Gerhart 1974, Allen 1969 and Ogada 1969). Whereas the re-use the of harvested crop as seed is not recommended because the 'hybrid power' of the seed disappears with the first planting, many of the smallholders in North Maragoli and in our sample survey attempt to re-use the hybrid seed rather than buy new stocks because they face to financial constraints.

Additional obstacles to an optimal harvest include the poor quality of hybrid seeds, the shortage of seeds and the late delivery of seeds to the buying centres. These were important causes of low maize yields and production in the years 1991-94 in many parts of Kenya, and North Maragoli was one of the areas affected (Daily Nation 14 July 1992, Sunday Nation 22 May 1994, Kenya). The body responsible for the production of hybrid seeds has consistently produced low quality seeds from 1989 to 1994, forcing the government to import maize or to rely on relief food (Daily Nation July 14, 1992). Government management of hybrid seed is poor, considering the fact that no explanation has been provided for the incessant shortages, late delivery and the poor quality of hybrid maize seeds. These examples highlight how farm-household management is influenced by a variety of exogenous factors beyond the farmer's control. The government and its policies play an important role in the administration, management and supervision of the institutions and personnel involved in agricultural innovations (see Sections 2.2, 7.4 and 7.6).

Another problem is the ever escalating price of hybrid maize seeds and other farm inputs, partly due to liberalization and Structural Adjustment Programmes that have been introduced piecemeal since the early 1980s (Kenya 1986). The stringent measures associated with these programmes have resulted in government budget cuts and the abandonment of subsidies on farm inputs and has led to high prices. Many small-scale farmers in North Maragoli can no longer afford to

Table 4.6
The Purchase of Certified Hybrid Maize Seeds in 1991 (source: North Maragoli Sample Survey, 1992).

Amount of Kshs for Purchase of Seeds	Frequency	Percent	Cumulative Percent
Kshs 00 - 00	65	36.1	36.1
Kshs 01 - 50	47	26.2	62.2
Kshs 51 - 100	27	15.0	77.2
Kshs 101 - 200	23	12.8	90.0
Kshs >201	18	10.0	100 %
	n = 180	100 %	100 %

purchase hybrid maize seeds and other necessary inputs. Hybrid maize seed prices have increased by over 250%. Therefore, some farmers resort to buying less than what is really needed on the farm. A multiple regression analysis shows that the use of hybrid seed may explain approximately 28% of the variation in maize production (Multiple R .53173 and R Square .28273). Table 4.6 illustrates the purchase and use of certified maize seeds by the farmers sampled in the research area.

In theory farmers required 120 shillings in 1991 to purchase hybrid maize seeds for an average maize plot of 0.4 ha. However, only 68.50 shillings were spent which is equivalent to 57% of the expected expenditure. This demonstrates that a number of farmers are financially handicapped when it comes to purchasing hybrid maize seeds. Table 4.6 gives a good insight into the situation in the field. Thirty-six percent of the households visited did not buy or plant hybrid seeds in 1991 but, instead, re-used their harvested maize or a corrupt (not pure) local variety. Twenty-six percent spent less than 50 shillings on the same investment. This suggests that they bought between 1 kg and 4 kgs of hybrid maize seeds.

Many farmers combine both local seeds and hybrid seeds when planting. This poses a problem for the management of the crop in terms of weeding and harvesting since the two grow at different rates. The local seeds mature faster than the hybrid seeds. The indigenous maize that mature earlier is eaten as green maize (green plucking) as it matures at a critical time - the time when there is a food shortage in many households. When selecting the maize that has matured first, farmers and children risk trampling (crushing) on the other maize crop thus affecting total production.

(c) Crop husbandry Practices

Another important factor in explaining low maize yields and production is poor crop husbandry. Hybrid seed varieties require crop husbandry practices that differ from the traditional varieties if their full yield potential is to be realized. Hybrid maize seed as an innovation was introduced in package form, involving fertilizers, pesticides and insecticides. The presentation of this package revolves around a set of crop husbandry practices and normally the following detailed instructions should accompany every package of maize seed. (Rundquist 1984, Gerhart 1974, Acland 1971 and communication with Provincial (Western) and District (Vihiga) agricultural staff- Kakamega, Vihiga):

(1) Land Preparation: this should be made well in advance of planting to ensure a ready seed-bed clean of weeds at the onset of the rains.
(2) Time of planting: planting should be done at the beginning of the rains, or shortly before.
(3) Choice of Hybrid : the right hybrid variety with respect to altitude and rainfall should be chosen. Below 1600m Hybrid 622 and 623, 1600-2000m Hybrid 613 and 611, 2000-2300m Hybrid 611B and 613B, above 2300m Hybrid 611B and 611C.
(4) Population and Spacing: a high, but not excessively high number of plants should be grown, this is achieved if planting is made in rows with a distance of 100 cm between rows and 25 cm between plants.
(5) Planting : two seeds should be planted in every hole and a later thinning should be made when the plants are 15 to 20 cm high.

(6) Fertilizers : these should be used at two times, first at planting and then later as top-dressing when the maize is knee-high. 100-150 kg per hectare double Superphosphate and 200-300 kg per hectare Ammonium Sulphate Nitrate or 200-300 kg per hectare Calcium Ammonium Nitrate are required
(7) Weeding : in addition to having a clean-bed early weeding is important and weeding should be a continuous process keeping the fields clean of weeds until the maize flowers.
(8) Stalkborer Protection : in order to prevent stalkborer, insecticides should be used on the growing maize

Many of these practices are not followed by the small-scale farmers in our study area and this contributes to poor performance on many farms. Therefore, insufficient knowledge and poor use of the agronomic practices related to maize is a big disadvantage in maize production. Farmers will have to spend more cash and time on improving farm management in order to get all the operations carried out on time and to follow crop agronomic requirements.

During the fieldwork it was observed that planting density, planting distance and inter-planting practices varied greatly as far as maize was concerned. Farmers practice mixed cropping and the availability of seed determined the extent of mixtures. Often there were too many mixtures many resulting in competition for food nutrients and a retardation of plant growth. This explains the enormous range in yields (see also Luning n.d, Kakamega mission report).

(d) Fertilizer Use
The use of chemical and organic fertilizers to replace lost soil nutrients and a combination of land improvement techniques are required to enhanced agricultural production. Many farmers do not buy fertilizers for their maize crop but use N.P.K instead. This is provided for tea cultivation and is not suitable for maize cultivation (Kakamega, 1992).

The apparent under-utilization or incorrect use of chemical fertilizers contributes to the low yields. In an earlier study it was thought that farmers were not yet fully aware of the benefits of fertilizer use (Moock 1973, 1980). Nowadays many farmers know how important fertilizer is but their major constraint is that they lack money to buy it and do not know how to use it correctly. The cost of a 50 kg bag of fertilizer went up to between Kshs 1,250 and Kshs 1,500 in 1993 and 1994 (KGGCU, KNTC, stockists and farmers). Multiple regression analysis results show that 22% of the variation in maize production can be explained by the use of fertilizer (Multiple R .47417 and R Square .22484).

(e) Lack of credit and general poverty
Many farmers are of the opinion that lack of finance contributes to poor management not only in maize cultivation but in all other farming sub-systems as well. However, given the dismal harvests and small sizes of the maize plots, it is almost certain that most of these farmers would not be able to repay credit from maize production.

The poverty of many farm-households influences decision-making, lack of investment and lack of confidence in the modernization of the farming system in this area very much. Some farm households are so poor that all that is important to them is obtaining food for the day. A few well-to-do farm families are able to invest in maize farming. Household income plays a role in investment in that those with higher incomes invest more in maize management. A multiple regression analysis shows that household incomes explain close to 23% of the variation in maize production (Multiple R .47967 and R Square .23009).

(f) Green plucking

Due to poverty and low food production, many farm-households are without adequate food for most of the year and this results in green plucking which further reduces the production that might otherwise be expected.

(g) Measurement and multiplication from a small acreage to a per hectare basis

Measurements of the irregular and very small fields (or parts of it) lead to large errors. The majority of the maize fields are between 0.01 ha and 0.4 ha (the small-plot bias) in size.

The present findings clearly show that for many households maize production is a questionable means of providing enough staple food to sustain them for the whole year. Production and yields are low because of poor management at the farm-household level and poor institutional management when it comes to the supply of seeds, fertilizers and extension.

4.4.3 Other Food Crops

Bananas are the second food crop grown. There are several cultivars available providing sweet bananas, and varieties for cooking and for other uses. A total of 19.3 ha (10%) of the total farm size (200.5 hectares) in our sample population was under banana cultivation. This means an average of 0.1 hectares per household. Bananas are important both as a source of carbohydrate and as a fruit. Because they fruit throughout the year, they are especially valuable not only as a family food but also in income generation. Their leaves are a good source of food for livestock and also provide wrapping, thatching and mulching material. Banana production in this area is affected by pests, diseases and poor husbandry practices. Some of the pests include the banana weevil and nematodes. For a long time the weevil and nematodes were controlled by the use of agro-chemicals. This has proved ineffective because the pests have developed resistance, pesticides were used non-selectively and were harmful to the environment and the user, were prohibitive in prices and growing increasingly expensive (Acland 1971 & ICIPE, no date). Use of chemical pesticides, therefore, cannot be considered a sustainable solution. Emphasis should be on good husbandry which is all important for high yields in banana production. However, on many farms bananas are grown using poor methods. Our rough estimates show a low average yield of approximately 500 bunches/ha in 1991 against a possible yield of between 1,000 to 2,500 bunches/ha/annum. However, the reader is warned that the bananas fields surveyed in North Maragoli were never pure stands.

Beans are an important leguminous crop and are interplanted with maize. The farmers reported that in the last few years production has been insignificant. They blamed the low production on pests and diseases, but it was noticeable on many farms that management of the crop was also inadequate. Our sample estimate was 80 kg/ha in 1991 against a minimum possible yield of 220 kg/ha and a maximum of 1,100 kg/ha/annum (Acland 1971).

Finger millet and sorghum were the main food crops before the introduction of maize. Finger millet was liked because it could be stored without the use of insecticides for a longer period than any other cereal (Acland 1971). However, both finger millet and sorghum have a lower yield capacity than maize, and require more labour at all stages of cultivation, particularly during the preparation of the seedbed, weeding, bird-scaring, harvesting and threshing. These tasks have made these crops unpopular and explain the preference for maize as a staple crop. The hectareage under these crops was found to be insignificant.

4.5 The Commercial Crop Sub-system

The main cash crops grown in our study area are coffee, tea and French beans. Tea and coffee were only introduced after the acceptance of the Swynnerton Plan in 1954. Coffee was the first cash crop brought in this area and arrived in 1955. Both coffee and tea were introduced and registered as men's crops while women were left in their traditional sector of food production. A few farmers in North Maragoli started growing tea in the early 1960s. French beans were introduced as a women's cash crop in the early 1980s (Sabatia and Wanjala, 1994).

Since Independence, the Kenya Government has encouraged the growing of coffee, tea and other cash crops by small-scale farmers as a method of redistributing incomes and diversifying cash crop production for the export market (Sessional Paper No. 1 of 1965 on African Socialism).

4.5.1 Tea Cultivation

Tea, covering 30% of total farm-household land, is considered by the North Maragoli farmers sampled to be the most important cash crop, followed by coffee (7%) and French beans (5%) by the sample farmers in North Maragoli. On average the tea plots measure about 0.3 hectares. Approximately 46% of the farmers have between 0.3 to 0.4 hectares planted with tea (see Table 4.7).

Table 4.7
Land Allocation to Tea by Sample Farmers (source: North Maragoli Sample survey 1992 (n=180)).

Size Tea Plots in hectares	Frequency	Percentage	Cumulative Percentage
0.0	21	11.7	11.7
0.1 - 0.2	46	25.6	37.2
0.3 - 0.4	82	45.5	82.8
0.5 - 0.6	20	11.1	93.9
> 0.7	11	6.1	100
	n = 180	100%	100%

Small-scale Farming in North-Maragoli 93

According to the Kenya Tea Development Authority (KTDA), an average yield of 8,600 kgs/ha/annum of green leaf is achieved when tea is grown under good management in this area (Jaetzold & Schmidt 1982, KTDA 1992). Farmers in North Maragoli can easily produce between 6,500 and 10,000 kgs/ha/year if the right husbandry practices are used (Tea Extension Officers, Western Province Region Office, 1994, Okelo and Oluoch 1992). In this study, the lower limit of 6,500 kgs/ha/year has been taken as a yardstick for the analysis of our participants' green leaf tea production. Table 4.8 presents data for tea production from the sample farms.

Approximately 40% of the farmers produced less than 500 kgs of green leaf in 1991. The total production for the whole sample population was 193,530 kgs of green leaf in 1991, averaging 1,217 kgs per household and 3,264 kgs/ha. This figure is low compared to an average national

Table 4.8
Total Tea Production by Sample Farmers 1991 (source: North Maragoli Sample survey 1992 (n=180))

Tea Production in Kgs per hh	Frequency	Percentage	Cumulative Percentage
00	28	15.6	15.6
01 - 500	44	24.4	40.0
501 - 1000	38	21.1	61.1
1001 - 1500	25	13.9	75.0
1501 - 2000	19	10.6	85.6
2001 - 2500	13	7.2	92.8
2501 - 3000	6	3.3	96.1
>3001	7	3.9	100 %
	n = 180	100 %	100 %

Table 4.9
Annual Tea Income (Bonus in 1991) to Farm Households (source: North Maragoli Sample Survey 1992 (n=180)).

Income in Kshs per hh/annum	Frequency	Percentage	Cumulative Percent
00	28	15.6	15.6
01 - 1000	30	16.6	32.2
1001 - 2500	34	18.9	51.1
2501 - 5000	44	24.5	75.1
5001 - 7500	27	15.0	90.6
>7501	17	9.4	100
	n = 180	100 %	100 %

production of 6000 kgs/ha and is also low when compared to other small-scale tea growing areas in Kenya (Section 3.4.1 and Table 3.10). Low production is also reflected in the meagre earnings from tea of the individual households (Table 4.9).

Seventy-five percent of the farmers in the sample survey earned less than 5,000 shillings from tea cultivation in 1991, giving an average of Kshs 3,444 per household. This is a very low income considering the cost of production (see Table 4.10). On a one hectare basis, a farmer in North Maragoli earned approximately 10,500 shillings per ha/annum when his/her counterpart in Embu earned approximately Kshs 63,000, in Kirinyaga 94,000, in Murang'a 69,000 and nationally 59,000 per ha/annum in 1991/92 (KTDA 1992). The poor production and low incomes are explained by a number of constraints.

Table 4.10
The Economics of Tea Production in Vihiga District: Gross-Margins/ha/year in 1991 (source: KTDA (1992) Kakamega Regional office.)

Expenditure	Kshs	Income
Plucking 8611 kg of green leaf X Kshs 1.50	12,916.50	1st payment: monthly payment 8611 kg green leaf X 3.00
Weeding: 2 times a year each at Kshs 1000 or 8611 bushes X Kshs 0.24	2,066.60	Kshs 25,833.00
Pruning Kshs 0.30 per Bush after 3 years 8611 bushes X Kshs 0.10	861.10	Second payment: End of year payment (bonus) 8611 kg of green leaf X Kshs 2.75 Kshs 23,680.25
Fertilizer costs/labour 17 bags X Kshs 345	5,865.00	Gross Income 49,513.25
Government Tax 6% of Gross Product Kshs 0.345 per kg	2,970.80	
KTDA cess 0.38 per kg of green leaf	3,272.20	
Total Expenditure	27,561.00	
Net Income (return)/ha	21,952.25	

4.5.2 Problems in Tea Cultivation

Problems that contribute to low production, poor quality and low incomes can be examined at two levels, i.e (a) the household level and (b) the institutional level.

Household level problems
At the household level, the problems identified are mainly related to poor management practices due to insufficient knowledge of crop husbandry, labour constraints and poverty. Many farmers' tea husbandry practices do not conform with the ones recommended. To a large extent this appears to be caused by a lack of capital which prevents many farm households from investing in the crop. This was aggravated by the fact that the males earn the tea income (bonus) while the women do the farm work (see Chapters 5 and 6)

(a) Plucking:
Tea farming is labour intensive; plucking takes up more than half the total time and cost involved in its production. Many farmers incur heavy losses because of poor labour allocation at the household level, especially for plucking. It was observed that many poor households who lacked financial support could not employ hired labour for tea picking. This resulted in much unplucked tea and low production.

On many farms, farmers found it difficult to maintain the best standard and quality of plucking of '2 leaves and a bud' because of unskilled pluckers. Poor plucking was reflected in leaves being crushed in the pluckers' hands or during transportation to the tea buying/collecting centre. On more than 50% of the farms tea was plucked only once or twice a month instead of the recommended four rounds. Poor quality tea was the result and, consequently, a low income. This was due to poor plucking schedules and labour constraints.

Another problem relating to plucking of which farmers are painfully aware is the way of paying hired labour. Farmers pay hired labour a daily rate, including meals, with no regard to the weight or amount of leaf plucked. It was quite apparent in the field that on many farms hired labourers were very poorly supervised. No wonder that in North Maragoli very little work is done per day and casual workers increase the number of working days at will. This inefficient method of payment contrasts strongly with the method of payment used in other tea-growing districts where hired labourers are paid per kilogram of green leaf plucked and no meals are served (Okelo 1992). It has been observed that such employees work faster and produce more kilos because of the incentive to make more money.

(b) Pruning:
Pruning is regarded as the main job of the men in tea cultivation. Pruning is carried out poorly because farmers do not prune as recommended. Some farmers prune tea after four to six years. It was found that many of the young men contracted to prune tea do not have any experience nor any training in tea pruning and tea is cut badly, exposing the stems to diseases and to drying up. A certain height has to be maintained when pruning tea bushes in order to accelerate canopy formation. It was observed that because of the lack of experience of the pruners, they pruned very low which is unsuitable because it retards growth and hinders high production. On a

number of farms it was observed that pruning was done at the wrong time which retards growth and reduces production.

Farmers are encouraged to retain the pruned vegetation as mulch to increase soil fertility. However, due to a shortage of firewood and little time to collect it or a lack of money to procure firewood from the local markets, most farmers use the potential mulch as firewood for cooking.

Another problem that negatively affects production is the low density of tea bushes per unit of land on some farms. Many gaps are found which need infills to increase both density and production. Gaps appear during the initial planting when some seedlings die off. Farmers rarely fill up these gaps, citing unavailability of seedlings and financial constraints.

(c) Fertilizer Use:
Low and incorrect use of fertilizer is typical and is caused by a low rate of acquisition, prohibitive prices on the local markets and a lack of correct information about its use or distorted information. Table 4.11 shows the purchase of fertilizer by farmers for tea cultivation.

On average Kshs 764 per household growing tea was spent on the purchase of fertilizer in 1991. The Kenya Tea Development Authority (KTDA) supplies tea farmers with fertilizer on credit at a subsidized rate and in 1991 this was Kshs 342.50 per 50 kg bag. Table 4.11 illustrates that 24% of the farmers did not purchase fertilizer, and 26% spent less than 500 shillings, which suggests that they bought less than one and half bags of fertilizers in 1991. The 77% farmers who spent Kshs 1000 or less on fertilizer were only able to obtain three bags or less from the KTDA. The fertilizer is specifically meant for tea but in fact ends up being used on all crops. The three bags of fertilizer are insufficient for average tea plots measuring 0.3 hectares or more since fertilizer should be applied to tea at least twice a year. A plot of 0.3 hectares requires a minimum of 5 bags of fertilizer in one year according to the recommendations of the KTDA.

The shortcomings observed in relation to fertilizer use in the study area are as follows: (i) not all farmers put up application requests for fertilizers from the K.T.D.A; (ii) those who submitted

Table 4.11
Purchase of Fertilizer for Tea Cultivation (source: North Maragoli Sample Survey 1992 (n= 180)).

Amount of Kshs for Purchase	Frequency	Percentage	Cumulative Percentage
00 - 00	44	24.4	24.4
01 - 500	47	26.2	50.6
501 - 1000	45	25.0	75.6
1001 - 1500	15	8.3	83.9
>1501	29	16.1	100.0
	n = 180	100 %	100 %

application forms ask for an inadequate number of bags; (iii) fertilizer is sometimes delivered late and at the wrong time of year; (iv) many farmers strew fertilizers on the tea crop without properly first weeding, resulting in considerable wastage as part of the fertilizer is then used up by the weeds. Fertilizer is also often spread too thinly in order to cover the whole tea plot and to minimize fertilizers costs; (v) fertilizers received from the KTDA are also frequently used for maize and other crops, reducing the quantity available for tea; and (vi) fertilizers obtained on credit from the KTDA are often sold to other farmers when there is a dire need for money.

(d) Weeding is poorly executed on many farms. Two weedings in a year is the recommended minimum, otherwise once every two months. On about 60% of the farms weeding was performed once a year, on 20% twice a year and on another 20% once about every two years. Use of wrong implements and methods in weeding caused and strengthened the destruction of roots, prompting root rot and the withering of bushes on a number of farms. Such examples show that at the household level the management of tea plots is associated with a variety of problems which explains low production and the inferior quality of the tea produced.

Institutional Problems

(a) Tea Payment:
As already mentioned, farmers receive two types of tea payments. Prices fluctuate depending on the quality and prices set at the tea auction. The amount of money per kilogram paid to farmers varies from one factory to another because the quality of tea varies according to the area in which it is grown and the processing factory. Higher prices are obtained in tea-growing areas east of the Rift Valley (Leonard 1991) where farmers have been able to produce and maintain high quality tea. Small-scale farmers to the east of the Rift Valley were incorporated into tea farming earlier than the smallholders to the west of the Rift Valley. This early start seems to have given the smallholders to the east more experience in the management of the crop which explains the high quality green leaf and better yields produced there. The cooler and thus more suitable weather for tea-growing in the east of the Rift Valley also influences the quality of the crop favourably. Price incentives seem to have worked better and favoured farmers in the eastern zone encouraging them to maintain the production of high quality tea: a sharp contrast with farmers from the western zone. The quality of tea from the latter region has been consistently low.

(b) Administration:
The farmers complained about the administration at the tea leaf buying/collection centres. Some of the clerks at these buying/collecting centres are not strict with weighing records. Female farmers, especially the illiterates, complained that they are cheated on the weight of the green leaf delivered. Some of the clerks appear to be bribed by certain farmers (see Chapter 6). This type of corruption affects delivery systems of other crops such as coffee and sugar cane as well. Non-punctuality and the inefficient working habits of clerks at the buying/collecting centres lead to harvested tea loosing weight and quality. Other complaints about the clerks' behaviour concern to their poor tea evaluation skills and the late submission of application for tea inputs. Inefficiencies in the transportation system are also a reflection of administrative shortcomings. These problems all contribute to low production.

(c) Processing Facility:

Neither Kakamega nor Vihiga District has a tea factory. All the tea plucked in these districts is transported to Chebut factory which is situated about 40 km away near Kapsabet town in Nandi District as shown in Figure 3.1. The factory is far away for the farmers and therefore expensive as farmers have to pay transportation costs. Because of the distance involved, the time allocated to plucking tea is restricted to make it easier to collect and transport it to the factory before dark. This restrictive plucking time schedule encourages under-plucking and tea spoilage on the farms. Sometimes, after the tea has been weighed, the transportation lorries do not collect the tea because of mechanical problems, the shortage of lorries and other breakdowns. Fermentation and loss of quality is not only incurred because of lack of transport and long waits in the hot sun but also by poor packaging on the lorries. Sometimes tea from this area is rejected at the factory and the loss is passed on to the farmers. It is not surprising therefore that this area is rated as producing low quality tea, and this has led to perpetual low prices being paid for tea from this zone.

The farmers have pleaded with the government and the KTDA to construct a factory in Western Province. A government pledge in 1991 to build a tea factory in Vihiga District appears to have been shelved until money becomes available.

It is now government policy that each leaf buying/collecting centre has to meet the cost of tea operation. At the moment, the cost involved in collecting and transporting a kilogram of green leaf to the KTDA Chebut tea factory is Kshs.0.99 for Kakamega and Vihiga farmers. Compared to the uniform transportation charges of Kshs.0.38 in areas which are situated close to a the factory, this means that farmers in Vihiga District are taxed heavily (KTDA 1992). These institutional and organizational problems are in addition to the problems identified at the household level and again help explain poor management and low yields in North Maragoli.

4.5.3 Coffee Production

Arabica coffee is grown because this is an area of comparatively high attitude. Coffee has been performing very poorly, however. The average amount of land allocated to coffee is very small in many households (Table 4.4). Only 14.7 hectares, (7% of the total land of the sample population) is under coffee. The average size of plots under coffee is 0.1 hectares. Coffee was the first cash crop to be introduced here and one might have expected it to command a bigger area. However, the plots under coffee production have decreased in size since the early 1980s. The limited size of land planted to coffee is in fact indicative of its poor performance and farmers' disinterest in it (Table 4.12).

Coffee production is determined by management at the farm-household level. It is estimated that on a one hectare farm, a production of 1,000 kg/ha to 1,250 kg/ha of clean coffee per annum is possible with good crop management, and the correct use of adequate fertilizers and other inputs (World Bank 1979 and Acland 1971). The average production for Vihiga District was between 200-300 kg/per/ha of clean coffee in 1991 (Vihiga District Development Plan 1994). Production amongst our sample population was also low (Table 4.13).

More than 60% of the farmers sampled either had no coffee or had coffee with no production (non-producing). The total production for the sample population was 9,391 kgs with an average

Table 4.12
Land under Coffee Cultivation in 1991 (source: North Maragoli Sample Survey 1992 (n=180)).

Size of land in hectares	Frequency	Percentage	Cumulative Percentage
0.0	67	37.2	37.2
0.1	87	48.3	85.6
0.2	20	11.1	96.7
0.3	4	2.2	98.9
0.4	2	1.1	100.0
	n = 180	100 %	100 %

Table 4.13
Coffee Production 1991 Among the Sample Farmers (source: North Maragoli Sample Survey 1992 (n=180)).

Coffee Production in kg per hh	Number of farmers	Percentage	Cumulative percentage
00 kg	111	61.7	61.7
01 - 100 kg	46	25.5	87.2
101 - 200 kg	13	7.2	94.4
>201 kg	10	5.6	100
	n = 180	100 %	100 %

Table 4.14
Coffee Income in 1991 for the Sample Farmers (source: North Maragoli Sample Survey 1992 (n=180))

Income in Kshs.	Frequency	Percentage	Cumulative
00 - 00	111	61.7	61.7
01 - 250	58	32.2	93.9
251 - 500	9	5.0	98.9
>501	2	1.1	100
	n = 180	100 %	100 %

of 52 kgs per household in 1991. Average production was 639 kgs/ha in 1991. Consequently, the average coffee incomes are also low (Table 4.14).

An average income of Kshs 74 per household was achieved by the sample population in 1991. When those growing coffee but not producing it are subtracted then an average earning of Kshs.130 per household in 1991 was obtained and Kshs 645/ha. Most of the farmers (99%) earn less than Kshs 500 a year from coffee farming. Given the inputs required for this crop, farmers do not benefit from its cultivation. On a one hectare plot an investment of approximately Kshs 12,000 is anticipated and a net income of Kshs 15,000/ha/year. Many farmers are of the opinion that coffee is occupying space which could otherwise be utilized for another more remunerative activity, especially because of the scarcity and the high value of the land in North Maragoli.

Reasons for low production and incomes:
Coffee as a crop was totally neglected on most of the farms in our sample. Weeding was poorly done or was hardly carried out at all. Pruning had been neglected for a long time. Other practices like pest and disease control were hardly referred to. Only about 2-3% of all the coffee farmers in Vihiga sprayed coffee during the period of our survey. Fertilizer application was almost nil as the coffee co-operative society in the location was almost non-operational due to lack of funds.

The problems diagnosed in the coffee industry in this area are: (i) poor extension delivery system resulting in poor crop husbandry; (ii) disorganized and non-functional credit system; (iii) disorganized co-operative societies; (iv) coffee factories that are expensive for farmers to maintain; (v) corrupt coffee officials; (vi) low payments due to low prices and late payments which clearly sap farmers' incentives to produce and; (viii) the continuation of the practice where males earn the coffee income while women provide the labour required for its management. This contributes to the disincentives working against good management. On some farms women have withdrawn their labour and the result is poor production; (ix) the 5% presumptive tax on gross earnings by farmers irrespective of the fact that farm proceeds had reached an all-time low is punitive; (x) restrictive legislation which bind farmers to tending their crop even if the proceeds fall below that of alternative agricultural investment; and (xi) the imposition by the Kenya Planters Co-operative Union (KPCU) of a development levy on farmers for the expansion of the union's milling capacity when actual national output was falling. Some of these problems were also identified by Grosh (1991) and Leonard (1991) as the causes underlying the deterioration of coffee production in Kenya.

Women's input in coffee management
An additional and important problem associated with coffee growing and management in North Maragoli is the ownership of land. Since the man is the head of the household and therefore the decision-maker, it has been the policy since coffee was first introduced to small-scale farmers that coffee be registered in the name of the head of the household and the owner of the land. The person in whose name coffee was registered is also the person who earns the income from the crop and/or applies for coffee loans and materials.

This has meant that men have continued to earn from this crop although it is a known fact that women are the ones working and providing most labour on many coffee fields. This was found to have discouraged many women from sticking to good management in weeding, pruning, picking and transporting of the crop. Some of the husbands who earn the income live away from home and at the time of loan application these men are not present to place in applications for farm inputs. This was one of the reasons why few farmers were applying for or receiving coffee inputs. If the crop had been registered in the women's names or in the names of both wife and husband, it would have been easy for women to apply for inputs as they were always on the farm.

The coffee officers were of the opinion that if women were to receive part of the earnings from the different cash crops, especially coffee, the management and production of coffee might improve. Most women are demoralized and those who still work on coffee plantations do it unwillingly because some husbands dictate that women should work and force them to. Most of the coffee farms are inter-cropped with bananas, napier grass, beans and maize.

4.5.4 French Beans

The French beans project was started in 1982 by Hortiquip Co Ltd. This project was initiated specifically for women farmers. The main objectives of the project were to: (i) increase family incomes specially for women because French beans are a crop that matures rapidly and a quick income earner; (ii) create employment for family labour (the company also encourages school children to register as members to get inputs for the cultivation of the crop). The crop is easily grown three times a year. However, the company encourages farmers to cultivate it twice because it is labour intensive just like tea and coffee. French beans are suitable for this area because the crop does not require a large amount of land. French beans help maintain soil fertility through their nitrogen fixing capacity. There are, in practice, always some rejected beans and these provide proteins and help to reduce the malnutrition which exists in some households. Finally, French beans are a foreign exchange earner as the crop is exported tinned and fresh to France and Finland.

Women, school children and also a few men have responded positively to the cultivation of French beans. It takes from between 45 to 50 days from planting to harvesting the crop, and if the weather and soil conditions remain favourable, harvesting can go on for three months. The company loans farmers inputs in 'kind': 1 kg of seeds and 2 kgs of fertilizers. The company has its own extension staff who visit the farmers to instruct them on management aspects. When the beans are about three to four weeks old the company sends its extension staff to spray them against possible diseases and pests. The Hortiquip company rails the beans to Njoro where they are canned, packed in 400g tins and exported. The farmers receive their earnings at the end of the season when all the produce has been harvested and delivered to the company. The company deducts Kshs 410 (1992) for the inputs from the total income per farmer. Those farmers who provide good management and devote ample time to harvesting, earn close to Kshs 2000 per season.

The company stated that it is impressed with the women's performance because the plots under French beans are well managed. It maintained that women were more reliable to deal with as far as loan disbursement and repayment were concerned for the simple reason that women were always there in the home in contrast to men who are more mobile. It was found that women attentively follow the instructions from the company's extension staff. Extension staff can work quite efficiently because they are dealing with smaller numbers and they are well paid for their work. This invalidates assumptions and allegations that maintain that women cannot listen to male extension officers without their husbands being present. The French bean experience seems to illustrate that women can be good managers, and are willing to learn new technologies as long as they have control over the crop and its earnings.

The main problem the company and women experienced with the growing of French beans is the destruction caused by hail stones, strong winds and the dry season. On average women earn between Kshs 300 and 5000 per annum from the crop. The variation is caused by differences in management. Women complained that some husbands threatened to stop them cultivating French beans if they did not share the money they earned from the crop with them.

4.6 The Livestock Farming Sub-system

Livestock-keeping is a long tradition among most of the people of North Maragoli and the Luhya group that occupy Western Province. Meat, milk and eggs provided by livestock serve as important sources of high quality protein to complement diets that are based on starchy crops like maize, bananas, millet and cassava. Cattle are important in a few homes for traction and in many households for manure. Cattle also serve as a safety valve because they are sold when money is needed in cash to meet family obligations. Socially and culturally cattle are important for dowries and other traditional functions and ceremonies such as funerals and weddings.

The main type of cattle kept by farmers are of the Zebu type. In the past, when the availability of land was not a problem because there were communal grazing lands and many other uncultivated pieces of land, households had many animals and many different types (cattle, sheep, goats, poultry).

The present land tenure system combined with population pressure creates problems when it comes to grazing. Land has been privatized and there is less access to neighbours' farms and uncultivated lands. Pressure on land and shrinking farm sizes have accelerated the decline in the numbers and types of livestock kept. The emphasis in North Maragoli and in many small-scale farming areas in recent years has been on keeping fewer but higher yielding grade cattle or improved cattle (cross-breed) where it is financially possible. Zero-grazing is now emphasized as a strategy for commercial farming. This employs a minimal amount of land, is intensive, high yielding, generates high income and minimizes the spread of cattle diseases.

A few farmers (20%) who have been able to accumulate capital have purchased cross-breeds and a few have acquired dairy grade cattle. However, many farmers are not following the recommended zero grazing practices. Out of the 180 households visited less than 5% had constructed the recommended zero grazing units.

Many farmers (80%) still keep the local Zebu, and amongst this group there is average of two cows per household, mainly used for milk production. In many homes, the animal was tethered within the compound and fed from the same place. Some farmers referred to this as zero grazing.

Table 4.15
Livestock Population among the Sample Households 1992 (source: North Maragoli Sample Survey, 1992.).

Type of livestock	Total number of livestock	Number of farmers	Percent of farmers
Local Cattle	271	164 out of 180	91
Improved Cattle	16	8 out of 180	4
Grade Cattle	13	8 out of 180	4
Goats	35	18 out of 180	10

Today, common grazing grounds are mainly along the main road where there is in fact very little grass. Table 4.15 shows the population of livestock in the farm households of the sample survey.

As we have already seen cattle belong to men. Yet approximately 70% of the labour required in livestock-keeping is provided by females. This is a complete shift from the old days when close to 90% of the labour and care for animals was provided by males. The labour required for cattle husbandry in many homes consists of planting napier grass, providing grass and other feeds, milking, watering and cleaning up the shed. However, when it comes to making decisions to sell or buy cattle, this is exclusively a male prerogative. More than 50% of the female respondents declared that the male's failure to share incomes from livestock with them contributed to the neglect of livestock on a good number of farms. Women neglect livestock because of the mounting pressure of work that faced them both on the farm and in the homestead. Most of the local cattle kept look malnourished and very thin. Poor management has resulted in a low milk production in many households.

Problems in Livestock Management
The range of problems include (i) lack of capital to invest in livestock and replace local breeds with cross-breeds or grade animals; (ii) the high cost of grade cattle for many farmers. One good grade cow was sold for between 20,000 to 30,000 shillings in 1994 (Kakamega District Development Plan 1994); (iii) poor management and administration of artificial insemination (AI):

The Ministry encourages farmers who are unable to purchase grade cows to upgrade the local breed, using artificial insemination or a good grade bull. The artificial insemination has suffered from a series of problems and hence its impact has not been felt in the last few years. The service ran short of funds and it is poorly staffed, making it difficult to provide the services required at the right time. As an example, the Division of Sabatia (where North Maragoli is to be found) with about 14,600 farm households only has one veterinary doctor. Due to a shortage of transport and staff, services are provided from Vihiga and staff drive along pre-set routes; farmers who wish to have their cows inseminated gather them at small roadside cattle crushes, where the inseminators stop to provide services and collect fees once in a while. Farmers living close to the road benefitted more than farmers far away from the road. The artificial insemination service has also been very poorly managed and administered. There are numerous cases where the officers have given this service but it failed due either to mis-timing, poor storage or infertile sperms making it necessary for the farmers to call the officers on many occasions and thus spending much money.

(iv) insufficient biomass during the short dry period as a result of small farm size; (v) poor disease and pest management; (vi) expensive and scarce commercial animal feeds. These are needed to enhance milk production, especially for grade cattle and cross-breeds; and (vii) poor management of pastures and napier grass plots with limited perception of the usefulness of fertilizer. These problems, in addition to specific household conditions, have resulted in poor livestock management and low milk production.

Poultry
Local birds are kept in all households. Ownership and the management of poultry is the responsibility of females and children. During the last five years there has been high incidence of diseases among the birds. Newcastle disease appears to the most frequent. Its spread is encouraged by the free movement of these birds, and the buying and selling of chicken on the local markets from other districts. Measures to control the spread of Newcastle disease by quarantining poultry from districts where the disease breaks out is not enforced. As a result many households sell off all chickens infected by the Newcastle disease and, in this way, encourage it to spread.

The impediments preventing women from adopting improved poultry are :- (i) lack of finance; (ii) inadequate and non-motivated field staff; (iii) expensive drugs which are at times unavailable.

4.7 The Tree Production Sub-system

As already mentioned, this area is situated in what was formerly the dense mountainous tropical forest (rainforest) zone. However, because of high population density and shrinking farm size, most of the natural vegetation has been cleared to give way to settlement and the cultivation of crops. But, surprisingly, the higher the population pressure the larger the amount of land devoted to woody biomass (Bradley 1984) as in the case of North Maragoli. With a heavy population, the proportion of 'man-made' as against natural woody biomass also increases.

North Maragoli's physical environment, like many parts of Vihiga District, is characterized by very many deep valleys. Typically these valleys have marshy bottoms which are unsuitable for many crops. Most of these valleys are planted with woody biomass, 70% of which consist of eucalyptus trees. The eucalyptus trees that are planted are for multipurpose use with a major emphasis on poles and frames for construction purposes and wood for sale. Other trees found in the area include a few remnants of indigenous trees, fruit trees (mango, avocado and pawpaw trees) planted in the homestead or on agricultural land and planted hedges (kie apple). Tree planting activities are mainly carried out by men and the concept of tree ownership (male dominated) is effectively sustained through well-manipulated cultural practices (taboos). As a result women hardly participate in tree planting activities.

In many homes trees are not specifically planted to supply fuelwood to the household. Fuelwood is considered to be a by-product after the tree's economic value has already been derived. Fuelwood procurement is predominantly a woman's task, and fuelwood collection outside the individual's farm is no longer possible. Women are therefore finding it necessary to engage in practices that will enable them to procure fuelwood either on their own farms or from the market. Through agro-forestry programmes, women are able to establish their own trees.

The Ministry of Environment and Natural Resources, Department of Forestry and some Non-Governmental Organizations (NGOs) - namely the Kenya Energy Non-Governmental Organization (KENGO), the Kenya Woodfuel Development Programme (KWDP) and the Kenya Woodfuel and Agroforestry Programme (KWAP) - have been sensitizing the rural population to the issue of agro-forestry and tree planting. These Non-Governmental Organizations and extension staff from the Department of Forestry are working through women groups, the media, and chief's 'Barazas' to promote agro-forestry (Dietz et al 1994, Bradley 1984 & Chavangi 1984). The awareness has been created and the people understand the importance of planting trees for timber, fodder, fruits and firewood, to maintain soil fertility and reduce soil erosion, and to providing more shade, especially in the home compound. Trees and agro-forestry plants cultivated are Leucaena Leucocephala, Sesbania Sesban, and Markhamia platycalyx, Kie apple, Lantana-camara, Cypress and Euphorbia tirucali (Bradley et al 1984). Some of these agro-forestry plants are now planted together with food crops.

Farmers have observed that eucalyptus trees draw considerable water from the soil and make them harder. Many of the swampy areas and small streams now dry up during the short dry season, creating a problem of water collection and retention. However, the eucalyptus trees are valued in all households because they provide an income. Although every household has eucalyptus trees, there is still a shortage of fuelwood in many households because males consider the economic value of trees to be more important than the household's need for wood for cooking. What is noticeable, therefore, is an intra-household conflict over the use of trees.

4.8 Agricultural Labour

We have already identified four farming sub-systems at the farm household level. These four sub-systems are present on all the farms selected for this study. Sufficient labour is essential for the effective and efficient operation of the four sub-systems.

In the North Maragoli farming system there are conflicting and coinciding activities within the household that require labour throughout the year. Serious competition for labour is visible as far as tea, coffee, French beans, food crops and cattle management are concerned, labour is also needed for domestic work and non-agricultural activities. Food crops and French beans are cultivated twice a year while tea and coffee are permanent crops. In total, all these sub-systems require a sizable labour input in order to reap satisfactory returns. All the adult women engage in farming activity on their own small holdings, apart from their domestic tasks of fetching water and firewood, (often at great distances from their homes) housekeeping, preparation of meals and childcare. The farmers conceded that family labour is not always available and it is not easy to control family labour. It is the adult female in every farm household who is the main source of labour. Given the situation of limited family labour, good husbandry requires that the farm households will have to spend money on hiring labour to ease labour bottlenecks.

Regardless of trends towards diversification and intensification, and the relative labour surplus associated with high population, women and poor farmers are seriously affected by labour shortages. The homogeneity in cropping systems prevalent in this area results in a labour peak from March through June and another smaller peak during the second cropping season from August to October. It was observed that there is a labour peak at times when food shortages are most severe. Most of the poor farmers are obliged to make trade-offs between working on their own farms and hiring out their labour for cash to buy food. These farmers have no choice but to neglect their own farms, thereby endangering the future food security of the household, cash crop production and incomes.

The existing relationships among members of the household have not been properly taken into consideration in the introduction of agricultural innovations by external agencies. Gender issues and intra-household dynamics are important in the structure and process of work patterns, capital accumulation and investment, off-farm employment, delegation of responsibilities and household status. Conflicts among members of a household are one of the causes of poor management and low agricultural production. It is important to find out the major causes of household conflicts as these relate to agricultural production. We have found that access to and control over resources such as land, livestock, trees and agricultural sales and incomes is highly conditioned by gender. The differentiation in gender relations and gender division of labour in relation to access to and control of resources affects farm management and the overall functioning of the household farming system.

Off-farm activities directly affecting the farm-household system in this area include the migration of members of the household. Some of the migrants are able to remit the needed incomes and others are not. On many occasions remittances are used for direct consumption and investment in education which is the most outstanding 'investment' in many of the sample households. The migration of members of the household reduces labour supply to the household. In many of the households, this labour is not compensated for. This is mainly due to the fact that in some households there are no remittances from the migrants and in the others the remittances are too small and irregular to be relied upon. Other activities found within a farm-household system that require labour and reduce labour for the farm are homestead household chores, child care, marketing and leisure. Sickness also makes inroads into the labour time available. Many of these activities are not considered as time and labour consuming variables and are not planned for in terms of labour time.

It is often assumed by policy makers and the agricultural staff that each household has sufficient labour. This explains the silence on the issue of financing labour in small-scale farming and the absence of government involvement in finding solutions to the labour constraints in this sector. It is taken for granted that all household members have equal access to farm incomes and have family savings that can be utilized to hire labour. The farming systems approach and the disaggregation of the household opens up a whole field of differentiation in production units, consumption patterns, decision-making patterns and forms of allocating of labour at the household level. These differentiations affect the management, production and eventually the incomes generated by the different sub-systems.

Labour analysis in many households reveals that women are the ones who are permanently around and who supply almost all household and farm labour. They are occasionally assisted by children on non-school days. Time allocated to different household and farm tasks appears, in practice, to be very fragmented because of the multiplicity of responsibilities to met each day. The labour problem is one of the reasons for poor management and low agricultural production on many farms and instances of late planting, and late and haphazard weeding are widespread. At times in the year when tasks overlap (labour peak period) it was observed that farmers would first tried to complete food crop activities before turning to the cash crops. Tea plucking is affected most as it has to compete with food crop labour.

Approximately 90% of the farmers complained of shortages of farm labour and that this affected a variety of crops. Many families do not generate sufficient income from their land to be able to afford even the low wages paid to agricultural workers in the area. The vicious circle of low production and low incomes repeats itself year after year. Farmers complained that the required wage level (Kshs 30 per day and one meal in 1992/93) was too high for them. It should be noted here that it was found easier to measure labour input by using the amount of money spent instead of hours of work per week as there was no systematic pattern of labour utilization and our survey was but a one-time visit interview. Table 4.16 shows expenditure on hired agricultural labour for the sample population.

The cheapest hired agricultural labour would cost approximately Kshs 600 per month (1991-1994). This represents one hired labourer assisting with all the farm work generated by a household. The total amount of money paid to hired labourers by the 180 farmers in sample survey was Kshs 88,400 in 1991. This was an average expenditure of Kshs 490 per household in

Table 4.16
Expenditure on Hired Agricultural Labour 1991 (source: North Maragoli Sample Survey 1992).

Amount Spent on Labour 1991 in Kshs	Number of Respondents	Percentage of Total Sample	Cumulative Percentage
00 - 00	104	57.8	57.8
01 - 500	26	14.4	72.2
501 - 1000	19	10.6	82.8
1001 - 2000	17	9.4	92.2
>2000	14	7.8	100.0
Total	n = 180	100 %	100 %

one year. The average amount spent on labour per month per household was only Kshs 41. At the time of this study this sum was only sufficient for a household to hire labour for two days a month. It is evident from the table above that farm households in fact spend very insignificant amounts on labour. In reality, in 1991, 58% of the sample households did not use hired labour but relied entirely on family labour. As a result, farm operations were quite often carried out late, done inefficiently or left out completely. Labour shortage partially explains poor management and low production in all the sub-systems in North Maragoli and Vihiga District.

Other problems related to labour are: (i) low levels of labour productivity caused by absence or severe lack of physical capital and experienced management. This resulted in inefficiency which could be seen in both the food and cash crop sub-systems discussed before (see Sections 4.4.1 and 5.5.2); (ii) poor worker motivation due to the marginalization of small-scale farming as seen in weak incentives and low incomes; and (iii) institutional inflexibility (see Todaro 1994). Institutional changes such as the reform of land tenure, credit, and the creation and strengthening of independent, honest and efficient administrative services may improve the motivation and productivity of farmers (especially women) in small-scale farming areas.

4.9 Agro-Support Institutions in North Maragoli

The success of the farm-household sub-systems depend very much on agro-support services. These are mainly supplied by the Government through different ministries but chiefly through the Ministry of Agriculture, Livestock Development and Marketing. Agro-support services in North Maragoli mainly consist of agricultural extension and credit. Such services are meant to assist farmers to cope with the farm management situation and to improve agricultural production. Indeed such services are a must and important for all farmers regardless of gender and social status.

4.9.1 Agricultural Credit

Money is required in agricultural production for the purchase of inputs and the payment of hired labour which is essential for good farm management. To raise productivity, domestic or household savings must be mobilized to generate new investments in physical capital goods and to build-up the stock of human capital (for example, managerial skills) (Todaro 1994). However, owing to the low margin of savings, most farmers are unable to accumulate capital. More than 90% of the farmers in the sample complained of lack of capital (money) as one of the constraints to improving management through farming investment. This implies that there is a need for other ways of funding agricultural production in small-scale farming. The most important sources of credit are the government and the remittances from working relatives. However, credit facilities for small-scale farmers are very limited and poorly distributed. The main reason cited by those concerned with credit facilities was that farmers were unable to offer acceptable security. Consequently, small-scale producers are caught up in a vicious circle: a low level of income leads to a low rate of capital investment which leads again to a low level of agricultural productivity. Thus the level of income remains low which explains the never-ending story of rural food shortage and poverty.

The Agricultural Finance Corporation (AFC) is the main government parastatal body charged with extending agricultural credit to farmers. Other groups who lend money to farmers are commercial banks, agricultural co-operatives and special rural development programmes or projects. The AFC extends three basic types of loans to farmers. Since the late 1960s farmers in North Maragoli and in other small-scale and sometimes large-scale farming areas are supposed to benefit from these loan facilities:

1 **Long-term loans** for land purchase, the purchase of dairy cattle and for permanent improvements of land. These loans are normally given to farmers with a minimum of five hectares; a land title deed as collateral is required and, where possible, an additional security (for example, certificate of registration), or salary. These long-term loans account for an estimated 43% of all farm credit in Kenya.
2 **Medium-term loans** for farm development, the purchase of machinery, and to establish crops. These are, in theory, loans available to all farmers. They also require land title deeds as security. These are estimated to account for 31% of the total farm credit in Kenya.
3 **Short-term loans**, such as Seasonal Farm Credit to cover costs in planting maize, wheat, pyrethrum and tea, and for the purchase of other farm inputs such as fertilizers and pesticides. These are estimated to account for about 20% of total farm credit.

We found that the majority of small-scale farmers in this area do not have access to AFC loans. Farmers complained that AFC discriminated in giving out loans. We found that the main impediment to credit in many households is the small size of the farm which is far below the size of the unit considered acceptable by the AFC and the commercial banks. Twenty percent of the farmers in our sample survey who had tried to get AFC or commercial loans had been asked for security in addition to a land title deed, such as fixed assets in the urban or rural areas. Only a limited number of farmers had these. Women are more disadvantaged because they cannot fulfil any of the conditions because they do not even own land.

Some small-scale farmers in our study area fear applying for loans because of the experience of other farmers, who took loans and used their land as collateral. When these farmers defaulted on their loan repayment, their land was auctioned. In our sample survey only five (3%) out of the 180 farmers had taken loans for land improvement between 1989-1991 from the AFC. Apart from the AFC, a few farmers had benefitted from commercial bank loans (2%) and loans from teachers co-operatives (5%), but these loans were not specifically for agricultural improvement (Table 4.17).

Agricultural credit institutions pointed out that at times loans have sometimes been misused and diverted to uses other than agriculture. Misuse of loan money by some farmers has dissuaded agricultural lending institutions from directing loans to smallholders whose holdings may not be worth the amount of the borrowed sum.

Apart from the above mentioned government sources of loans, the small-scale farmers in our study area have access to credit through internationally sponsored programmes:-
(1) the Integrated Agricultural Development Programme (IADP) and (2) the Smallholders Production Service Credit Programme (SPSCP). The IADP was sponsored by both the Kenya Government and the World Bank, while the SPSCP was partially sponsored by the Kenya Government and USAID. Very few farmers have benefitted from loans from these organizations.

Table 4.17
Sources of Loans taken by Farmers 1991 (source: North Maragoli Sample Survey 1992 ROSCA : Rotating Savings and Credit Associations).

Source of Loan	Respondents	Percentage
Comm. Bank	4	2.2
AFC	5	2.8
Teachers Co-op	9	5.0
Farmers Co-op	12	6.7
Revolving Funds (ROSCA)	95	52.8
None	55	30.5
Total	n = 180	100.0

A farm size of at least 1.2 hectares and a land title deed were among the conditions that had to be fulfilled before farmers qualified for these loans. Farmers had to have been active members of a co-operative society for a minimum of three years and should have attended farmers training courses. Farmers were also obliged to sell their produce through the society to repay the loans advanced. The loans were meant to improve maize, beans and groundnut cultivation (i.e food crops). Here again farm size was the limiting factor and the returns to the farmer and the society were too small to warrant giving loans. Moreover, the amount of food crops harvested in the study area is usually in sufficient to feed the household, let alone sell through the society. Only two out of 180 farmers interviewed had benefitted from these loans. They were government employees with other sources of income and bigger farms. Red-tape and bureaucratic procedures make these sources unaccessible.

For many rural households an important means of getting money is by forming of women's, men's and self-help associations (groups). These groups function somewhat like co-operative societies and are known in the international literature as 'ROSCA' (Rotating Savings and Credit Associations). These groups are informal and function with limited capital. One of their main objectives is to contribute money each month for use by members of the group and in this way a 'revolving fund' or 'Merry-Go-Round' is created (see also Chapter 8). The monthly contributions are made in rotation until all the members of the group have had their turn and then the cycle begins again. We found that the money from this source is mainly used to pay school fees and purchase of household goods, roofing materials but is rarely used for farming investment.

Another important source of money is remittances from migrant relatives, especially from migrant husbands. Many farmers stated that remittance money is usually earmarked for paying school fees, buying of food, building a house, or purchasing additional land. It is clear that very little is budgeted for farm management. It is also clear that agricultural credit and other sources of money do not contribute either to improving farm management and agricultural production as they should. In a nutshell, poor management and low production in North Maragoli and in other similar small-scale farming locations are closely associated with low farm investment due to the limited capital resources available.

4.9.2 Co-operatives in Production and Marketing

Since Independence, the objective of Co-operative Unions has been to strengthen small rural Co-operative Societies such as those found in North Maragoli (coffee co-operative societies, dairy co-operative societies). At the location level, the co-operative society is supposed to be the most important organization. It assists the smallholder obtain the advantages of economies of scale. The mother of all the co-operative societies in Vihiga District is the Lunyerere Coffee Co-operative Society, established in 1955. This society's premises are situated to the west of the district and because of distance and congestion it was unable to serve all the farmers in the district effectively. Therefore, co-operative societies have been established in almost every location and nowadays Vihiga District has 48 co-operative societies, mainly dealing with coffee cultivation. North Maragoli is served by the Mudete Coffee Co-operative Society and the Sabatia Dairy Co-operative Society.

The tasks of co-operative society officials, working together with the Ministry of Co-operatives, include: providing credit and inputs to farmers; processing and marketing, (especially coffee); providing extension to farmers; and taking care of the administration, supervision and management of the coffee factories. However, as elsewhere in the country, the co-operative movement has had its own problems. Among these in this area are: (i) inefficient and incompetent officials, many of them technically illiterate and hence incapable of running the factories and advising farmers; (ii) mismanagement of the co-operative funds, loans and inputs when these become available; and (iii) corruption and inequality in the way farmers are treated. Many of the co-operative societies in North Maragoli have not been able to fund coffee farmers and other enterprises, and this has resulted in the total collapse of coffee farming. The co-operative officers complained of the unwillingness of the farmers to learn and to co-operate with them. However, they confirmed that this unwillingness is a result of poor incentives and low payments from the cash crops. Many farmers want the government to allow them to uproot the coffee trees.

In contrast, tea seems to be a success story. This is attributed to the fact that the KTDA deals directly with farmers, while there are about five different organizations dealing with coffee growing, processing and marketing. As pointed out by Leonard (1991), from the very beginning every aspect of tea production was to be financed out of growers' sales, even the agricultural extension. The one exception was the necessary and extensive development of feeder roads, which the government agreed to provide. Tea was therefore spared the many taxes that coffee growers have had to pay in return for their extension and road services. This trade-off between more comprehensive responsibility in return for no taxes also added to the KTDA's autonomy by making it self-sufficient in most of resources. The organizational approach adopted by the KTDA was a typical late colonial one, stressing control as it sought to incorporate African peasants into the market. Many have credited the profit incentives, built into the KTDA system for its success though Leonard suggests that the KTDA organizational design reflects the colonial officers' lack of faith in the workings of the market. They feared that the many services needed for commercial tea production to succeed would not come together simply through market pressure. Thus they created a single, vertically integrated organization to provide all the planting materials, extension, transport, processing, and marketing necessary. The KTDA, therefore, was constructed with monopoly and monopsony powers and multiple controls over grower behaviour (Leonard, 1991:127-128). The success of tea is also linked to transportation arrangements and regular monthly and end of year payments.

4.9.3 Agricultural Extension

A major responsibility of the Ministry of Agriculture, Livestock Development and Marketing is to promote agricultural and livestock development so as to (1) conserve the basic national resources of land and water and maintain them for efficient productivity; (2) ensure adequate and reliable food supply; (3) support the agricultural processing industries by assuring a flow of agricultural and livestock raw materials; (4) promote the production of exportable commodities while increasing national food stocks; and, (5) create employment opportunities in the agricultural sector to meet the needs of an ever increasing population (Ong'ondo, 1976). To achieve these aims, the Ministry works with farmers through the agricultural extension services who convey innovations in agriculture to the farmers themselves.

Extension officers provide advise on efficient and more productive methods that can be used in crop and animal husbandry and also provide information about the right inputs and marketing strategies for particular produce (Okuku 1993). The agricultural staff are categorized according to qualification. The lowest category staff interacting directly with farmers are referred to as Agricultural Extension Agents or Frontline Agricultural Extension Agents or Junior Technical Assistants (see Figure 4.5). Many of the Frontline Extension Agents have had a few months of training at an agricultural college. For the rest they are trained on the job or once every fortnight before they convey information to the farmers. Extension staff in Kenya have tended to be single sector and mono-crop in their orientation (Sterkenburg 1987). The zoning of some areas for sugar cane, coffee, and tea means that small-scale farmers in areas without these crops are not included in extension programmes and do not get the right attention from extension officers neither do they receive encouragement to venture into these particular commodities (personal communication with the DDO, Kakamega).

It has been suggested in District Development Committee meetings that extension services should be organized on the basis of teamwork so that the beneficiaries receive the impact of different messages simultaneously and relationships between the sectoral packages can be harmonized. This may, however, be easier said than done unless sectoral policies and statements as well as training packages for all staff in line Ministries undergo innovative changes to achieve the real integration of extension services. Even within the same Ministry of Agriculture, Livestock Development and Marketing field training programmes cannot be organized simultaneously to targeted beneficiaries in the same area because the subject matter and issues to be covered would take different amounts of time to cover. The different departments of the Ministry of Agriculture try to be independent of each other even though they work with the same clientele. The extension approach is disaggregated and causes confusion to farmers at times. Even at very localized levels and in short multi-sectoral seminars or workshops there still appeared to be a clear compartmentalization of approaches and groups targeted within the same meeting.

Selected farmers are targeted to receive information on crops from the crop officers and information on livestock from livestock production and veterinary officers. Marketing and co-operative issues are hardly touched. In the final analysis, the integrated approach envisaged by these programmes is not realized because no linkages have been developed between production, markets, incomes, health facilities, nutrition practices and the commitment of all farmers to the strengthening co-operatives.

Another problem facing extension is the inability of the Government to increase the scale of field extension to maintain a reasonable farmer/extension officer ratio. This is a particular problem in areas such as our study area which has a dense population. Despite the rapid expansion of extension staff since Independence, the field staff/farmer ratio is still low and most farmers cannot get appropriate advice when they need it. Sabatia Division, of which North Maragoli is a part, had the following extension staff for a total of 14,692 farm households: one Extension Divisional Agricultural Officer who is the overall head of the station and under her was a Divisional Horticultural Officer, a Divisional Co-operative Officer, a Divisional Livestock Officer, a Divisional Soil Conservation Officer, a Divisional Crops Officer, a Divisional Farm Management Officer and a Divisional Veterinary Officer. Each of the Divisional Officers was supposed to have two assistant officers who work at location level (1:1837 farm households). Below the officer were the technical assistants or frontline workers and the junior technical assistants, each serving one sub-location (1:530 farm households) (see Figure 4.5). This division was understaffed by 8 assistants and this created considerable work for the few officers employed. The lack of transport was seen as a major hindrance to an effective extension: the division has only one vehicle and on many occasions it could not run because of a lack of servicing and fuel. The assistants to the Divisional Officers need motor bikes to make their work easier. At the moment there are only two motor cycles in the division, and these are specifically for livestock extension and for soil and water conservation projects. Bicycles may be cheaper but the steep slopes of the area do not favour them.

4.9.4 Types of Information Transmitted and Target Groups

The Ministry of Agriculture has arranged that communications concerning technological innovations and practices are delivered via monthly training seminars for field extension staff at Divisional levels and fortnightly, one-day training sessions for the lowest level or frontline extension staff. The agricultural extension agent or frontline extension staff selects eight farmers as representatives. The selected farmers are called 'contact farmers' because they are the farmers that interact with the extension staff. The method for disseminating agricultural messages is called 'Training and Visit' ('T&V'). It simply means that frontline extension staff are trained and they in turn visit the 'contact farmer' to convey news about new messages or innovations. It is hoped that by using this method there is a trickle down effect, so that the agricultural messages passed on to the 'contact farmer' are passed on by him to other farmers in his neighbourhood. The contact farmer's farm is considered to be a government demonstration farm, and therefore whenever the frontline officer visits all the farmers in the neighbourhood should gather on this farm to watch farming demonstrations or to learn about innovations. The observation was made that farmers were not informed of these meetings. Contact farmers also never make any effort to pass on extension information to other farmers in their neigbourhood. It was quite obvious that the selection of 'contact farmers' is unfair to farmers with a poor resource base. It was observed that more 'progressive' and predominantly male farmers were selected (see Chapter six, Mzee

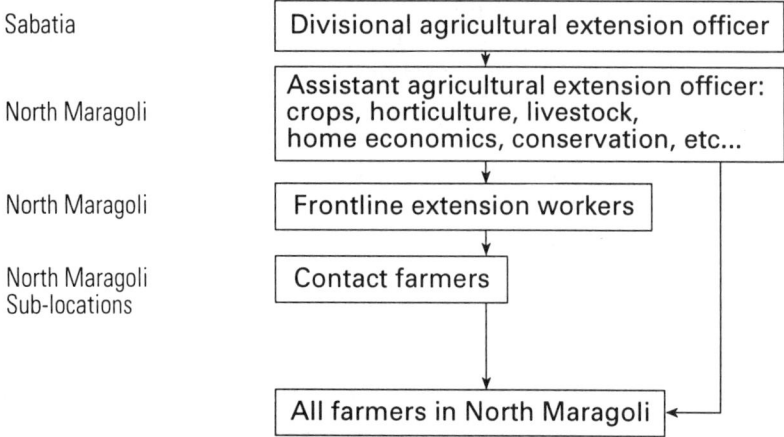

Figure 4.5
Structure of Agricultural Extension in North Maragoli (source: Study Results).

Musasa). Progressive farmers are farmers who are better placed than others in the neighbourhood in terms of finance, education and leadership skills.

Ideally, the T&V method should ensure a four-way traffic between research station, agricultural and livestock extension officers, frontline technical staff and the farmers and vice visa. Farmers reported that the content of the messages tend to be repetitive, apparently indicating that there is poor contact with research stations, no new messages and a lack of familiarity with or absent recent relevant publications. In the field it was apparent that only the 'contact farmers' receive these messages while the rest of the small-scale farmers were given little attention. Each Sub-location only has eight contact farmers among an average of 530 farm households. Only about 2% of the smallholders receive general agricultural messages directly. It is not realistic to expect contact farmers to take time to pass on messages to other farmers in the neighbourhood. Some of the reasons for why messages and technologies fail to be diffused are: (1) non-payment of the 'contact farmers', and a consequent lack of incentive; (2) lack of time to disseminate messages; (3) jealousies and ill feelings towards 'contact farmers' and allegations that the government only seems concerned with the well-to-do in the community. Frontline extension staff are seen as being the guest of certain people in the village and therefore the rest of the smallholders do not bother with his presence; (4) lack of administrative and organizational skills among frontline staff reflected in their failure to get neighbourhood farmers to assemble on the farms of contact farmers; (5) resource differentiation which contributes immensely to the disinterest of poor farmers.

Farmers complained that some frontline extension staff appear to know very little about actual farming and are strangers to the localities where they are stationed. Many of them are not aware of the needs and constraints of the small-scale farmers. Frontline extension staff do not visit 'contact farmers' as regularly as might be expected. Language plays a large role as far as the poor communication of the extension messages is concerned. Some of the officers use Kiswahili as a medium of communication. Many of the people, especially women do not use this language. Farmers are therefore unable to understand messages or ask questions. Some technical terms cannot be translated into Kiswahili or Luhya and so the frontline extension agent uses the English language. Worst placed are frontline officers who come from another ethnic group. The rate and regularity with which non-contact farmers learn from the 'progressive' farmers is either non-existent or very low.

Our findings indicate that the Ministry of Agriculture's measure of the short and long-term impact of staff and 'contact farmers' during demonstrations is expressed as a function of the number of farmers attending. No reference is made as to farm outputs as a measure of results. The messages mainly emphasize fertilizer use, and crop spacing. No effort is made to discuss the integration of traditional with new farming methods and relate this to the capital resource base of the farmers concerned.

Farming demonstrations by the Ministry are organized on the fields of a few individual farmers and at the Farmers' Training Centre (FTC) - Bukura for this area. Model farms operated by 'progressive' farmers or FTC farms are above standard in terms of inputs and intensive care. Ordinary farmers may not be able to afford this. Appeals to the ordinary farmers to follow the example of progressive farmers have yielded little positive result. The dilemma that confronts critical agricultural officers is whether to visit the already well-educated and 'overtrained' contact farmers, the non-contact interested farmers or the laggards or resource-poor farmers who really need a push but may not command the resources or the courage to learn from the 'progressives'. Although it is a common practice to deliver packages already tested in research stations concerning farming methodology, improved seeds, fertilizers, concentrates, fodder crops and pesticides for example, the tendency to develop information packages without prior consultation with farmers or methods developed from farm trials carried out over a number of years in other areas has been noted as a problem.

The Ministry's current technical packages delivered in both Kakamega and Vihiga Districts through extension networks were found to be unco-ordinated and beyond the purchasing power of most smallholders (Okuku, 1993). Extension officers give greater attention to the needs and problems of large-scale farmers than to the requirements of small-scale farmers. This tendency appears to have been strengthened by the emphasis on quick results in the various donor-funded programmes in the primary production sector. Programme investment in the poorest of the poor has been discouraged because of their slow response and poor results.

Government officers suggested that extension services have also been selective in that they favour the ease with which they can communicate with the educated farmers who can easily understand the pamphlets and discuss issues in the English language which is the language in which the staff are trained. Most new things are learnt by farmers in the verbal medium during 'barazas' (meetings) with practical demonstrations. Hand-outs are rare and if available these are in English or Kiswahili and thus incomprehensible to most rural farmers in North Maragoli. Staudt (1976) already found that female-headed households tended to be discriminated against as far as receiving attention from extension officers and donor-funded agricultural programmes are concerned. Extension officers generally expect women to be conservative, traditional and to resist or have difficulties with new crop and livestock techniques. Some extension agents at the local level still avoid contact with women farmers and groups, claiming cultural restraints. An earlier study based on Kakamega District observed that women farmers often resort to learning about agricultural innovations from other women through informal channels and networks (Fogelberg 1990).

These forms of selective targeting by extension services hampers people's participation. In many parts of the district it has been noted in 'barazas' that a majority of farmers neither acknowledge the presence, assistance nor the routine work of the extension services. This may be the result of the combined effect of extension being increasingly thin on the ground and officers becoming increasingly desk-bound because of the constraints of transport and lack of finance. As a result, most of the officers remain unfamiliar with the actual problems facing farmers in particular areas.

Approximately 90% of the farmers in the field survey expressed dissatisfaction with the extension services in our study area. Some farmers were surprised and did not believe that there were field extension officers who were supposed to keep farmers informed about the most recent technology or new development in all fields of farming. Such farmers had never gone to contact farmers' fields. Some farmers felt that there was discrimination in the way contact farmers and those attending courses at Farmers' Training Centres were selected. These farmers believed that mainly progressive and well-to-do farmers were selected. However, the officers are of the opinion that only farmers who can afford to spend money and are good in farm management are selected in order not to waste time. Some contact farmers are also disillusioned because some of them have not been visited by extension officers for a long time and have not received the promised inputs. Moreover, farmers expressed dissatisfaction with the 'rawness' (young and less knowledgeable) of some of the graduates fresh out of agricultural training institutions who are neither experts, farmers in their own right, nor show great interest in farm-work. Some farmers maintained that the exchange of ideas between the local farmers is more important than information received from young frontline officers.

The major extension constraints can be summarized as being the limited technical know-how of frontline extension workers; limited opportunities for regular consultation between extension workers and agricultural and livestock researchers; problems in communication between poor unmotivated farmers and extension agents; and the conservative dependence of farmers on extension agents to the point where the failure of a recommended technique or crop is blamed on extension agents (Bahemuka 1983).
This leads to the observation by Fuglesang (1984:1-2) that local farmers have often developed more subtle solutions to their problems using their intimate indigenous knowledge of their environment. Their survival strategies often appear to go against the rational modernistic advice and economic incentives suggested by non-farming trainers and experts. We agree with Fuglesang's suggestion that the idea that people cannot think for themselves and must be guided in the elementary aspects of farming needs to be discarded and replaced by a willingness amongst extension officers to learn first from the farmers about the local farming environment. This is partly emphasized in the farming systems approach. The irrelevance of the extension messages, often compounded by their non-innovative character and repetitiveness, may explain

Small-scale Farming in North-Maragoli 117

why the farmers are reluctant to acknowledge assistance from the extension services. Unrealistic advice about how to achieve better production and higher incomes is given to local farmers regardless of the inflated costs of inputs, the poor level of farm technology and the complications that usually arise from delayed payments for produce delivered. The situation is made worse by delays in the delivery of farm inputs such as fertilizers and seeds.

4.10 Conclusions

This chapter has made it clear that the farm management of most of the farming sub-systems is poor and the overall agricultural production in North Maragoli is low. Poor farm management and low agricultural production is influenced by problems at household, local and institutional levels.

Household level problems:- Farm management is poor to modest in many of the farm-households. Poverty and backwardness are widespread and these prevent farmers from investing in labour, purchasing farm inputs and using hired labour. At the same time there is a severe shortage of household labour. Indeed, farm tasks cannot be performed on time or are performed inefficiently due to labour shortages and scarcity of money. Many farmers have little or no education. This contributes to poor husbandry practices in the management of those exotic crops requiring specialized knowledge and management practices.

Local level problems:- North Maragoli Location and Vihiga District have the highest population densities in Kenya. This high population density has contributed to increased farm fragmentation and decreasing farm size. There is, therefore, a competition for land and only small plots are allocated to each farming sub-system. Such intensive use of land calls for good planning and management. However, these two factors are missing. Due to poor farm management and low agricultural production at the household level, the location and district is food deficient and the cash crops produced are of poor quality. Thus farm incomes are depressed and many households are impoverished. Local co-operatives dealing with coffee and the coffee factories themselves are inefficiently and poorly managed. This has contributed to disincentives against coffee farming.

Institutional level problems:- Good farm management, especially of the main crops (maize, tea, coffee and French beans) and improved livestock production depend on the efficiency of the agro-support services extended to farmers. As already noted, these crops require specialized husbandry for successful management and production. However, this study has made it clear that agricultural extension is inadequate as far as many farmers are concerned and this explains the poor management of tea, coffee and hybrid maize crops in particular. However, farmers need credit in order to be able to buy farm inputs. As we have seen above, many farmers have no access to credit and this impedes the use of adequate farm inputs. Other problems involve distribution and production. It has been shown that low quality seeds are produced and sold to farmers, sometimes there is shortage of inputs or these are delivered too late and so contribute to untimeliness in agricultural activities. Tea is singled out as performing slightly better than the others but because there is no processing facility in the district or province, the quality of the produce is often degraded due to spoilage during transportation over long distances.

Notes

1 An earlier study by Luning (1987) found farm sizes in the range of 0.1 to 1.2 ha with an average of 0.8 ha per farm family.

2 "A hybrid is produced by crossing different varieties in two-steps thus reaching a new variety which contains the genetic qualities of the so-called inbreed lines. In the process of crossing, deliberate selections are made of the inbreed lines with respect to the particular qualities one would like to have in the final hybrid. In making this selection it is possible to produce seed varieties adapted to specific environments. A primary focus in the breeding is to adapt seed varieties to the wide differences in altitude and subsequently, differences in rainfall and temperatures" (Rundquist, 1984:94).

3 Kshs 1,250 instead of Kshs 450 per 25 kg and Kshs 550 instead of Kshs 120 for 10 kg bag (KNTC and KGGCU price list 1994).

4 There are two types of tea payments to farmers. One is the monthly payment and the second is the end of year payment which is referred to as 'Bonus'. The end of year tea payment is usually the largest of the two.

5 As already been pointed out, these officers are recruited according to qualifications. The Divisional Officers are qualified and have a Diploma in agriculture or livestock from Egerton, followed by those with certificates from Farmers' Training Colleges such as Bukura. The rest are school drop-outs who are trained on the job. The officers in contact with farmers are those who are trained on the job and do not have an overwhelming exposure to technical training. This creates a problem in the trickle-down effect of messages. It is quite possible for the messages to be distorted in this long chain of top-down approach of conveyance.

6 Farmers' Training Centres (FTCs) have been used since the colonial time as institutions for training farmers by demonstrations and teaching. During the course of the study, this researcher visited Bukura Farmers' Training Centre which trains farmers for both Kakamega and Vihiga Districts. The researcher had an opportunity to follow some of the courses offered to agricultural staff and to the farmers and finally perused the curriculum used. Farmers in 1992/93 were expected to pay 60 shillings as a registration fee in order to attend organized courses. The courses that FTC offers are: (1) general management including soil conservation; (2) horticultural farming; (3) general crop production; (4) livestock courses e.g poultry course, dairy course, bee-keeping, pig-keeping; (5) coffee; (6) home economics; and (7) women group leadership courses. Each of these courses takes between one to two weeks. Each course is offered once or twice a year. The FTC has a capacity for only 52 farmers at every session. Only four participants can be selected from each Division. Therefore, for Divisions like Sabatia wherein North Maragoli falls, four farm households represent approximately 15,000 farm households whenever these courses are offered. This clearly is a very small number and is unlikely to have an impact in a division or location. The farmers who participant in these courses are selected by frontline extension/technical assistants in liaison with divisional staff. The farmers selected must of course have an interest in the courses and must have money for registration. This automatically qualifies the 'contact farmers.' More than 80% of the participants in these courses are male farmers often the same farmers appear in most of the courses. Women are recruited for the home economics and management and women leadership courses. This clearly marginalizes the role of women and emphasizes their traditional role.

5

Farm Management Situations and Types of Household in Small-scale Farming in North Maragoli

5.1 Introduction

In the previous chapter we defined the farm-household system in North Maragoli. Subsequently, four sub-systems were described and analyzed in detail. It was shown that in our North Maragoli farming system, yields, production and incomes in most of the sub-systems are low due to multifarious management problems and therefore do not satisfy many household needs.

The purpose of the present chapter is to clarify and analyze the farm-household system from the point of view of household types and from a gender perspective in order to gain a better insight into farm management situations and agricultural production. Three types of households have been identified and are analyzed in this chapter: male-headed households, female-managed households and female-headed households:

(i) Male-headed households - a household where a husband is present throughout the year and is directly involved in farm management decision-making, supposedly with his wife. (Staudt 1985)
(ii) Female-managed households - a household where the female manages the household and the farm in the absence of the husband who has migrated to work elsewhere. He may send money/goods to the household and he also receives goods (food) from the household. Once in a while he comes home; he helps the wife in farm management decisions either through writing, or sending messages through friends and relatives or when he comes home. Thus, the husband has a full-time occupation other than farming but continues to be a part-time farmer by virtue of the fact that he owns the land, contributes some labour himself and/or pays for hired labour (Safilios, 1988 a).

There are cases where the husband is in and out of the home very often. In this study, however, if the man was away from the household for more than three months between March and August when farming operations are most intense, then the household was considered being female-managed. If the man was away for less than three months, however, then such household was considered as male-headed.

(iii) Female-headed households - a household where the female head is a widow, a divorcee, a woman who has been abandoned by her husband or a single woman who has acquired her own land. In this case, the woman is entirely responsible for the household and farm management, and thus makes all key agricultural decisions.

This chapter will address the following research questions:-
1. What differences can we observe between and among the three types of households, with

special reference to the quality of farm management in terms of farming investments, farming techniques, crop production and farm incomes?
2. What are the main factors that negatively affect the adoption of farming innovations, farm investments, crop production and farm incomes in the three types of households?
3. What are the main problems experienced by women farmers in overcoming these negative factors?
4. What coping strategies are employed by the various households?

In an attempt to answer the above questions, various aspects of the characteristics of the sample households will be analyzed. The most important factor to establish is whether gender differentiation is an important issue in explaining of farm management and agricultural production within the farm-household system. If a differentiation does exist, then there must be reasons for the variations. In relation to this the topics discussed are as follows:

(i) Land tenure system and farm size
(ii) Maize and tea farming sub-systems
(iii) Livestock and milk production farming sub-system
(iv) Tree production farming sub-system
(iv) Gender division of labour and labour use
(v) The role of extension in the three types of households
(vi) Coping strategies, general strategies for the three household types and specific strategies for women
(vii) Analysis of household incomes
(viii) Production and living conditions of selected farm households
(ix) Problems affecting the sample farms
(x) Conclusions derived from the household analysis.

5.2 The Selection of the Respondents

The analysis of farm management situations in relation to types of households is based on an in-depth study of 33 households selected from among the 180 respondents in the general survey. Apart from a representative distribution over types of households, the other criteria considered in the selection of the cases were:

(i) a few 'contact farmers' to study the question of agricultural extension and its diffusion of information to all the farmers, (ii) a few progressive/rich and a few poor farmers to compare farm management differences, (iii) a few farmers in full-time employment but living in the village to compare with the majority of the farmers who are not in formal employment and, (iv) households along the main road and others in the interior with no access roads, to evaluate agro-support services and input delivery systems.

All the heads of households had farms ranging in size from 0.2 to 4.6 hectares. Of the 33 households selected, 13 were male-headed, 11 were female-managed and 9 were female-headed households. All the small-scale farmers selected were visited on a monthly basis for a period of twelve months and information was gathered using a checklist. Data collection included information on: (1) tea farming operations, for example, number of tea bushes, farm size under tea, extension services provided for tea by KTDA, information on weeding, pruning, plucking,

Farm Management Situations and Types of Household

labour allocation, acquisition and use of fertilizers, tea production and incomes; (2) subsistence farming with an emphasis on maize. Investment in subsistence farming, for instance the use of improved seeds (Hybrid seeds), the use of fertilizer, actual farming operations, labour allocation, production and farm incomes; (3) livestock management - Types of livestock kept by the different types of households, feeding and dipping of livestock, milk production and income from sale of livestock and milk; (4) tree production sub-system; (5) general extension visits and advice; (6) loan acquisition; (7) other sources of income; (8) farming investments; (9) general labour allocation; and (10) general and specific household problems and household coping strategies.

5.3 Characteristics of the Selected Households

Among the members of the households studied, about one half were children and young people under 15 years of age. These were mainly school-going children and youngsters and it could hardly be said that they were available for farm work. Their labour was perhaps only available during the weekends and in school holidays.

The total population of the selected 33 households was 247. Forty-three percent of the total household population fell in the age category 0 to 19 years. Two percent of the household population consisted of elderly people over 60 years. This implies that approximately 45% of the population was made up of dependents. This creates economic and labour constraints on those who work or provide for the livelihood of the households. However, the research findings indicate that females heading households were still actively participating in farm management even at the advanced age of 60 plus because they lacked social and economic support and often had many young grandchildren under their care. Fifty-five percent of the population consisted of active working adults (Table 5.1 and Figure 5.1). In the age category 20 to 59 years, approximately 50% of men had migrated out of the area, joining what can be referred to as off-farm employment. Table 5.1 shows age and population distribution among the households of the in-depth case studies.

Total male household population is 117 (47%) and the total female population is 130 (53%). This pattern is similar to the patterns observed amongst the Vihiga and North Maragoli population in general (see 3.3.1 & 4.2).

Table 5.1
Case Studies Population Distribution by Age and Sex (source: North Maragoli In-depth Case studies, 1993)

Type of	00-14		15-19		20-29		30-39		40-49		50-59		60-69		Total	
Household	M	F	M	F	M	F	M	F	M	F	M	F	M	F	M	F
Male hh	22	15	08	10	13	12	06	07	04	05	04	03	03	00	60	52
Female mh	11	10	04	08	06	09	02	11	01	04	00	03	00	01	24	46
Female hh	05	03	04	05	13	06	09	06	02	04	00	05	00	03	33	32
Total	38	28	16	23	32	27	17	24	07	13	04	11	03	04	117	130

North Maragoli-Case Studies

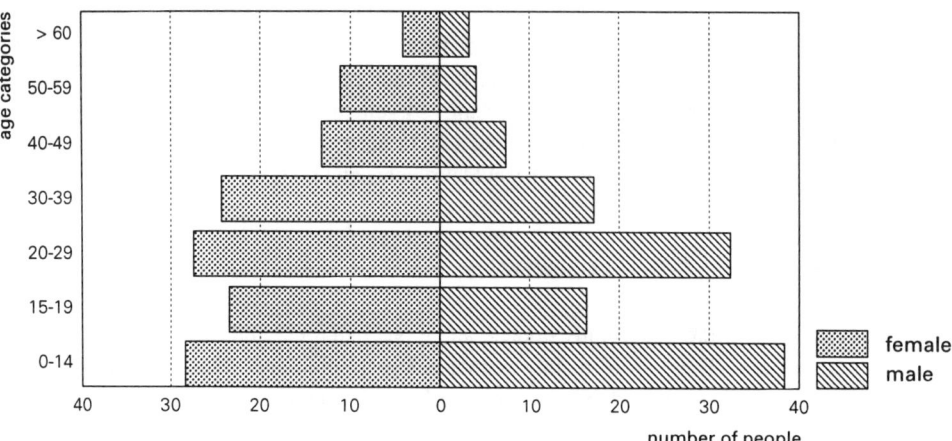

Figure 5.1
Population Pyramid by Sex and Age (source: North Maragoli In-depth Case Studies, 1993).

The average age for heads of male-headed households was 50 years. These were mainly men who had retired from off-farm employment and had settled into peasant rural life, a phenomenon common amongst retired urban workers. The rural home represents the forces of reproduction, production and the ancestral burial place. It is the place to return to after retirement (Gugler 1991, Anderson 1990). It appeared that many of the retired men provided only limited labour on the farms although they make decisions and give instructions to the members of their respective households. Some of these men seemed disillusioned. Alcohol was identified by many members of households as one of the social problems that caused family and household conflicts and to some extent negatively affected farm management and agricultural production.

The average age for the **de facto** head in the female-managed households was 43 years. Female-managed households were found to consist of the younger and more active farmers. The average age for **de jure** heads in the female-headed households was 55 years. These women are mainly widows, or women who had been deserted by their husbands for longer than five years. Age is an important parameter in agricultural production because it determines the number of working hours that can be undertaken. Age may also determine efficiency in the utilization of resources, investment, and access to the technical information necessary for cash crop cultivation and improved livestock keeping. This is particularly the case among those farmers who have very few years of schooling.

Education is an important parameter, especially today when new scientific knowledge is increasingly being applied to agriculture. Anker et.al (1983) in their analysis of Kenya and Vihiga state that inputs, expenses and net income vary with farmers' education. They found that the amount of purchased inputs per acre increases directly with the level of the farmer's education and that the increased use of inputs was reflected in significantly higher levels of net income per hectare. More educated farmers exhibit a greater degree of economic efficiency. It was found during this study that farmers with a basic education (8 years of schooling and more) were able to read farming instructions in booklets, posters, pamphlets or on packaged inputs such as hybrid seeds, pesticides and fertilizers or instructions on how to use different farming equipment. They could also easily communicate with extension officers.

Upper primary is the minimum level of schooling necessary for a farmer to be able to follow and understand farming instructions, especially when those introductions concern new crops such as tea, coffee, French beans, and hybrid maize which need specialized management skills. In many households, the children who might be able to read for their parents are away from home most of the time and farming until someone is found who can read the instructions. Instructions are often in English and/or Kiswahili which some primary school children may find difficult to interpret correctly into the local language. Table 5.2 shows the levels of education for the household heads on the sample farms.

Table 5.2
Levels of Education for Household Heads (source: North Maragoli In-depth Case studies, 1993).

Type of household	No Education	Lower Primary 1 - 4	Upper Primary 5-7-8	Secondary	High School	Total
Male hh	1 (8%)	3 (23%)	5 (38%)	3 (23%)	1 (8%)	13
Female m	1 (9%)	5 (46%)	3 (27%)	2 (18%)	0 (0%)	11
Female h	6 (67%)	2 (22%)	0 (00%)	1 (11%)	0 (0%)	9
Total	8 (24%)	10 (30%)	8 (24%)	6 (18%)	1 (3%)	33

Twenty-four percent of the farmers had no formal schooling and 30% had lower primary education which is considered inadequate because such farmers were unable to read and interpret farming instructions correctly. Approximately 55% of the farmers can be considered illiterate. There is a considerable difference in education levels between different household types.

The level of education was lowest among the female-headed households where 67% of the heads had no education and 22% had only lower primary education. None had upper primary education and only one female in this household type had secondary education and was the only member of the group in formal government employment. Part of the reason for this low level of education could be age. Female-headed households are mainly managed by older women over 50 years of age. Many of these female heads were children during the colonial period, when there was no wide spread access to education. Education was for the select few, the sons of chiefs or religious converts attached to church mission stations. During the colonial period formal schooling for girls was restricted or unavailable (Lavrijsen 1984, Moock 1976).
In female-managed households, 9% of the respondents had no education, 46% had lower primary and 45% had upper primary and more. Education among male-headed households was

found to be well distributed in all the categories with 69% of the males having had upper primary education, secondary education and more. There are many reasons why women have a lower education than men: families have limited financial resources and make a choice in favour of boys, early pregnancies, and a tradition which considered males to be breadwinners and the heads of family while females are seen as housewives. Thus there is a considerable differentiation between the three types of households in our study area. Does this differentiation affect household decision-making and the farming system?

Illiteracy handicaps women in farm management. The recommended crop and livestock management practices are not followed, and more often than not are taken for granted. It was apparent that women heads of household sometimes bought the wrong seed and other inputs which were not suitable for their region because they could not read. Under-utilization and the incorrect use of fertilizer is one of the problems associated with illiteracy. Cases were found where CAN was applied while planting maize when it is in fact a top dressing, or NPK which is meant for tea was applied when planting maize. Such incidences caused by ignorance and illiteracy lead to low harvests. These illiteracy-related problems are more common in female-headed households than in the other two types of households.

Women's disinterest in and non-attendance of the chief's 'barazas', seminars and workshops is partially because of the result of illiteracy as they cannot follow the proceedings in such meetings, especially when in English and/or Kiswahili are used which is most often the case. Illiteracy, if combined with lack of extension messages, lead to poor crop and livestock management. The marginalization of women in education in this area appears to be affecting agricultural production. The disparity in education received by men and women is becoming more pronounced as the structural adjustment programmes whereby the government is decreasing its education subsidy are implemented. The implication of these programmes is that parents must pay more for the education of their children. Many poor rural households (especially female-headed households) cannot afford the fees and other expenses associated with schooling. This eventually leads to gender discrimination when there is shortage of money as many households prefer to educate boys. Thus the disparity in education between men and women may remain a permanent feature in small-scale farming areas, reinforcing the dominance of women in agricultural production while men, who enjoy better education, have opportunities of migrating in search of employment. At the same time, the low status of women's education may continue to present problems as far as the management of farm-household systems are concerned and this is particularly so in the more intensive commercial and improved livestock production sub-systems.

5.4 Land Tenure and Farm Size in Relation to Gender and Types of Households

As mentioned in Section 2.2.1, the main system of land tenure in the zones of high agricultural potential is individual freehold or private land tenure. Although this system was established to encourage farmers to consolidate holdings under individual rather than collective ownership and to introduce more profitable crops and technology, the plan set the precedent for post-colonial land tenure policies that legitimized differential access to land between men and women and the unequal land distribution. The land tenure system is totally biased against women (Davison 1988, Saito et al 1994).

A critical issue for smallholders throughout Kenya is the shortage of good quality farming land. Increasing population pressure and the fragmentation of holdings have sharply reduced the cultivated area per person (see 3.3.1). For women, the situation is more critical. Faced with uncertain tenure and the decreasing size and quality of plots to farm, they have an exceptionally difficult task in maintaining levels of output and household food security (Saito et al 1994). It is therefore not surprising that farms in North Maragoli are registered in the names of men. Two-thirds of male heads had land titles and only one female had land registered in her name.

The land tenure system marginalizes women and creates problems for agricultural production because decisions on land use are made by men. Women are marginalized because of the patrilineal system that is prevalent in many parts of Kenya and particularly in North Maragoli. Women's land rights are dependent on their relation to males, usually a father, brother, husband or a son (Takyiwaa 1989). Among the Luhya people, women do not inherit land in their own right. Cases where husbands sell the land unilaterally leaving small amounts of poor unproductive land or no land for female management were reported. Conflicts over land were reported especially in polygamous marriages or in cases of desertion. It was observed that insecurity about land ownership and use contributed to the poor farm management prevalent in female-headed and some female-managed households. More than 45% of the females in the survey suffered from landuse insecurity. Land is a very sensitive issue in many families as there is obvious competition for use as can be seen in cases where there are grown-up sons who demand their own share and right. Such competition and conflict leads to family instability and therefore to poor farm management.

As already discussed in Section 4.2, farms in this area are becoming progressively smaller and smaller because of a growing population and the farm fragmentation that accompanies this.

Table 5.3
Farm Sizes among the Case Study Samples 1993 (source: North Maragoli In-depth Case Studies, 1993).

Household type Farm size in ha	Male-headed		Female-managed		Female-headed	
< 0.4	4	31 %	5	46 %	1	11 %
0.5 - 0.9	2	15 %	1	09 %	4	44 %
1.0 - 1.5	6	46 %	4	36 %	2	22 %
1.6 - 2.0	0	00 %	1	09 %	0	00 %
>2.0	1	08 %	0	00 %	2	22 %
	13	100 %	11	100 %	9	100 %
Total Average	13.3 ha 1.0 ha		8.6 ha 0.8 ha		13.5 ha 1.5 ha	
Total for all 35.4 ha	Average for all 1.1 ha per hhh					

However, we want to find out whether there is a difference in farm sizes between the three types of households shown in Table 5.3.

The size of the farm a household operates is influenced by a several of factors. Three are particularly important in North Maragoli, i.e type of household, age of household head (life cycle) and sources of income related to the level of education and/or off-farm employment and wage levels. The farmers above 50 years of age had slightly larger farms of one hectare or more (the terms large and small farm size are relative terms for the North Maragoli situation). Many farms of more than one hectare had not been subdivided. This phenomenon was more prevalent in female-headed households where the husband had died before the farm could be subdivided. It was also found in male-headed households where the heads were over 50 years of age and had not sub-divided the land for their sons. The female-managed households tend to comprise of the younger generation and most of them had smaller parcels of land. This small farm size may be one of the explanations for the high rate of male out-migration.

Respondents who had at least completed secondary education and were in fairly well paid jobs (for example teachers, civil servants) had farms of more than one hectare. All of them had used part of their earnings or had acquired loans from co-operative societies to procure more land. This appears to be the norm for most men from this area. Only one female in the female-headed household category had secondary education and she was a primary school teacher. Her farm was larger than one hectare because she had used the teachers' co-operative society to get loans and this had enabled her to buy more land. In male-headed households, four out of seven heads, with farms larger than one hectare, had secondary school education; three were teachers and one had been working with the Kenya Grain Growers Co-operative Union (KGGCU). Many of the absent males in female-managed households did not appear to have well-paying jobs but most of them claimed that they were still accumulating savings to buy additional land later on. In this area, land has been a thorny and disturbing issue since the time of migration and settlement, and during the colonial period. The amount of land available continues to shrink, yet many rural peoples' livelihood is pegged to it. Immediate attention must be given to dealing with inequality (prejudice) in land ownership and the use of land on a gender basis.

5.5 The Household Type and its Effect on the Maize Farming Sub-system

The introduction of hybrid maize seeds means that farmers are obliged to follow certain agronomic practices (see Section 4.4). Farmers who do not follow these practices risk lower yields than those farmers using seeds from local varieties (Allen 1969, Rundquist 1984).

Per household, the average area under maize for the total in-depth sample is 0.3 hectares. There were slight differences in the average size of land allocated to maize in the three different types of households: female-headed households had 0.4 hectares, male-headed households 0.3 hectares and female-managed households 0.2 hectares (Table 5.4).

Maize production occupies approximately one third of the total amount of farm land owned by our in-depth case study respondents, the same fraction as in the sample survey. The averages in

Table 5.4
Farm Size under Maize Production (source: North Maragoli In-depth Case Studies, 1993).

Type of household	Total farm for maize in ha	Average per hh member -ha	% of total group land	Sample Total	Survey Ave
Male-headed hh	4.4 of 13.3	0.3 ha	33%	21 ha	0.35
Female-managed hh	1.8 of 8.6	0.2 ha	21%	20 ha	0.33
Female-headed hh	3.5 of 13.5	0.4 ha	26%	24.5 ha	0.41
Total and average	9.7 of 35.4	0.3 ha	27%	65.5 ha	0.36

Table 5.4 demonstrate that female-headed households have, on average, and just slightly the largest land area under maize cultivation per household as compared to the other two households. For North Maragoli, these small differences may carry some weight in explaining variations in management and production. Female heads asserted that they plant a larger plot with maize because they are more concerned about household food security. They claimed this is because they have no regular income from other sources such as remittances, and it is therefore important for them to try to produce as much as possible from the family holding.

Most of the land in North Maragoli is being used permanently and hence plots in maize cultivation remain under crop for a long time. Close to 90% of the farmers pointed out that plots under maize have been under this crop for the last ten years or more with hardly any form of rotation to improve the soil. The maize crop on many farms is inter-planted with beans but the bean crop has been failing over the last five years because of pests and diseases.

Farm preparations were inadequate on many farms. Poor and simple implements (traditional hoes) are used especially on female-headed farms. More recent and better hoes (jembes) were recorded on both male-headed and female-managed households. Tractor and oxen plough were used on a few female-headed farms (15%). Many maize plots were only ploughed once and then planted. This inadequate preparation encouraged the growth of couch grass and striga weed which are common in this area and very destructive to the maize crop. Inadequate land preparation was more striking on female-headed farms and the reasons given were lack of money and shortage of labour.

Many farmers (approximately 80%) explained vividly the importance of planting hybrid seeds especially during the long rains season (main crop season). They were convinced that hybrid seeds were best if one wanted a high yield. Yet, the use of hybrid seeds appeared average and sometimes inadequate. The quantities of hybrid seed bought by small-scale farmers, especially in female-headed households, were insufficient for the farm size. Only 45% of the respondents planted pure hybrid maize seeds during the main crop season of 1993. More hybrid seeds were planted on male-headed household farms. As discussed in Section 4.4, during the recent years (since 1990), farmers have been sold wrong and sub-standard (low quality) hybrid maize seeds or commercial maize which was packed as hybrid seeds. This had a marked effect and resulted in low production (Daily Nation, 7th July, 1993, Communication with one of the hybrid seed farmers). During the short rains season most farmers in the three types of households did not use hybrid maize seeds. Local maize is used as seed because it matures fast and the second crop is less significant as it contributes to about one third of the total maize production.

The farmers ignored information on spacing and plant density. On some farms the plant population density was very low and this was when the hybrid maize seeds had been used. It indicated a shortage of hybrid seeds in the household. When the local maize or previously harvested maize were used as seed, between four to six plants are found in one hole. This did not create a situation conducive to growth because the plants are competing for food nutrients in soils that are already over-utilized, and as a result development is slowed down and outputs are low. Female-headed households displayed many of the features associated with low maize management. Female-managed households rated better than male-headed households. The question of food production being the portfolio of women was clearly displayed in male-headed households where despite the presence of males in the homes, women were still the ones carrying full responsibility for management and production. Indeed, providing and feeding of the household was still the responsibility of women in many male-headed households.

The use of fertilizer on maize was found to be inadequate throughout the area. Only 42% of the in-depth study respondents used sufficient artificial fertilizers for their maize crop. Of these, 54% were female-managed, 31% male-headed and 15% female-headed households. The rest of the farmers (58%) either did not use fertilizer or applied inadequate amounts taken from the fertilizer meant for tea and which apparently is not suitable for maize. Compost and manure production is limited because many families only own an average of two cows. Cow dung is used for plastering floors and very little is reserved for making manure. Only a limited amount of the waste (garbage) generated by the households is worthwhile collecting and using in compost preparation.

Maize was weeded only once on many farms (about 74% of the farms), especially in female-headed farm household and only 26% of the farmers weeded twice. Top dressing was not used (91%). A small percentage (9 %) of the farmers practiced top-dressing and this was confined to male-headed farm household.

Given the above piecemeal and dismal management conditions, maize yield per hectare in this area were much lower than the expected 6,000 kg/ha on the lower scale and food deficiency is common in most households. Poverty and hunger in many of the households, especially in female-headed households, lead to critical green plucking of maize cobs which further reduces the prospects of a good harvest.

Given the plot sizes in the three types of households, good management of the maize plots would produce an average of 20 bags (1,800 kg) on farms that are male-headed, 13 bags (1200 kg) in female-managed households and 27 bags (2,400 kg) on female-headed households during the two maize growing seasons based on a yield of 6,000 kg/ha (Section 3.4.1 and 4.4). An adult consumes on average between 200 to 250 kg cereals a year. Since many families averaged of six to ten persons with an adult equivalent ratio of say six on the higher side, the above production levels would be reasonably sufficient food in male-headed and female-headed households for a whole year.

Despite the small average size of plots (0.2 ha) reserved for maize production by female-managed households, their average yield of maize was higher than in the other two types

Table 5.5
General Summary Table on Maize Production in North Maragoli (in 90 kg bags) (source: North Maragoli In-depth Case Studies, 1993).

Types of Households	August 1992	November 1992	1992 tot harvest	August 1993	November har 1993	Total 1993
Male-headed hh	55 bags	29 bags	84 bags	55 bags	28 bags	83 bags
Production in kgs	4,950 kg	2,610 kg	7,560 kg	4,950 kg	2,520 kg	7,470 kg
Average yield kg/ha	1,125 kg	593 kg	1,719 kg	1,125 kg	573 kg	1,698 kg
Female-Managed hh	33 bags	11 bags	44 bags	29 bags	11 bags	40 bags
Production in kgs	2,970 kg	990 kg	3,960 kg	2,610 kg	990 kg	3,600 kg
Average yield kg/ha	1,650 kg	550 kg	2,200 kg	1,450 kg	550 kg	2,000 kg
Female-Headed hh	21 bags	12 bags	33 bags	15 bags	8 bags	23 bags
Production in kgs	1,890 kg	1,080 kg	2,970 kg	1,350 kg	720 kg	2,070 kg
Average yield kg/ha	540 kg	309 kg	849 kg	386 kg	205 kg	591 kg

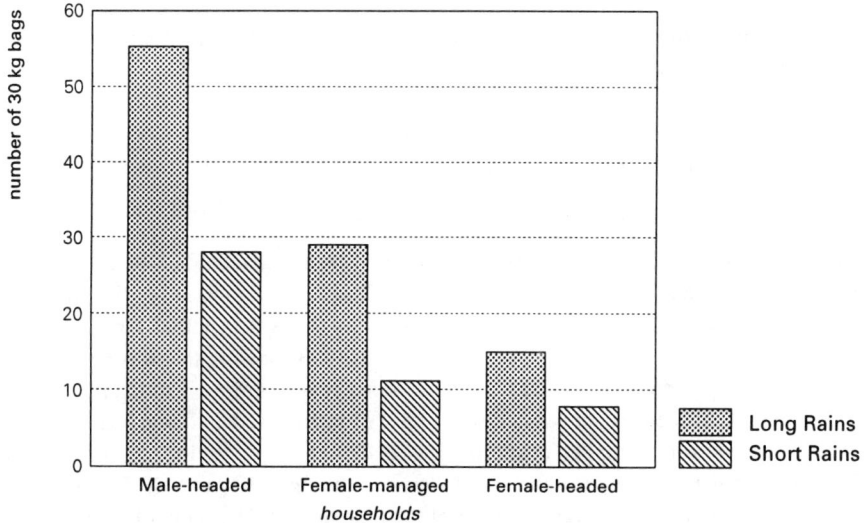

Figure 5.2
Total Maize Production, 1993 (source: North Maragoli In-depth Case Studies, 1993).

households (2,200 kg/ha in 1992 and 2,000 kg/ha in 1993). This average yield was in fact above the average yield expected from this type of household. Because the family size is small, there was not much green plucking. Maize yields from male-headed farms were 1,719 kg/ha in 1992 and 1,698 kg/ha in 1993, whilst female-heads' plots yielded 849 kg/ha in 1992 and 591 kg/ha in

Three type of household in North Maragoli

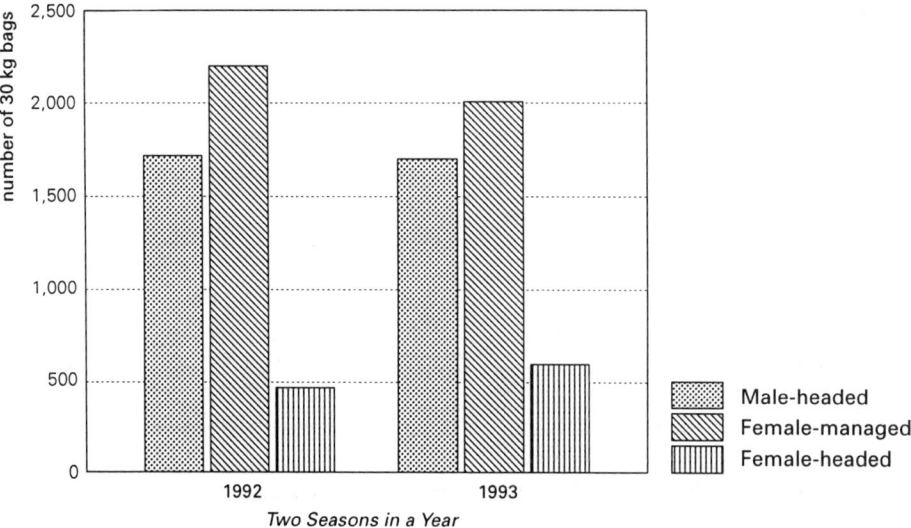

Figure 5.3
Annual Average Maize Yield per Hectare (source: North Maragoli In-depth Case Studies, 1983).

1993 (see Table 5.5 and Figures 5.2 and 5.3). However, we can only make firm conclusions on these differences if longitudinal research over at least five years is conducted.

The yield for all the in-depth case studies computed from the data in Table 5.5 is 1,494 kg/ha in 1992 and 1,250 kg/ha in 1993. This evidently shows that production is low when measured against 6000 kg/ha/year for this region.

Among the male-headed households maize field sizes varied from between 0.03 to 1.5 ha. The largest farm under maize was 1.5 ha and yielded a total of 8 bags (720 kg instead of the potential 9,000 kg, i.e 8% of the expected production) in both seasons of 1993. Another farmer with a maize field of 0.2 ha produced a total of 9.8 bags (882 kg) in 1993 showing that it is possible to come rather close to the potential yield. It is too simple to say that farm size is to blame for low production. The biggest differences are caused by the level of inputs used and crop husbandry in the fields. The farmer with the smaller maize field used hybrid seeds, fertilizers and weeded twice while the farmer with the larger maize field was sick during 1993, and all the money in this household was used to pay medical bills. Most of the time the wife was either taking care of the husband in hospital or at home and therefore very little attention was paid to the farm. This highlights how inter- and intra-households situations affect management and production experiences on the different farms.

The fact that female-managed households produced more per hectare than male-headed households may indicate the important role of women in food production in small-scale farming. It may also imply that of the three groups of women in this study (female managers, female heads and females in male-headed households), female managers have better resources, higher investment and a freer hand in decision-making as far as maize production is concerned. Females in female-managed households had more years of schooling than the other two groups of females in male-headed and female-headed households. This may partially explain the better and more slightly efficient management of maize in female-managed farm household. Most females in male-headed households were controlled by the males and thus the inputs and labour used in maize production in this type of household were subject to limitations. The emphasis in these households was more on the commercial farming sub-system because it earned more money for the male head of the household. Female-headed households seemed to suffer more from both social, financial, health and emotional constraints and this affected their production. These factors seemed to have less effect on the other two types of households. Socially, some women have difficult relations where land and other properties were concerned, and felt unwanted within the community and lacked security.

Some of these women had many health problems which prevented them from working on the farm. Other general problems noticed were frequent theft and green plucking (see Chapter Four). Female-headed households were more affected by these problems than the other two types of households.

The reasons given above illustrate why there is a difference in maize production between the different types of households. The many problems mentioned here lead to poor maize yields and production in North Maragoli and many parts of Vihiga District. Many farmers blamed the poor management on their fields on lack of money, the escalating prices of farm inputs, the late arrival of farm inputs and the unavailability of inputs, sub-standard hybrid maize seeds or the wrong seeds being sold to farmers, the inaccessibility of loans, a shortage of labour, lack of proper technical information from the extension staff, over-utilized soils, and declining farm sizes. Farmers were of the opinion that the failure of the rains was only partly to blame because this failure only occurs once in so many years while the other problems were an ever-present reality. The role of women cannot be overemphasized because on all the maize farms visited, most of the labour (approximately 90%) was provided by women. Thus management may improve if some of the constraints mentioned could be dealt with.

In conclusion, it has been shown that despite the many odds women have to fight against, female-managed households manage to produce higher maize yields per hectare compared to male-headed households. It has also been emphasized that in the different types of households women are responsible for food (maize) production and the feeding of the household members. This emphasizes the important role women play in farm-household food production. Female-headed households performed very poorly in the study area because of the multitude of problems that confront this household type both from within and from outside. Maize production is very low and the problems observed and enumerated by farmers are: (a) poor farm management; (b) gender differentiation; (c) lack of money to buy inputs; (d) faulty and low quality hybrid seeds produced and sold to farmers; (e) poor and incorrect use of inputs; (f) shortage of labour in some households; (g) theft of maize from the farm; (h) green maize plucking; (i) low plant population density for hybrid maize; (j) high plant population density for local maize; (k) health problems and time spent on social and cultural functions; (l) poor education and lack of detailed technical information due to poor agricultural extension channels and (m) poverty.

5.6 The Impact of Household Type on the Tea Farming Sub-system

This section looks at the tea farming sub-system in the three types of households. Tea is considered to be the main cash crop for most households. Table 5.6 shows how the different types of households have allocated land to tea cultivation.

Table 5.6
Distribution of Plot Size Under Tea by Household Type in 1993 (source: North Maragoli In-depth Case Studies, 1993).

Households	Total size tea plot/hht in ha	Ave tea field per hht in ha	% total farm size	Sample size	survey ave	1992 % tot
Male-headed	3.86 of 13.3	0.3 ha	29 %	20.4	0.38	29.2%
Female-managed	4.22 of 8.6	0.4 ha	49 %	20.9	0.39	33%
Female-headed	2.20 of 13.5	0.3 ha	16 %	18.0	0.33	26%
Total/average	10.28 of 35.4	0.3 ha	29 %	59.3	0.37	29.6%

Tea takes up 29% of the total land and occupies an average plot size of 0.3 ha for all the respondents in the in-depth case study. In the larger sample survey (180 farmers) tea covered 30% of the total amount of land (see Tables 4.4 and 5.6). The plots under tea are about the same size in the three types of households. The female-managed households have put slightly more land under tea cultivation, 49% of their total land at an average of 0.4 hectare in the in-depth studies and 33% in the sample survey. The managers in female-managed households explained that their husbands are eager to accumulate more wealth as fast as possible. Once these young couples inherit land from parents, they sub-divide the farm into almost two equal halves: one half for tea and one for food crop cultivation. Their tea was still relatively young, while in both male-headed and female-headed households the tea was mature and well established.

The younger households appeared to understand the benefits that cash crop farming could bring. Emphasis is on cash crop farming with the hope of buying food from the market with remittance money. The female managers confirmed that their spouses made decisions about allocating larger fields to tea cultivation. Cases were found where all the family's land was under tea while additional land was bought or hired for food production (18%).

In fact, one would expect that male-headed and female-headed households would have bigger tea plots than female-managed ones because of the age difference of the respective heads. The younger the couple the smaller their farm because of shrinking farm sizes at every subsequent sub-division. However, males in employment put priority on accumulation of wealth through commercial farming and the purchase of additional land when money was available. This may partially explain the slight differences in tea plots.

Female-headed households disclosed that they had not added additional tea bushes since the

Table 5.7
Population Density of Tea Bushes (per Hectare Within the Three Types of Households) (source: North Maragoli In-depth Case Studies, 1993).

Types of household	Density of tea bushes per ha	Average tea bushes per hh member	% of total tea bushes - survey
Male-headed hht	9,000 bushes/ha	2,700 per member	42%
Female-managed hht	6,300 bushes/ha	2,400 per member	30%
Female-headed hht	6,000 bushes/ha	2,000 per member	28%

death of their spouses. Some of them (60%) asserted that part of the tea plots had been transferred to the sons in the households, thus greatly reducing the size of the plot under tea. A few (30%) remarked that the late husband's kin had acquired part of the tea holding.

As discussed in Section 4.4, the number of tea bushes is one of the factors that may influence the production of tea. On a one hectare plot it is possible to have 12,000 bushes (Acland 1971, KTDA 1971). The number of tea bushes varied greatly between the three households (Table 5.7)

Male-headed households had more tea bushes per hectare (9,000 bushes/ha) than female-managed and female-headed households whilst female-managed ones had 6,300 bushes/ha and female-headed ones 6,000 bushes/ha (Table 5.7). Female-managed households where a larger area was under tea had less bushes than the male-headed ones. Some of the reasons given for this difference are: female-managed farms have more open spacings and their tea is relatively younger when compared to the farms of the other two types of households (most of it had been planted in the late 1980s and early 1990s whilst in the farms of the other two types of households tea had been planted in the 1960s and 1970s). The age of the crop is important for establishing the tea bush population density. Variations in the population of tea bushes was observed between these three types of households and among the members of the same household and this directly influenced production from the farms. Other reasons given for low bush population density were poor pruning methods causing bushes to wilt, the lack of weeding which accelerated the spread of diseases and pests, and unprotected land that is vulnerable to erosion and runoff during heavy storms thus decimating tea bushes.

Other determinants of tea production are labour input for weeding, pruning and plucking. Shortage of labour for these activities was a constraint, especially in female-headed households. Despite the fact that some farmers had a larger number of tea bushes, for example, there were farmers with 3,000 bushes, production was still low because no weeding had been done because of labour shortage. If tea bushes are well managed, farmers may pluck four rounds in a month and an average of one kilogramme of green leaf per bush in a year. North Maragoli farmers pluck on average 0.5 kilogram of green leaf per bush/year compared to other areas such as Murang'a 1.2 kilograms, Kericho 0.8 kilograms, and Kirinyaga 1.5 kilograms of green leaf per bush per year (KTDA 1992).

In some female-headed households, production was as low as 0.2 kilograms of green leaf per bush per year, and there were quite large variations between and among households. This ranged from between 0.1 to 1 kg of green leaf per year. Averages of above 0.6 kg were realized on male-headed tea farms, 0.5 kg on female-managed and 0.3 kg on female-headed tea farms. On average three rounds per month were plucked in male-headed households, two rounds in female-managed households and one round in female-headed households. The reasons for these differences lies in the different tea management practices in the three households.

Tea Production and Incomes in the Three Types of Households

Tea Production

Tea yields and production are dependant on a number of variables. Following the correct agronomic practices and the right use of inputs is essential for high production, and high incomes per hectare. An average tea plot in North Maragoli requires a minimum of two weedings per year, and a minimum of two applications of fertilizer per year at three to five bags for a 0.2 to 0.3 ha plot. Such a plot should have a minimum establishment of 2,500 bushes, and a maximum of about 4,000 bushes (Table 5.7). We have shown above that management varies from one household type to another with female-headed households displaying the lowest management standards. Table 5.8 presents average tea yields in the three types of households investigated.

As already mentioned, an expected yield of 6,500 kg/ha can be obtained in this area. The average yield for all respondents is 4,342 kg/ha/year. If compared with the average for Vihiga District of 4,125 kg/ha/year, the yield in North Maragoli is higher. In fact, tea yields and production in North Maragoli is the highest for Vihiga District as a whole, yet compared to other tea growing areas and what is possible given the agro-ecological conditions in the area, yield and production are still low (Section 3.4.1).

Yields vary as we have seen from household type to households type. It is apparent that female-headed farms have the lowest average production per hectare while male-headed farms have the highest average yield per hectare. The yield on male-headed farms is 14% higher than on female-managed and 30% higher than on female-headed farms (see Table 5.8 and Figure 5.4). This was also observed during the sample survey (see Table 5.9).

Table 5.8
Average Tea Yield in Three Years in the Three Types of Households in 1993 (source: North Maragoli In-depth Case Studies, 1993).

Households	1990/91 Ave/ha/yr	1991/92 Ave/ha/yr	1992/93 Ave/ha/yr	Average in 3 yrs/ha
Male-headed	4,738 kg	4,364 kg	3,848 kg	4,317 kg
Female-managed	3,878 kg	3,444 kg	3,839 kg	3,839 kg
Female-headed	2,970 kg	2,700 kg	3,450 kg	3,040 kg

Three types of households in North Maragoli

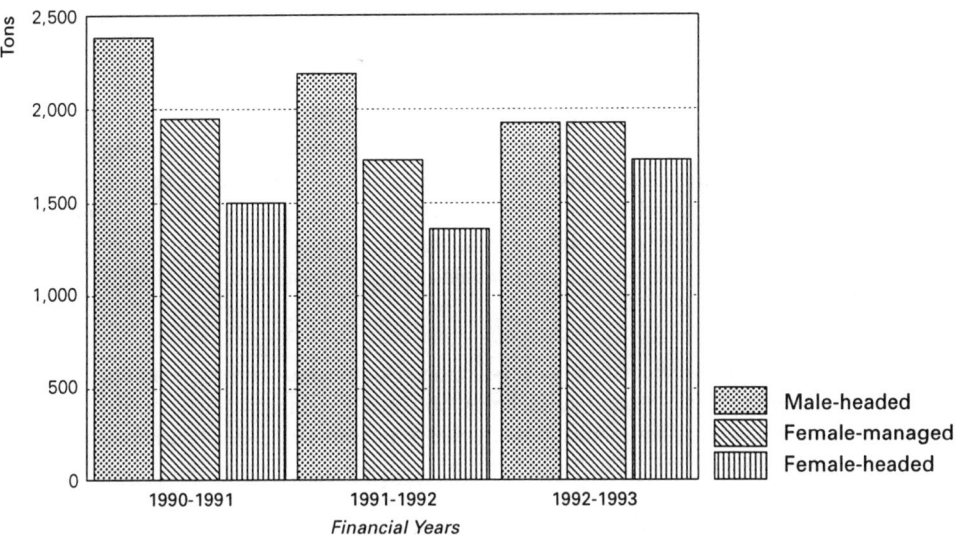

Figure 5.4
Mean Annual Tea Yield per Hectare (source: North Maragoli In-depth Case Studies, 1993).

The total monthly tea production in 1993 also shows that male-headed households produced more than the other two households in most months (see Figure 5.5). Male-headed households have higher monthly and annual tea production in contrast to the female-managed and female-headed households.

Tea production increases during the rainy season and declines during the dry months. But production deteriorated in female-managed and female-headed households during some months of the rainy season as displayed in Figure 5.5. This was attributed to a number of factors. Tea

Table 5.9
Summary of Tea Production and Incomes 1991 (source: North Maragoli Sample Survey 1992).

Household type	Total production in kg	Yield/ha in kg	Total income in Kshs	Income/ha in Ksh
Male-headed-(60)	82,423	4,040	270,790	13,274
Female-managed (60)	61,583	2,947	189,467	9,065
Female-headed-60	49,525	2,751	159,583	8,866
Total (180)	193,531	3,364	619,840	10,453

production was low in March because of the apparent competition for labour between tea and food crop production. The heads of these households spent more time weeding food crops than plucking tea whereas in male-headed households more hired labour was used during this period. Production was low in all households in May because tea was destroyed by hailstones in 1993.

The harvest of long rains food crops such as maize and the planting of the short rains food crop are carried out in August. More time in the households was allocated to harvesting the long season crop and planting the short season crop and hence the production of tea decreased in this month. However, as the tasks related to food crop production diminish, more time and labour is devoted to plucking tea and this explains the peaking up of production in the months of June, July, and September in both male-headed and female-managed households (Figure 5.5).

Tea Incomes
Tea incomes follow the same pattern as production. Households with high production also achieve high incomes. Male-headed households obtained the highest incomes per hectare, followed by female-managed households (less by 11%), and female-headed households received the lowest incomes (less by 31%) (see Table 5.10 and compare with Table 5.9).

As mentioned in Section 4.4, a hectare of tea on average earns farmers Kshs 59,000 per annum (KTDA 1992) taking the figure for Kenya as a whole. Other areas such as Embu, Kirinyaga, Kisii, and Murang'a for example, earn much more than the national average (see 4.5). When the figures for North Maragoli are compared to national tea income per hectare and tea earnings per hectare for other small-scale tea growing areas, farmers in North Maragoli earn very low incomes (see Tables 5.9 and 5.10).

Per household type, North Maragoli

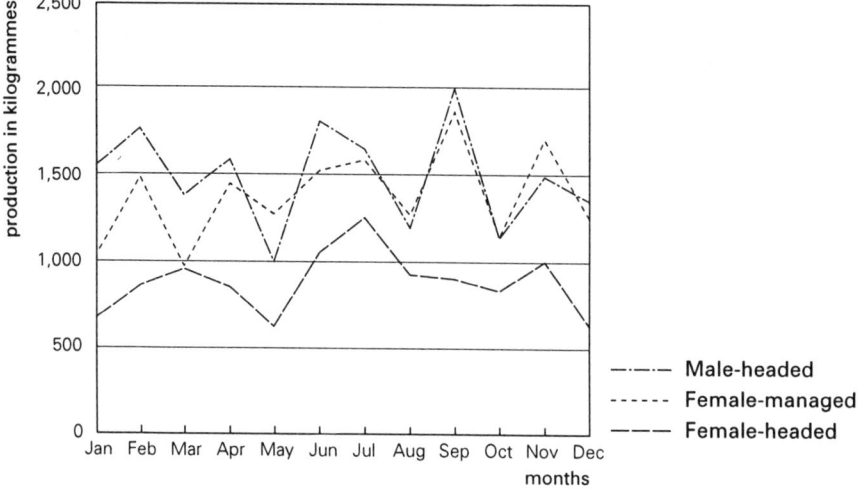

Figure 5.5
Monthly Tea Production, 1993 (source: North Maragoli In-depth Case Studies, 1993).

Table 5.10
1993 End of Year Tea Incomes (in Kshs) in the Three Types of Households in 1993 (source: North Maragoli In-depth Case Studies, 1993).

Households	Ave income 1990/91/ha	Ave income 1991/92/ha	Ave incom 1992/93ha	Ave income in 3 years	% of the ave 59000/ha-Kenya
Male-headed hh	12,843.95	13,457.60	33,719.10	20,006.80	34% less 66%
Female-managed hh	9,361.90	11,120.80	32,741.15	17,741.80	30% less 70%
Female-headed hh	7,495.00	9,346.00	30,669.30	15,836.80	7% less 73%

There are again differences in earnings between the three different types of households. Male-headed households receive the most income, followed by female-managed households. Female-headed households earn the least from the tea farming sub-system (see Tables 5.9 and 5.10). Farmers in North Maragoli have been earning less than Kshs 4.50 per/kg of green leaf. However, in 1992/93 the price of tea shot up on the world market and farmers in North Maragoli earned Kshs 9.50 per/kg of green leaf. This explains the steep raise in average tea earnings in 1993 (Table 5.10). Male-headed households earn higher incomes per month and per year (see figures 5.6 and 5.7). These higher incomes are related to better accessibility to factors of production,

Per household type, North Maragoli

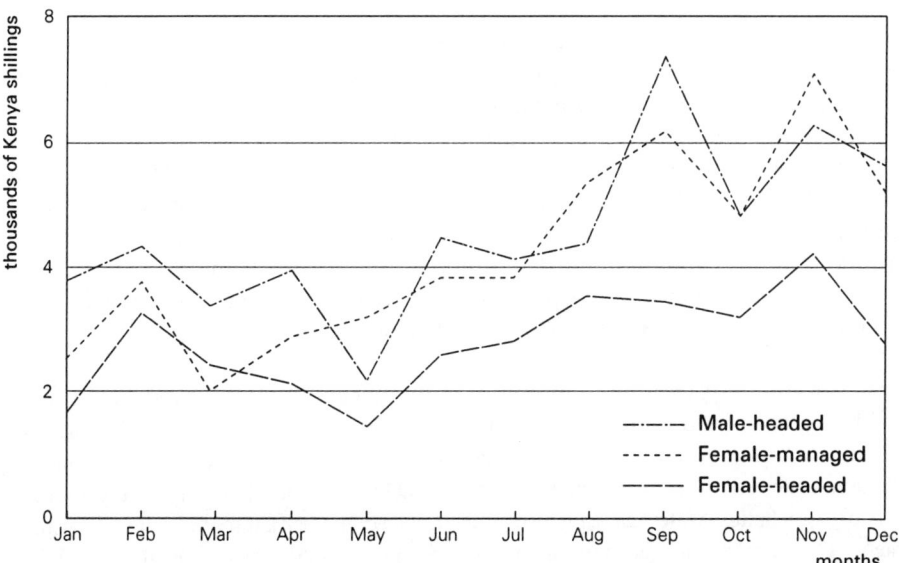

Figure 5.6
Monthly Tea Income, 1993 (source: North Maragoli In-depth Case Studies, 1993).

Three types of households in North Maragoli

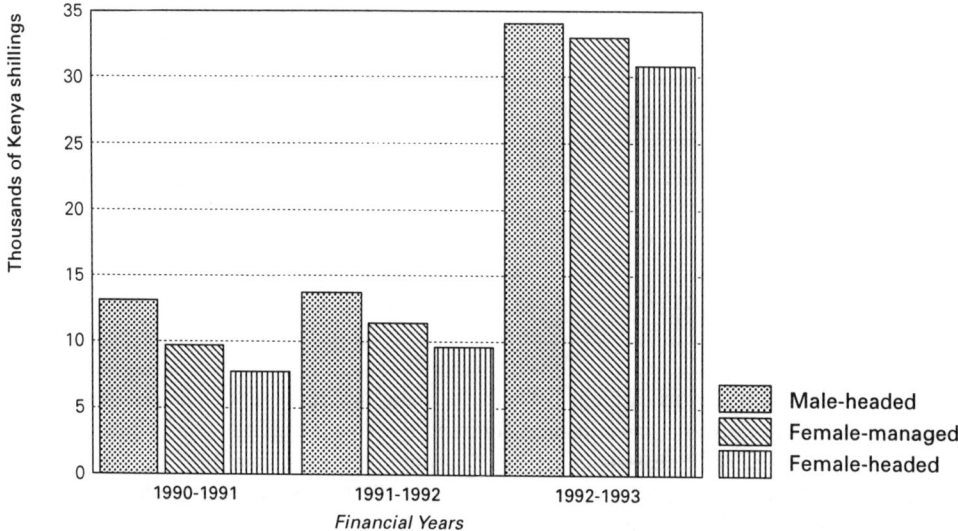

Figure 5.7
Mean Annual Tea Income per hectare, 1990-1993 (source: North Maragoli In-depth Case Studies, 1993).

better management due to the higher use of extra labour and high production. Given the higher incomes, this type of household enjoys a slightly higher socio-economic status than the other two types of households.

An investigation into the distribution of tea incomes indicates that the 'end of year' (bonus) income in both male and female-managed households is collected by the males. In female-headed households, approximately 70% of the heads collected the income whereas in 30% of the cases the bonus was collected by the sons. The income earned by female-headed households was shared out between the female head and the sons. In the male-headed households, monthly tea income was collected by males or was directly accredited on their bank accounts. In the female-managed households, women collected and used the monthly tea income but the end of year payment was collected by their husbands. Farmers have been encouraged to open bank accounts because of recurring theft on pay days. Opening of bank accounts was more widespread in male-headed households. The other farmers received cash money on pay days.

Our findings indicate that there exist differences in the amount of tea produced and incomes generated by tea among the three types of households. These differences are based on gender differentiation and farm management. Male-headed households enjoyed better access to factors of production, better education, more money, and were able to use hired labour. As a result their tea husbandry was better. Male-headed households produced and earned more in the tea farming sub-system. The female-managed households took an intermediate position in the production and earnings from tea. The female-headed households had the lowest production of and earnings from tea. Generally, women experienced many problems and these problems appeared more compounded in female-headed households. The presence of a male in a given household seems to ensure a larger chance of access to the inputs necessary for good management and also represents security within the community.

To summarize, the problems encountered in tea production include: (a) gender differentiation and biases appeared to play a significant role in all processes in the tea farming sub-system. It influences, as discussed, decisions on plot size, labour allocation, investment, production and incomes; (b) low tea bush population density, and a lot of open spaces in the tea fields; (c) low and poor use of fertilizer leading to low quality tea; (d) tea collection and transportation problems; (e) no factory in the province; (f) low tea price in the whole area acts as a disincentive to farmers with exception of the earnings for 1993; (g) conflicts about ownership of tea and subsequently tea earnings; (h) scarcity and shortage of labour; and (i) control over resources and earnings.

5.7 Livestock Sub-system and Milk Production in the Three Types of Households

In addition to the maize and tea farming sub-systems, the livestock sub-system is also important. In the past (1940s and early 1950s) it was possible for farmers to own more cattle because the farms were still reasonably large. Each household had an average of 10 hectares of land and there were some communal grazing lands. Farmers then owned on average heads of 20 cattle. Today, there is hardly any (communal) grazing land left. Each household has to devise strategies of feeding its cattle. Approximately 90% of the farmers now keep one or two local zebu cattle. A few farmers (12%) sold off their cattle during this study to settle school fees and now keep one or two heifers. There was no significant difference observed in the number and types of animals kept in the three types of households.

More livestock was sold in male-headed households than in the other two types. The presence of the male facilitated decision-making and the actual sale of livestock in male-headed households. Female-managed households had to wait for such a decision to be made by the husbands. Female heads of households rarely sell their livestock because in most cases these animals have been inherited from the dead husbands. If such cows have to be sold, the female-heads have to consult either their brothers in-law, their fathers in-law or their sons. All these people are entitled to a certain amount of the price the animal fetches. Conflicts over inheritance and the ownership of cattle crop up frequently when a husband die. If a man dies leaving many cattle then these are shared out between his sons, leaving one for his widow. If the deceased man owned only one cow, then it is sold and the money shared out to the next of kin who may include the wife, sons and brothers. Approximately 75% of female-headed households preferred not to sell the cattle as it would be difficult for them to buy a milk producing cow.
As already discussed in Section 4.6.1, livestock management was generally low in many farm-households. The feeding of cattle depended on the availability of napier grass established and the green roughage available on each farm. Farm measurements revealed that farmers reserved very small parcels of land for napier grass. In male-headed 4.5%, in female-managed 4.1%, and in female-headed households 3.0% of the total land was under napier grass cultivation. The napier grass plots measured on average less than 0.1 hectares. Some farmers claimed that they found it difficult to get the right seedlings and knowledge about planting napier grass. There was therefore a

perpetual shortage of grass which contributed to the poor feeding of the cattle. The establishment of napier grass was poorest on female-headed household farms (a problem of shortage of labour again). Many cows looked emaciated mainly due to inadequate care. Water was an added problem in households where it had to be collected from a point far away from the homestead.

Local zebu cattle are assumed to be quite resistant to many diseases but it is recommended by the Department of Livestock Development that animals should be dipped four times a month. This helps to eradicate ticks that cause such tick-borne diseases as Anaplasmosis and East Coast Fever. Dips have been constructed in every location for this purpose. The rate of dipping was low in many households leading to tick infestation in cattle. The farmers claimed that it was very expensive to dip cattle. Those with grade cattle or cross-breeds frequently spray these animals against ticks in their homes. In male-headed households cattle were dipped once a month, in female-managed household dipping was carried out once every two months and in female-headed household once in three months. Labour and money shortage were blamed for the low dipping rates.

Many farmers have no access to livestock extension services because these services suffer from shortage of staff, and the narrow and selective approach of the delivery of extension. The animals are rarely vaccinated against diseases. Many farmers are not able to detect when animals are diseased. Only 18% of the respondents were visited once in 1993 for livestock extension. Extension visits vary between households, being highest in male-headed households (31% visited twice), 18% of the female-managed households were visited once, and were lowest in female-headed households (11% visited once). This shows the infrequency of these services to farmers and the differentiation that exists along gender lines.

The prevailing poor livestock management conditions have contributed to poor milk production. Farmers with local zebu cattle produce between 1 to 2 litres of milk per cow a day as opposed to a possible 5 litres and more with good management. Farmers with grade and cross-breeds produce between 10 to 15, and 8 to 12 litres of milk per day respectively as opposed to about 30 litres with good management. Many farmers pointed out that the problems associated with livestock management include expensive inputs, lack of capital, shortage of grass and scarcity of labour. Women were again responsible for livestock management when already over-burdened with other farm-household system tasks. Livestock were neglected because women did not have enough time for them. Milk production and the income from milk varies between households (see Figures 5.8 and 5.9). Male-headed households again had a higher milk production and income than the other two types of households.

The livestock sub-system fulfills social, cultural and economic functions within the North Maragoli farming system. Environmentally, its by-products (cowdung) serve as a source of energy and manure for organic farming, especially for farmers who find it too expensive to purchase chemical fertilizers. However, these by-products are inadequate.
Though the campaign for small-scale farmers to adopt zero-grazing started in 1980, it is spreading slowly due to the high capital investment required. Only 25% of the male heads, 10% of the female managers and 10% female heads had adopted zero-grazing with grade cows. There is room for further expansion of zero-grazing. In fact, North Maragoli is very suitable for zero-

Per type of household, North Maragoli

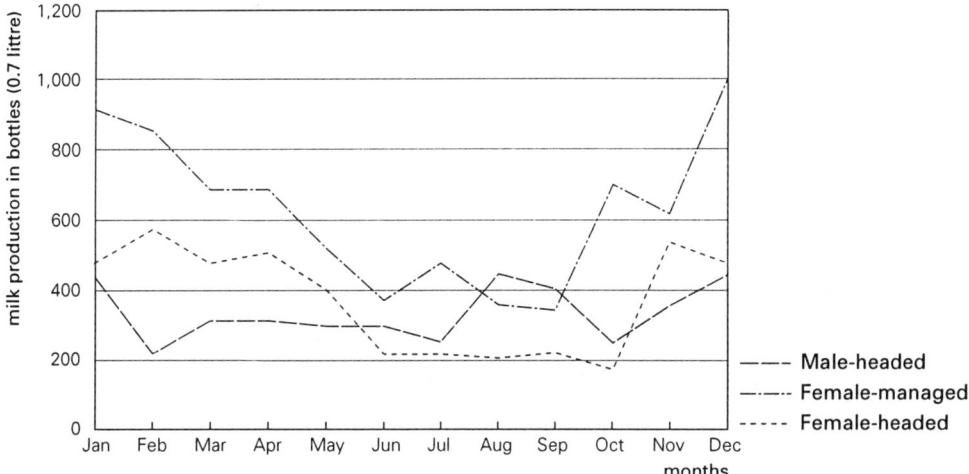

Figure 5.8
Total Milk Production, 1993 (source: North Maragoli In-depth Case Studies, 1993).

Per type of household, North Maragoli

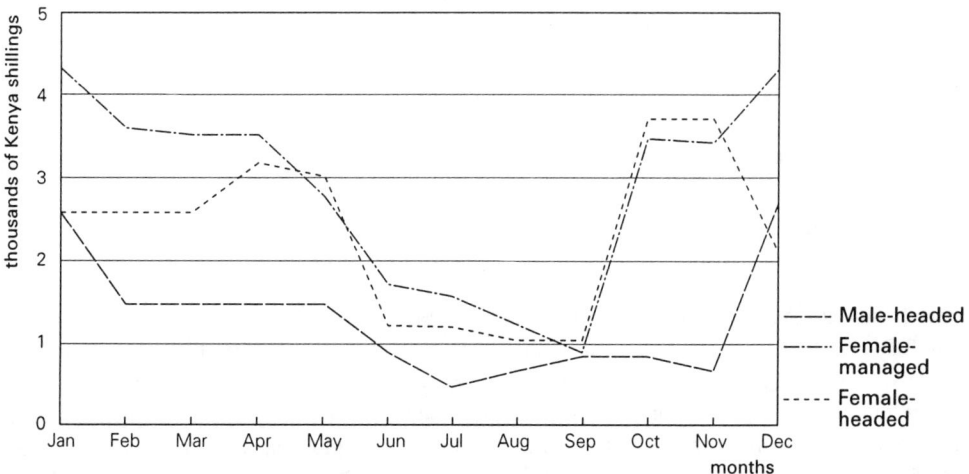

Figure 5.9
Total Milk Income, 1993 (source: North Maragoli In-depth Case Studies, 1993).

grazing because of favourable climatic conditions and the small farms. It has been alleged, quite correctly, that keeping one grade cow intensively under good management may be more paying than the present practice of keeping the local zebu cows in North Maragoli. Many farmers appeared to dream of saving money, or that the government should give them financial assistance or loans to buy grade cows.

It was found that the livestock sub-system, especially where farmers still own the local zebu, was more neglected than the food and commercial crop farming sub-systems. The disregard for providing good management is not intentional but circumstantial. Many households were unable to provide sufficient cattle feeds (grass and supplementary feeds) because of deficiency on the farms. Shortage of labour greatly contributed to poor husbandry. However, male-headed households performed better than the other two types of households.

5.8 The Tree Production Sub-system

The importance of the tree production farming sub-system has been emphasized in Section 4.7. The total area under trees in our sample is 5.3 ha which is 15% of the total land in our study area (see Table 5.11).

Table 5.11
Land under Tree Cultivation in the Three Types of Households (source: North Maragoli In-depth Case Studies, 1993).

Household type	Total land for trees/hht in ha	% of total land area under trees/hh
Male-headed hh	1.1 ave = 0.1	8.1 %
Female-managed hh	0.9 ave = 0.1	10 %
Female-headed hh	3.2 ave = 0.4	24 %
Total & average	5.3 ave = 0.2	15 %

Female-headed households had more land under trees than the other two types of households (see Table 5.11). Generally, women appeared to be more concerned with the preservation and conservation of the environment than men because degradation affects them directly as deforestation and depletion of trees means they have to walk long distances in search of fuelwood. Many women are neither free to nor allowed to exploit the woodlots since the trees belong to the men. Most of the woodlots had been established by the husbands. On many farms fruit trees (avocadoes, mangoes and pawpaws) are interplanted with food crops and they are also grown in the farm compound. The fruit trees are exploited both for home consumption and for sale and this was mainly the task of the women.

Trees are mainly timber for construction and sale in most households. Some of the trees provide feed for livestock, and others provide medicine that is used both for humans and livestock. In some homes, especially those of the female-headed households, women have to provide shelter for their families and it is therefore important to have their own tree resources as they cannot afford to buy them from others.

Several widows confirmed that many of the trees had been established before the death of their spouses and not many had been lost through felling for sale as the case was in male-headed households. An awareness that the environment should be protected by all means was greater among women than it was among men. This is associated with fact that the involvement of women in the farm-household system is greater than that of men and men appeared to be more concerned with earning money.

Sixty percent of the women in the sample stated that it was now possible for them to plant trees. This implies a slow change in attitude towards certain traditions which had forbidden women to plant trees. However, there are a few traditional trees which, by custom, are considered permanent property and have many symbolic meanings and taboos. Only men are allowed to plant such species. Some women used their sons and/or workmen to plant such trees for them. Women above 50 years of age have not changed their attitude towards the ownership and use of trees. They cannot plant trees by themselves but do ask male relatives or their own sons to plant for them. They still believe that it is unwomanly and against tradition to plant trees because trees are a permanent resource and permanent resources are owned by men. They also believe that if women plant certain trees this will bring misfortune on the family.

At the farm level trees that were managed by females yielded a higher production than those managed by males according to the in-depth survey. This is because trees on female farms are left to mature while men cut down their trees either for firing bricks or for sale when the trees are still quite young.

5.9 Gender and Labour Distribution and their Effect on the Farm-Household System

Female labour input in the small-scale farming sector in North Maragoli is greater than the male input and this increases every year as the out-migration of men from the rural areas also increases. Agricultural development in terms of the introduction of commercial farming often increases the amount of agricultural work in terms of hours per day and days per year, for both men and women. However, women's labour time has increased much more than men's because women now have additional tasks which were traditionally the men's farm tasks. With the introduction of cash crops, the adoption of stable-feeding and zero-grazing in addition to the cultivation of food crops, women find themselves with increasingly more farm work as men have continued to move out of agricultural production to join other sectors of the economy. Men are able to do this because they have better education and opportunities than do most women.

Crops like tea and coffee were specifically introduced as men's crops. It was assumed that women would remain in their traditional food crop cultivation sector. It was found during this study that women spent more time on farm work apart from their household chores. Women were found to cultivate all the crops, care for livestock, process and prepare all the food for the farm households, and were also responsible food distribution and transportation, and the sale of most cash crops. The production of crops entails preparation of the land, planting, weeding, harvesting and many other related tasks.

The gender division of labour by task has broken down and is no longer applicable either in North Maragoli or in other small-scale farming areas in Kenya. Women now do most farm operations themselves with supplementary help from the family and hired labour where affordable (Saito, 1992, 1994). From field observations and discussions with respondents it appeared that the gender division of labour reflects the social customs, norms, and beliefs which govern and circumscribe behaviours and that this was further enhanced by the gender ideology of the people. In the past the gender division of labour was applicable in this area when farms were still large and there was no education and formal employment sector; and so everybody had to work on the land to sustain the household livelihood. However, today's social and economic environment is very different.

With the education and migration of males, the assignment of specific duties to males is no longer feasible. Also the change in the land tenure system and the reduction of farm size, combined with the changing economic and political conditions have forced some sort of social restructuring in small-scale farming. Some of the tasks that were reserved for men no longer exist on a large scale. For instance, men were responsible for clearing new ground, producing surplus food for the household in case of famine and drought and for herding livestock. The clearing of new ground is no longer applicable as there is no spare land. Men's labour was assigned an economic value with the introduction of commercial farming and the possibilities of off-farm wage labour employment.

Women's labour is not economically valued. Women's labour is seen as a given and something that fulfills socially accepted functions for the survival of the household. It appears more natural to see women working on farms than men in this area. Often when men are seen working alone they are asked what happened to the wives. The fact that women strive to provide all the labour is an issue related to gender ideology and gender relations. Women who do not work hard on the farms are criticized and labelled by other women and community members as lazy and irresponsible. However, although today's farms are small, there are more farm activities than in the traditional setting and therefore more labour is required than before. It was observed that women are their own worst enemies as they marginalize themselves in some aspects. Women train the children and from a early age they burden the girls with a lot of work while boys do little or nothing.

In North Maragoli, work done by women on the farms i.e producing food crops, working on their husbands' cash crop plots and all the homestead and household activities they carry out do not appear to be considered as work and as time consuming both by policy makers, researchers and men because when the word 'work' is used it connotes someone working for a salary/wage. It is assumed that there is an enormous reservoir of labour in rural households. However, our findings indicate that labour constraints in the different types of households appear to be a big contributory factor to poor management, low production and low incomes. For instance, in smallholder tea production, men are responsible for pruning and in some households there are tasks that are shared by both women and men. In many households however, women perform these tasks single-handedly.

According to the theory of household economics, Moock states that the time of its members is the basic resource of households. The opportunity cost of this resource varies over time among household members of different gender, age and skills (Moock 1986). The management of the farm-household systems relies on the efficient use of this time and its allocation to the different sub-systems. An implication of the household economics theory is that time and cash are substitutables. Time can be 'sold' to generate cash or non-market goods and it can be 'purchased' by spending cash on time-saving inputs or on extra labour. This partially explains one of the reasons for the out-migration of men from the small-scale farming sub-sector. They believe that by selling their time more money can be generated which can be recycled back into agriculture. In almost all households, one or a few members are away selling their time in other employment and many children are in school. In any given household there was therefore a limited number of farm- working members and this was certainly worse in female-headed households. Farm-households had to spend money to purchase extra labour from outside the household in order to cope with the different farm tasks especially for tea cultivation (see Table 5.12). However, this was only possible where money was available and there were distinct differences between household types where spending on labour was concerned. Male-headed households invested more directly in labour. The expenditure recommended on one hectare of tea by Kenya Tea Development Authority (KTDA) was approximately Kshs 7,000/ha in addition to family labour in 1992 (Okelo and Oluoch 1992) (see 4.5.1, Gross margins in tea production).

Some farmers often compromised on their own crop and livestock management not because of lack of knowledge but because of lack of cash and time. Sometimes poor farmers had to sell their own labour for cash regardless of their own labour requirements. There was a constant competition for labour in the farm households. The problem experienced by many households was shortage of labour and therefore the need to purchase labour from outside. Yet, many of the households did not have the financial capability to 'purchase' labour, especially during the period of the year when labour was in most demand and both cash and food crops needed attention.

Table 5.12
Money (in Kshs) Spent on Hired Labour in 1991 (source: North Maragoli Sample Survey 1992).

Household type	Total Expenditure in Kshs	Household ave- Expenditure	Expenditure/ha of tea
Male-headed (60)	42,720	712	2,094
Female-managed (60)	28,740	479	1,375
Total (180 farmers)	88,400	491	1,491

First we will examine the issues of the availability and provision of labour in our case study sample. We will then consider the question of purchasing labour in the different household types.

Table 5.13
Average Time Devoted to Tea Production (Hours/Day/Week/Year) (source: North Maragoli In-depth Case Studies, 1993).

Type of household	Women			Men		
	Day	Week	Year	Day	Week	Year
Male-headed hh	4.30	13.30	360	2.30	9.00	246
Female-Managed hh	5	15.30	520	0.00	0.00	120
Female-headed hh	8	16	672	0.00	0.00	0.00

Table 5.14
Labour Time Allocation on Food Production (source: North Maragoli In-depth Case Studies 1993).

Type of household	Women		Main	Men		Main
	Day	Week	Season	Day	Week	Season
Male-headed hh	6.00	36	137	2	4	50
Female-managed hh	6.00	36	150	0	0	24
Female-headed hh	8.30	50	224	0	0	0

The above tables show the impact of the division of labour on the time allocation of women and men in North Maragoli. Field results indicate that men spent very little time providing both homestead and farm labour. Despite the fact that tea is supposed to be a man's crop, women never the less provided three-quarters (77%) and men 23% of the labour required. Women provide 95% and men 5% of labour needed for food production. Women provide 80% and men 20% of the total labour for both food and cash crop farming sub-systems. Our findings further indicate that cash crop operations use 79% of the total time available for both sub-systems. Most of the time available in the farm-households is spent on these two sub-systems, leaving very little time for the other sub-systems.

In an earlier study, it was established that women spent 13 to 14 hours a day working (Horenstein 1989). During this study it was established that women in North Maragoli worked between 14 to 16 hours a day. Women work long hours but often do not achieve much productivity. This is because of the conflict surrounding their use of time and decisions made on the operation of the different activities in the farm-household system. Women's attempts to reallocate their labour in the face of competing demands on their time posed serious agricultural productivity implications/repercussions. The competing activities requiring labour as observed in the field led to women rushing through different operations in an incomplete manner which resulted in inefficiency (allocative inefficiency).

Our findings conform with Bryceson and McCall's findings that women work longer hours than men in the rural areas. They pointed out that already decades ago the working day of the African woman was both longer than that of the man and was subject to less seasonal fluctuation. An earlier survey in 1939 amongst the Bemba records women doing approximately 6 hours of work during the peak and slack agricultural seasons, as opposed to men whose working hours ranged from 4 to 2.75 hours per day. Another survey conducted in the same region in 1980s showed women performing 30 to 40% more agricultural field work compared to men. In fact, approximately sixty years ago women's labour input is shown to have exceeded men's by between 50 and 120%. When all the different tasks women perform per day are put together, then it is estimated that women perform almost 300% more work than men (Bryceson and McCall, 1994:1-2). Our findings in Tables 5.13, 5.14 and 5.15 can be compared with the findings in Table 5.16 adapted from Bryceson and McCall, showing the differences in the gender division of work-time in Africa.

As observed earlier, changes in the division of labour between women and men often intensify the work of women and results in a loss of their economic independence and social status. Changes in cropping patterns and farming technology as evidenced in North Maragoli and

Table 5.15
Weighted Farm Labour in Hours (Derived from the tables 5.13 and 5.14) (source: North Maragoli In-depth Case Studies, 1993).

Type of household	Females	Males	Tot-Females	Tot-Males
Male-headed hh (13)	360 + 137	246 + 50	6,461	3,848
Percentage of farm	labour	weighted	63%	37%
Female-managed hh (11)	520 + 150	120 + 24	7,370	1,584
Percentage of farm	labour	weighted	82%	18%
Female-headed hh (9)	672 + 224	0 + 0	8064	0
Percentage of farm	labour	weighted	100%	0%
Total hours			21,895	5,432
Total No. of hours in all households		27,327	80%	20%

Table 5.16
Gender Division of Working Hours (source: Bryceson, D.F and M. McCall, Lightening the Load: Women's Labour and Appropriate Rural Technology in Sub-Saharan Africa. 1994:2).

Place	Length of Working Women	Day in farm Season Men
Zambia-Mpika, Mazambuka and Mumbwa Districts (Due & Mudenda 1983)		
Agricultural Work	8.5 hrs/day	7.4 hrs/day
Household Work	5.0 hrs/day	1.1 hrs/day
Working Day	13.5 hrs/day	8.5 hrs/day
Cameroons (McSweeney 1979)		
Agriculture and		
Domestic Work	7.3 hrs/day	5.0 hrs/day
Burkina Faso (McSweeney 1979)		
Agricultural and		
Domestic Work	8.6 hrs/day	3.4 hrs/day
Tanzania - Bukoba District (Kamuzora 1984)		
On-Farm Work	4.37	3.1 hrs/day
Domestic Work :- Food,		
Water, Wood, Childcare	3.37	1.16 hrs/day
Leisure	3.41	4.83 hrs/day
Burkina Faso, Kenya,	World Bank Study	
Nigeria, Zambia	(Saito et al. 1992)	
Farm Work	8.3 hrs/day	7.0 hrs/day
Non-Farm Work	6.0 hrs/day	1.7 hrs/day
Working Day	14.3 hrs/day	8.3 hrs/day

Vihiga District has relegated women to the lowest productivity branches of production, increased their working hours whilst augmenting the productivity and incomes of men (Boserup 1970). This is clearly demonstrated in North Maragoli. Production and income have been found to be generally low in female-managed and female-headed households. Studies have shown that (Anker et al. 1983), the amount of work, women do, is inversely related to household production and income levels, so that the poorer the household the higher the number of farm work hours of women. This is very true for North Maragoli. During this study it was found that women in poor households, especially the female-headed households, worked the longest hours of all three groups studied and they often work as wage labourers on other farms as well. Yet, production and incomes were lowest in this household type (Tables 5.5, 5.8, 5.9 and 5.10).

Labour is a considerable constraint and hampers farmers from following some of the crop and livestock husbandry practices suggested. The shortage of labour explains the 'under'-plucking of tea and inadequate weeding done for both maize and tea. It also explains some of the problems of livestock husbandry and tree production. The disorganization in the allocation of labour and labour shortage were more noticeable on female-headed household plots.

Hired labour is of increasing importance as production for market spreads and individualism increases (Upton 1987). The wages of hired labour makes up the largest single item of expenditure on most farms, despite the fact that hired labour generally provides less than 20% of the total farm work input. However, farm households must have ready cash to use hired labour. An analysis of the use of wage labour shows that 54% of the male-headed households, 45% of the female-managed households and 11% of the female-headed households used hired labour particularly for plucking tea and during the peak labour season of the year (long rains season). In a number of households labour was used on a temporary basis only when money was available.

Female-headed households had a more pressing problem in the provision of labour than the other two types of households. Most of these females conceded that they received very little or no support either physically, materially or financially from their children.

The tradition in this area is that boys and men have very little respect for women and so female-headed households find it very difficult to engage and supervise male labour. Yet, there are certain tasks that can only be done by men since specialized skills are required and until now these have been the prerogative of men and such jobs at times can be quite strenuous. Female-headed households observed that when they contract men to prune tea, the men sometimes delay the work or do a sub-standard job and yet still demand cash and food. Eighty percent of the female heads pointed out that these attitudes and behaviour cost them money, time, frustration and eventually lead to poor production. Men often delayed taking up farm jobs when contracted by females. This meant that farm operations were carried out late with a subsequent effect on production.

There are also many social functions that take up a considerable amount of time. Tradition dictates that during funerals people in the neighbourhood of the deceased are not allowed to work on their farms. Other social obligations include weddings, political rallies, prayer meetings and traditional visits to relatives. These, coupled with the inevitable many illnesses takes up a great deal of time that could otherwise have been utilized on farms. The female heads appear to have more health problems and complications because of their age than the other two types of households. This may be because of their age.

We noticed that in many households labour provided for farming operations by other members of the household was very unreliable. In a few male-headed households there was a problem of labour for tea production because women had withdrawn their labour because their husbands did not share tea earnings with them. Women claimed that they would use their time in producing food for the household, selling their labour on neighbour's farms or running small petty businesses on the local market. But this problem, where it existed, was overcome by the use of hired labour. In female-managed and female-headed households, women now keep the monthly income from tea and this has acted as an incentive for them to invest more time in the crop whenever possible. However, the end of year payment (bonus), which is the major earning from tea, is paid to the males in female-managed households. This creates many problems. It is the bulk of the tea payment and so part of it should be re-invested in the crop. Problems and conflicts between members of the household around these earnings and provision for labour indicate that the household cannot be treated as a single unit of production with uniform decision-making mechanisms and expenditure patterns.

Complications arise in the tea sub-system's management when a man dies before sub-dividing the land to his sons. Some tea plots (24%) had been neglected because of cases pending in the courts or with the chief on the sub-division of land between sons and mothers and sometimes between brothers. Conflicts also surfaced between sons and mothers as to who was the right person to manage tea and earn the income. Inter- and intra-household relations and conflicts caused many management problems, especially in the case of tea (the high money-earning crop).

Many farm households experience labour shortage and therefore the inadequacy of labour allocation to different farm-household sub-systems consequently leads to poor management and low production. It is clear that women provide most of the labour on the farms in North Maragoli and other small-scale farming areas in Kenya. Many of the households, especially the female-headed households were found to be resource poor and therefore unable to employ wage labourers. This implies that a solution to this problem is urgently required.

5.10 Fertilizer Use in the Three Types of Households

Many farmers attempt to use fertilizer. The yardstick for our analysis of fertilizer use is the fertilizer received from KTDA because it provides reliable information, and all tea farmers use this fertilizer. Farmers are encouraged to place application requests for fertilizer, supplied on

Table 5.17
Amount of Fertilizer Taken on Credit from KTDA by Household Type (for Tea) 1992/93 (source: North Maragoli In-depth Case Studies, 1993).

No of 50kg bags-KTDA	0 bags hh	tot	2 bags hh	tot	3 bags hh	tot	4 bags hh	tot	5 bags hh	tot	7 bags hh	tot	8 bag hh	t	10 bag hh	tot	12 hh	t	Tot
Male-hhh	3	0	2	4	2	6	2	8	1	5	0	0	2	16	1	10	0	0	49
Percent %	23.1		15.5		15.4		15.4		7.7		00.0		15.4		07.7		00.0		100%
Female-mhh	4	0	2	4	2	6	0	0	1	5	1	7	0	0	0	0	1	12	34
Percent %	36.4		18.2		18.2		00.0		9.1		9.1		00.0		00.0		9.1		100%
Female-hhh	3	0	2	4	1	3	2	8	1	5	0	0	0	0	0	0	0	0	20
Percent %	33.3		22.2		11.1		22.2		11.1		00.0		00.0		00.0		00.0		100%
Totals		0		12		15		16		15		7		16		10		12	103

credit, a year in advance. One of the conditions attached to applications for fertilizer is that the farmer must be a registered tea producer. Though farmers request for fertilizer, the amount they asked for often is insufficient for their tea plots. The reasons given for only requesting a few bags of fertilizer concern price and repayment. Farmers are afraid of the high monthly deductions involved in fertilizer repayment as this reduces what they can take home. Farmers only worry about the question of repayment and do not consider the fact that the more fertilizer used, of course if it is used in the correct way, the higher production will be. This is especially so because fertilizer is sold to farmers at a subsidized price by the Kenya Tea Development Authority. It was observed that the poorest households, especially the female-headed households submit requests for fertilizer which created a vicious circle of low yields, low incomes and continued poverty (see Table 5.17).

On average each male-headed household received four bags, female-managed three bags and female-headed two bags of fertilizer from KTDA in 1993. Yet, each farmer was expected to request for a minimum of five bags for an average tea plot of 0.3 ha. This evidently shows that less fertilizer than the recommended measure (amount) is being used on many tea plots. The following specific reasons for the limited amount of fertilizer used were mentioned.

In all the households 33% of the farmers did not apply for nor received fertilizers because some claimed that the information was not relayed to them in good time. In female-managed households 36% did not use fertilizer because the tea crop was registered under the husband's name and in many instances, at the time of applying for the fertilizers, spouses were away; no letters authorizing to submit applications for fertilizer had been given to the wives as the responsible managers in their absence.

Some of the households' (especially male-headed and female-managed) failed to use fertilizer because their were conflicts between husbands and wives, between brothers etc over the ownership of tea plots. Under such confrontational and unstable circumstances, application requests for fertilizers were not submitted.

Another problem observed was the incorrect use of fertilizer. Application of fertilizer at the wrong time and on unweeded tea bushes was prevalent in female-headed households. Farmers did not measure the fertilizer but used rough estimates and these were often on the lower side. Some farmers sold fertilizers to other farmers in order to raise money to solve pressing needs within the household and often the fertilizer was sold at a cheaper price than the KTDA price, at the expense of their tea crop. Generally, male-headed households used more fertilizers than the other two types of households. The use of fertilizers in female-managed households was intermediate and was deficient in female-headed households.

Insufficient and sometimes the incorrect fertilizers were used on maize plots. Approximately 42% of the farmers of male-headed, 46% of female-managed and 27% of female-headed households used some fertilizers for maize cultivation. The low production of both food crops and cash crops is partly affected by the low and incorrect use of fertilizers.

5.11 Farm and Household Incomes and Farming Investments in Relation to Gender

Good farm management and high agricultural production depends to a large extent on the investment made by the household in the farm-household system. High farm incomes depend on good farm management and high agricultural production, and this creates a vicious circle type of situation. It has been shown that female-managed and female-headed households face a number of constraints and that these limitations prevent them from providing good farm management and from achieving high agricultural production in the different farming sub-systems.

The main sources of farm income are tea, French beans, milk, livestock, and trees. In many households, it appeared that tea was the most reliable source of farm income since the farmer received monthly payments with a bonus at the end of year. Our field analysis shows that agricultural production varied greatly from one household type to another (see Tables 5.5 and 5.8). Considerable variations were observed even among members of the same type of household. This could be explained by differences in financial state, decision-making and managerial capability. The male-headed households achieved higher yields and production per hectare in the tea farming sub-system. Subsequently, male-headed households received more farm income than female-managed and female-headed households. This was obvious because male-headed households invested more fertilizers and labour in tea production than female-managed and female-headed households. Our analysis has also revealed that farm management in female-managed households was better than in female-headed households, and resulted in higher tea production. Farm incomes subsequently followed the same trends as production (see Table 5.18)

Total household income consisted of total farm income, remittances from spouse or other relatives, salary for those in both small-scale farming and non-agricultural employment, and incomes from petty businesses (trade). Again it was the female-headed households that earned the least.

Total farm incomes and total household incomes were higher in male-headed households and represented 43% and 46% of the total annual farm and household incomes respectively in North Maragoli. This could be explained by the fact that most males were involved in off-farm

Table 5.18
Total and Estimated Net Annual Farm and Household Incomes by Household Type 1993 (source: North Maragoli In-depth Case Studies, 1993).

Types of households	Total farm income/year in Kshs	Ave-farm income/year /hh- Kshs	Total house-hold income/ year in Kshs	Ave-house-hold income /yr/hh-Kshs	Index sub-sampl farm income	hh income
Male-hhh	208,542.00	16,042.00	475,038.00	36,541.00	100	100
Female-mhh	175,469.00	15,952.00	341,406.00	31,037.00	99	85
Female-hhh	102,717.00	11,413.00	214,341.00	23,816.00	71	65

employment or businesses which provided permanent monthly and annual sources of income for male-headed households. Some of the males who were not in permanent employment had stable businesses. This is in contrast to women who were not in employment and whose businesses were of a temporary nature. The question of money and free time were considered as the main determinants of business establishment and choice. Men appeared to have more money for themselves and more time since they contributed very little time towards household agricultural labour. Women's money however was used more for the running the household and women were also constricted by lack of time because of farm and household labour demands. This aspect demonstrated the success of male-headed household in accumulating higher total households incomes than female-managed and female-headed households.

The purchase of hybrid seeds, fertilizers, cross-breed cows or grade cows, labour and hire of tractor and oxen are the major monetary investments. Male-headed households invest more money than female-managed and female-headed households (see Table 5.19 and Figure 5.10). We can infer from Figure 5.10 that farmers spend more money in January and February, presumably for the purchase of seed and fertilizers, and for farm preparation and planting. Weeding is mostly done by members of the household and this is carried out during the month of March. The other peaks on the investment graph are mainly investments in the purchase of livestock and tea plucking. The pattern of investment in farming is similar to that in production and incomes with male-headed households spending more money on farming investment.

Farm investments consume a very small percentage of the farm income: in male-headed households the figure was 24%, in female-managed ones 17% and in female-headed households 17%. This may suggest that differentiation by gender and type of household greatly influences the amount of money invested in farming. Looking at the ratio of output to input and if we compare the farm income index and investment index (see Tables 5.18 and 5.19), it can be deduced that both female-managed and female-headed households are more efficient in the use of farm inputs than male-headed households (1, 1.4 and 1.4 for male-headed, female-managed and female-headed households respectively).

Table 5.19
Farm Investments by Household Type (source: North Maragoli In-depth Case Studies, 1993).

Type of household	Total invested by all farmers per hh	Average Kshs tinvested per hhh	Index sub-sample
Male-headed hh	Kshs 50,487.00	Kshs 3,883.60	100
Female-managed hh	Kshs 29,780.00	Kshs 2,707.30	70
Female-headed hh	Kshs 18,281.00	Kshs 2,031.20	52

Per type of household, North Maragoli

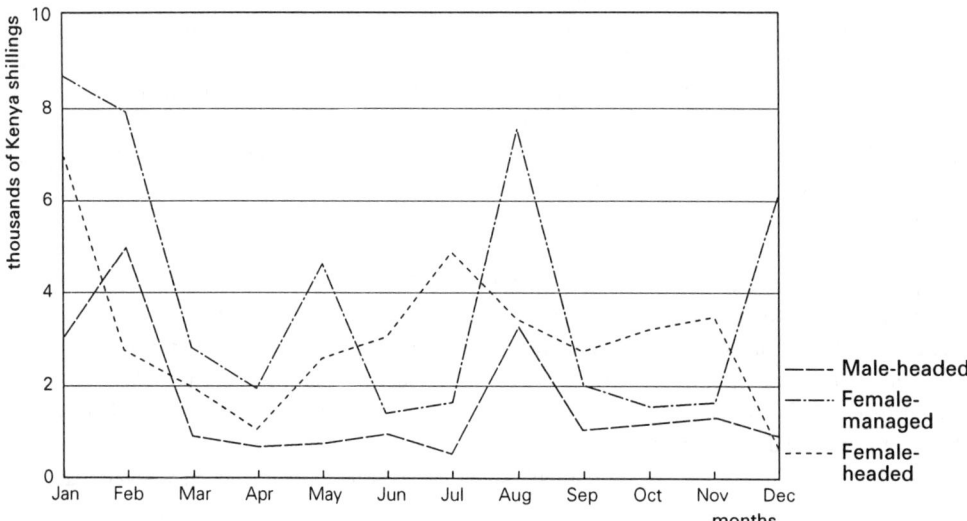

Figure 5.10
Monthly Investment in farming, 1993 (source: North Maragoli In-depth Case Studies, 1993).

5.12 Gender Differentiation and Access to Agro-Support Institutions

In this section we try to establish whether there are any differences in access to agro-support institutions in the three types of households. In section 4.9 we pointed out that not many farmers have benefitted from these services. The theory of the trickle-down effect does not seem to work in practice in the three types of households studied. The success in the management of crops and livestock that have been introduced to farmers such as hybrid maize, coffee, tea, French beans and zero grazing require efficient and effective agricultural extension and agricultural credit services. All the mentioned crops require specialized agronomic practices and appropriate and adequate inputs. Table 5.20 shows the number of times that farmers were visited by extension agents in 1993.

Male-headed households receive priority in extension services and this may explain the reason why they perform better than other households. They are followed in frequency by female-managed households. The farmers selected and visited by extension agents had better incomes, more education and were generally good farmers. They were the 'progressive' people in the village (small sample size).

The same pattern was repeated in the case of livestock extension officers. Four male heads (31%) were visited by livestock extension officers in 1993, compared to two female managers (18%) and only one female head (11%) (small sample size). Again here two factors were significant, the issue of status and location near the main road. Both the female managers and female heads did not attend any farming demonstrations and field days whereas six male heads (46%) did attend such events in 1993. Only one male farmer attended a farmers training course in 1993 and no female was given a chance.

Table 5.20
Visits by Extension Agents in 1993 (Except for Tea Extension) (source: North Maragoli In-depth Case Studies, 1993).

No. of times visited	Male-headed hh		Female-managed		Female-headed	
Zero	6	46 %	7	64 %	7	78 %
One	3	23 %	3	27 %	1	11 %
Two	0	0 %	1	9 %	1	11 %
Four	2	15 %	0	0 %	0	0 %
Five	2	15 %	0	0 %	0	0 %
Total No. of visits	21		5		3	
Farmers	13	100%	11	100%	9	100%

Many of extension messages and innovations are conveyed through the Training and Visit (T&V) method as discussed in Sections 2.2.3 and 4.9. Six of the male heads were 'contact' farmers and they also belonged to the 'water catchment and soil conservation committee'. Only two female managers were 'contact farmers' and none of the female heads was held this position. Many farmers learn by trial and error and some inquire from their friends or relatives in the neighbourhood for information. Some farmers seek agricultural information from shopkeepers who may be just as ignorant of new technologies/innovations as the farmers themselves, despite the fact that they sell the inputs.

Tea extension was found to be better organized than extension services to the other crops. All the farmers irrespective of gender were visited at least three times in 1993. However, the male-headed households were visited more frequently than the other two types of households because of the apparent friendship that developed between the extension worker and the male farmers. The tea field extension workers were males. Tea extension is better organized because KTDA is an autonomous parastatal, is free from government intervention and is at present still being well managed. It is also free of political pressure. KTDA staff also appear to be better motivated and supervised, and this results in better extension.

The problems associated with agricultural extension in Kenya centre around poor organization and the non-accountability of most of its staff. (see also Leonard 1977, 1990 & 1991). It has also been pointed out by Schwartz et al.(1992) that extension has been failing because of weak research-extension linkages, weak extension supervision and management, an inadequate focus on women, poor monitoring and evaluation and conflicts between the Departments of Agriculture and the Department of Livestock Development and other Ministries with activities related to farmers. Extension messages which are technically feasible, economically attractive and socially acceptable are urgently required. All these elements of the extension problem were conspicuous in this study. The neglect of women in extension was found to be overwhelmingly obvious.

Many farmers are insolvent and therefore unable to purchase seeds, other farm inputs and to contract labour as a result of lack of and/or shortage of money. The government has developed several methods which make it possible for farmers to get access to credit (see Sections 2.2.3 and 4.9) but many small-scale farmers have not been able to benefit from agricultural credit systems due to:-organizational and administrative limitations of the funding organizations, especially AFC; bureaucratic procedures that discourage farmers,; strict conditions of loan disbursement which many small-scale farmers fail to fulfil; and the centralization of the loaning institutions at provincial and district headquarters and their lack of knowledge of the needs of the different groups of farmers found in small-scale farming. It has been established that where there are possibilities of receiving loans, rural men benefit more than rural women because most women lack the necessary collateral. For instance, as was mentioned before, only one female in our survey had a land title deed.

As a result, only two farmers (one male head and one female head) received AFC loans in 1993. Four farmers (12%, three males and one female) had received loans from the teachers savings, credit and co-operative society because they are members of this co-operative had shares, colleagues to sign for them as guarantors and enjoyed job security. Many men among our sample think in terms of prestigious long-term investments, while women think more of consumption-type investments centering on food security for the family. In general an acute lack of money was observed in many households. To a great extent this lack of money negatively affected farm investments and household decision-making generally.

5.13 Coping Strategies and Non-Agricultural Activities in North Maragoli

Our discussion so far on food and commercial crops, livestock and tree production farming sub-systems has revealed that production and incomes are low. It has been illustrated that management is poor on many of the farms. Indeed, throughout the study, gender was seen to be an overriding factor: it either created or impeded access to and use of resources. The result of these differences were clear from the levels of management observed.

Poverty also contributed greatly to the low standard of farm management. Poverty was worst in female-headed households but as a whole more than two-thirds of the population in North Maragoli can be called poor because most of them were basically surviving from day to day, sometimes not even sure where the next meal was coming from, especially during the months of food shortage. Poverty was found to cause many conflicts in the family, and this, in turn, caused farm management crises and household instability.

Classifying of our households into three groups has enabled us to identify the reasons why male-headed households have relatively good farm management, production and incomes as compared to female-managed and female-headed households. Indeed, it has been shown that some of the justifications for the better performance of male-headed households are linked to the accessibility of production resources and support services, the ease at which they can be incorporated into the job sector and gender differentiation. Women fight against many odds to cope with daily life in this area. Farm production, and farm and total household incomes were lowest in female-headed households. Many farm households did not generate sufficient incomes from the commercial crop farming sub-system to satisfy their minimum financial needs.

Off-farm occupations and activities offer income opportunities for families without sufficient land available for an adequately remunerative employment of family members (Burger 1994). In North Maragoli, the situation is such that most of the farms cannot sufficiently fulfill the food and monetary needs of their household members. Some farmers have devised coping strategies by getting involved in a diversity of non-agricultural activities (also included under the off-farm activities in our farm-household system) to supplement farm income. Families claimed that they are involved in other strategies in order to diversify income sources and spread risks. Coping strategies mainly revolve around ways of earning or getting some income to satisfy household needs. However, income matters are confidential and sensitive. The most important point is that other methods of getting money have been devised in order to supplement agricultural production and the low income associated with it.

Remittances
For a long time, the most important source of income for has been and remains to be remittances from working relatives. The large proportion of male labour migration from North Maragoli can be associated with lack of genuinely adequate social and economic opportunities in this area. Many farm households have at least one member working off-agriculture (off-farm). The female-managed households are directly affected because partners are away. Many women commented that they did not know how much their partners earned. The question of different economic, consumption and production units within the same household arises here. The household has, in fact, been seen by some writers as a site of tension and conflict as well as of co-operation, as a site of inequality as well as mutuality (Beneria et al. 1992, Dwyer and Bruce, 1988, Dwyer et al. 1988). The description of distribution and expenditure patterns of income within the household makes it clear that none of the assumptions regarding the monolithic household/family are valid for our sample. Husbands and wives differ in their definitions about what constitutes the basic necessities of the family complex, their consumption priorities, the way in which income should be distributed, and the proportion to be allocated to the common fund if there is one. Only one household in this survey could be described as having one common fund: the wife and husband had a joint bank account.

A survey on remittances from spouses revealed that although wives depended on remittances, in a few households these were regular whilst in many households they were irregular and small. In some cases (42%) there were no remittances and the wives lived as deserted women because their husbands had a second wife in town and rarely visited and sent money to the rural home. When money was remitted by the spouse it was for a specific purpose. The few husbands surveyed when they came home on annual leave complained that the wages they earned were too low to fulfil the financial demands of the rural family. They pointed out that the cost of living in urban areas was very high and so they expected their wives to be able to manage the village farm and feed the family and meet other obligations.

Eighty-two percent of the female-managers received some money from their husbands in 1993. The money received varied from Kshs 250 to Kshs 9,800 in 1993 in the different female-managed households. The extent and amount of remittances were also dependent on inter- and intra-household relationships. More money was sent where a smooth and good relationship was experienced, especially with newly married couples. Also more money was received where the woman proved to feel more responsible in the management of the family resources. Female managers received remittances on average of Kshs 265 a month and Kshs 3,174 per annum. The spouses of the female-managed households worked mainly in Nairobi (55%). They were engaged mostly in low wage jobs. A small number worked as clerks in government service while a number worked in Nairobi's industrial area on semi-skilled occupations. About 30% were working in Eldoret, Kisumu and Kakamega. Those working in the Western Region of Kenya were employed in factories or as subordinate staff in government institutions (schools and health centres). Some of the spouses appeared underpaid and so rarely have money to remit to their rural families. Others earned good salaries but had neglected their rural households. Remittances are still a big source of money for many rural households. Approximately 15% of the female-managed households in North Maragoli were living on their pension, and had bought land in a settlement scheme where they lived with second wives.

Another source of remittances was working children or relatives. Many farm families spend considerable amounts of money on the education of their children and often they expect monetary assistance from the children when they take up a job. Many parents asserted that they received very little financial support from working children. In male-headed households three-quarters reported that they did not receive any remittances from working children or relatives. Only 23% of the male heads received some money, on average approximately Kshs 1,000 in 1993. However, the amount remitted varied between Kshs 600 to 8,000 in female-headed household in 1993.

In female-managed households, one-third of the farmers received financial support from their working children and this varied from Kshs 400 to 2,900 a year (average of Kshs 427 in 1993 per household). In some female-managed households this was a very important source of income but in other households this was not very important because the amounts involved were small, irregular and unreliable. It was again noticed that inter- and intra-household relationships determined the amount of remitted from the working relatives. As observed, the cost of living in the urban areas has risen in the last few years and this may be one of the reasons why some urban workers are not in a position to send remittances to their rural homes regularly.

Other sources
Many of the female farmers are involved in petty businesses on the nearby markets. They process and sell food-stuffs, hawking items and small merchandise. The women declared that most of the money earned from the petty business sector was utilized in the purchase of other consumption goods, mostly dry maize for grinding into maize, vegetables, salt, cooking fat, match boxes etc. Female-headed households are not involved (on a large scale) in these petty businesses because

of lack of starting capital. Sometimes all the money earned from petty trade is used to buy food and hence no more money is available for the continuation of the business.

These non-agricultural activities are normally carried on in the afternoons, after the women have worked on the farms. One of the respondents was involved in the making and selling of illicit local brews known as 'Busaa' and 'Chang'aa'. The female involved in this (illegal) trade was always harassed by the police and the local chief and therefore she was very suspicious of strangers coming to her home. The management on her farm was very poor because she spent most of her time selling beer or sometimes in police cells when arrested. However, she possibly made more money than the average farmer in North Maragoli. Other cases of families relying on the illicit beer trade- though not in the survey-were observed.

Respondents gave the impression that they made little money from these non-agricultural activities, although this is doubtful. The level of involvement and the time spent by many households indicated that this sector contributed some reasonable amount of income to the farmers. However, these activities took much time away from the time needed on the farm. This happened particularly when the women travelled to other areas to buy the commodities for sale. On official market days, (two days a week), most of the women involved in petty businesses spent the whole day at the market selling their wares and then farm work was neglected. Non-agricultural activities give rural people, especially women, much satisfaction because they feel involved in concrete income-earning activities and that they are making a contribution to the economy of the household. Household subsistence, especially the welfare of the children, is the most motivating force behind everything that women do. It appears that many households are prepared to try any option which might produce the money they need. Where farms were too small and production was very low, more time was spent on non-agricultural activities.

There are many other non-agricultural activities that take up women's time. These include household chores such as collecting water, firewood, cooking for the family and caring for children. It was quite apparent in the field that women used much time on these activities. Women's farming operations are affected by all these non-agricultural activities. Not enough time is invested in the different farming operations. This is made worse when women have babies or when one of the children or members of the family is sick. Women have to re-distribute their time on such occasions.

Men appeared to be involved in more lucrative off-farm activities. Some were involved in masonry and carpentry work, running a small shop, a butchery, tailoring and livestock trade. Men seemed to devote more money and time to trade and in doing so they earned more. This partly explains the differences observed in total household incomes.

5.14 Conclusions

The quality of farm management greatly affects production, incomes and the overall operation of the farm. Gender differences in the headship of the household plays a considerable role in the efficiency of the farm-household system. Male-headed households have better access to production factors and agro-support services than female-managed and female-headed households. Lack of access to these factors is fostered by gender-bias and affects farm management negatively. What our study has revealed is the poor performance of women in many of the farming sub-systems.

Social and cultural problems in small-scale farming: The present freehold and customary land tenure system, where men are the title holders discriminates against women's productivity. Indeed, in many farm-households women cannot make major decisions or initiate changes in land use. Conflicts over land ownership and the right to use it were apparent in some households and affected production process. The man in the household receives benefits from Government and the community on behalf of his household, but on many occasions he does not share these benefits with members of his household. This denies women the means to improve farming.

Economic aspects in the farming-system: The smallholder sector in North Maragoli is relatively under-developed because farming has not reached the stage of being income-earning intensive and therefore lacks the structural and institutional capacity to attract and absorb a large and rapidly growing young male labour force. Many interventions in agriculture by the government have contributed to the high prices of inputs and the low prices of the agricultural products. Agricultural production in this area is therefore not very reliable as a means of sustaining household livelihood without an additional source of money from outside the farm. To a large extent, this phenomenon has contributed to out-migration of males, creating labour bottle-necks in many farm households and eventually affecting the farming economy. Many households are poor and lack sources or means of income. The female-headed households were found to be the poorest of all the rural households. This led to low or no investments in the various farming sub-systems leading to continued poor production and low incomes.

Agro-support services: These were found to be gender-biased in favour of men, and inefficient. They had a limited impact on many farms. It appeared that there was no effective system of delivering knowledge and skills because agronomic practices for most crops are very poorly grasped and poorly used. This has to do with the agricultural extension services in this area which serve more men than women. The trickle-down effect or diffusion of information as anticipated to occur through the 'T&V' method in combination with the 'contact farmer' approach did not work to the benefit of many farmers. Women were particularly marginalized by this approach. Male-headed households seem to be preferred as 'contact farmers'. Indeed, even between the female-managed and female-headed households, there was a big difference as the latter were more neglected.

Agricultural credit was very limited. In fact, very few farmers - mostly men - had benefitted from this service. Where possible they used savings, credit and co-operative societies in their working places to get loans. Some had used land title deeds to get loans from commercial banks. Some of these loans had been used to buy more land or to invest in zero-grazing. Women in North Maragoli are marginalized because they do not have title deeds to use as collateral. There is also a lack of information on the credit system among the farmers. There is an apparent lack of an effective system for delivering financial and material support to the actual people involved in farming.

Labour and labour shortage: Labour was a big problem in many households. The concept of the division of labour according to gender lines in many farm-households was no longer in existence. Women provided most of the labour in all households. Women are therefore constrained to find adequate time to provide sufficient labour to all the sub-systems in the household-farm system. There was conflict of interest and this led to some sub-systems being neglected, consequently leading to poor production and low incomes. Male-headed households were better placed because many of these households had other sources of income and thus were in a position to hire additional labour to combine with household labour. Moreover, not much labour-saving technology is available to help meet labour peak requirements. The unpaid domestic and family farm labour is considered as private by the government. Therefore, no social system of incentives and rewards and penalties have been developed to encourage a change. This has lead to the invisibility and non-recognition of women's labour. In male-headed and female managed households, the individual woman confronted the individual man, and in the politics of the household the husband enjoyed the higher status and more power because he is the head of the household and the owner of the productive resources. We have shown that this affects agricultural production because of the conflicts, confrontations and controversies which develop particularly around cash crop earnings.

Deficiencies in marketing, infrastructure, price incentives and general agricultural policies as we will discuss in Chapter Seven have hindered the growth and expansion of small-scale farming. There is little participation by local people, especially of women, in the design, planning and execution of policies that affect them. The top-down approach that has been used since the colonial period does not consider household problems and characteristics. The scarcity of money and the low stanard of education among farm households, especially among women, negatively affects farming investments, and results in low production. Finally, it has been shown that female-headed households experienced more constraints and that these constraints lead to poor farm management, low agricultural production and low incomes. A big difference in terms of farm-household management, production and incomes was found between the three types of small-scale farming households. Male-headed households fared better than the others.

Notes

1 Producing the maximum-valued output possible, using cost-minimizing techniques of produc-tion and considering effective market demand (Todaro, 1994).

2 Gender ideology are the ideas which people in a society have about what behaviour is correct for the different sexes. The aspects of gender ideology may include questions as (i) what is believed to be a good girl or woman, a good man and boy? (ii) ideas about roles and tasks of women and girls and of men and boys (iii) perception of women about their own abilities and self image and (iv) women always struggle to overwork themselves to win praise from husbands and neighbours.

3 Ellis (1988) points out that the division of labour between the sexes is not 'natural' in the sense of being ordained by biological differences between them. Peasant households where women often do the most onerous physical tasks are living proof of this. It reflects social customs, norms, and beliefs which govern and circumscribe individual behaviour. For this reason it is misleading to refer to the division of labour

between women and men as the 'sexual division of labour' with its overtones of causation by the biological differences between sexes. An alternative is to refer to the 'gender division of labour' in which gender is used as a shorthand for the social meaning which is attached to the roles of women and men in different societies (Ellis 1988:166).

4 The working day starts at approximately 5.45 a.m and goes on up to about 10.00 p.m. Household work includes child care and upbringing of children, cooking, cleaning, washing, firewood collection, fetching water, marketing, house building and repair etc. Tea production includes, plucking, pruning, weeding, fertilizing, marketing etc. Food (maize) crop production includes, land preparation, planting, weeding, harvesting - two seasons in a year. Labour is also provided for cattle and many other tasks within the farm-household system and community social functions. In female-managed households, we counted the hours that a husband worked on the farm when he came home on 'leave' during the year. We therefore decided to leave out daily and weekly estimates as the husbands did not have a consistent farm working pattern.

6

Inter- and Intra-household Farm Management Situations: Case Studies

6.1 Introduction

This chapter selects a few of the households visited during the case study period to illustrate the differences in farm management and agricultural production, the reasons for the differences and the coping strategies as analyzed in chapter five. The household heads are divided into three groups using the life cycle approach: (1) Those where the household head is below the age of 45 years, i.e the young households, (2) those where the household head is in the middle age group between 45 and 55 years and (3) those where the household head is over 55. The households are categorized into three groups: male-headed, female-managed and female-headed households.

The state of the household does influence decisions-making processes and farm-management. Main features observed in the field that influence decision-making and farm management are: (a) the household type (b) standard of education (c) ownership, accessibility and control of the factors of production, (d) age of the household head (f) inter-and intra-household relationships, and (e) accessibility to money and pattern of expenditure by the members of the household.

The distributional patterns of income within the household make it clear that none of the assumptions regarding the monolithic household/family are valid in our sample as shown in Chapters 4 and 5. Husbands and wives differ in the definition of the basic necessities for the family, their consumption priorities, the way in which income should be distributed, and the proportion to be allocated for the common fund, if such exist. It was found that women were more concerned about consumption needs whilst men were interested in prestigious investments and non-consumption and leisure types of investment.

As already discussed, though women are not in control of household resources, they have the obligation and duty to manage household resources in order to feed, cloth, house, and educate the children especially those in primary school: many women in rural areas cannot afford to pay for secondary school education. This responsibility for making ends meet without control over resources is a source of constant problems and anxiety for poor women. It is they who have to devise household coping strategies, but it is their husbands who control access to major resources.

As Beneria (1992), has rightly observed and as we noticed during this study, relations of male domination and female subordination that characterize the household as a social institution have enormous implications for farm-household systems.

The coping strategies of some households undoubtedly lead to a deterioration in the position of women or the disintegration of households with men leaving (to seek work elsewhere) and sometimes starting new households in their new work locations. The increase of female-managed households is not a sign of emancipation from male power. In fact, in some cases the growth of female-headed and female-managed households to some degree induces further marginalization. This was illustrated with our examples of female-headed households who become more impoverished when the husband dies because of the social, economic, cultural and attitudinal norms binding rural households. In a society in which women as a gender are subordinate, the absence of a husband leaves most women worse off and more vulnerable to cultural, social and economic constraints. The core of gender subordination lies in the fact that most women are unable to mobilize adequate resources (both material and in terms of social identity) except by making themselves dependent on a man (Beneria et al 1992). Male migration on the one hand reduces the expenses of the household, but all too frequently reduces household resources and labour as well.

The number of female-managed households relying on insufficient and unstable remittances is significant. Cases of women in female-managed households being deserted by their husbands because the males are unable or unwilling to share their earnings is on the increase in both rural and urban areas and this explains the increasing number of households managed and headed by women. Approximately 60% of women in our female-managed households were as good as deserted because their husbands rarely sent remittances and visited very infrequently (once in a year or after a couple of years).

This Chapter forms part of the in-depth analysis of our North Maragoli study. It was found illustrative and revealing to use a few case studies to deconstruct and look inside the household to show the different types of households' dynamics and levels of farm management practices and agricultural production and how they were influenced by inter- and intra-household relations and factors. A brief background is given of the members of the specific households, an analysis is made of the farm-household system indicating the farm management situations and production performances as far as cash and food crop, livestock and tree production farming sub-systems are concerned, and finally an attempt is made to identify and describe the coping strategies and constraints faced by the farmers selected. Indeed, it is hoped that at the end of the chapter the reader will be able to single out household constraints to farm management which contribute to low production and to identify how gender plays a role in farm-household systems in small-scale farming.

6.2 Male-Headed Households

Baba Agesa

Background History
Agesa was born in 1944 in the Mudete Sub-location of North Maragoli. He is 50 years old and therefore belongs to category two of the life cycle categories i.e middle-aged household heads. He started school in 1952 at Digula primary school. In 1964, he sat for the Kenya Primary

Education Certificate (KPE) which was then the qualifying examination for secondary school. He did not pass the examination (8 years of schooling). He says that in those days, rural Kenyans went to school, especially upper primary and secondary school, when they were already quite old perhaps over 16 years of age, and that sometimes they were even married but still they continued with their primary education.

He got married in 1963; his wife also comes from Mudete Sub-location. They have nine children, six boys and three girls. The first two children have completed school (12 years of schooling). The first-born son teaches in a secondary school as an untrained teacher, while the second born was following a teacher training course at the time of our survey. Other children are still in various stages of education while a few are still too young to start school. The farmer claimed that all his earnings go towards paying school fees for the children, especially for the ones in secondary school where the amounts involved are very high. The fact that most of the money earned in this household is invested into the payment of school fees is an illustration of the importance attached to education by many farm households. Many farmers are willing to deny themselves all the basic things in life and put very little investment in farming so they can pay school fees.

To some extent, the education of children contributes greatly to the impoverishment of some households. Some farmers even sell part of the small farms to raise money for fees. Yet, there is no assurance of the benefits the household will receive from their children's education given to the children given the rigorous examination system and the scarce job opportunities in Kenya. Since most of the money is invested in paying school fees, the members in this household were barely dressed and only three-quarters of the house was complete in terms of roofing. They could barely have a square meal. On the many occasions that I visited them in the morning, ate either soft porridge or drank tea, lunch was soft porridge and the only real meal was supper when they ate 'Ugali' accompanied by green vegetables. This was the case for both children and parents. During the months of April, May and June when there is the most food shortages, they skipped the lunch porridge and only the children drank porridge in the morning. The man is permanently in debt because he has to raise money to pay school fees for his children. The farmer depends on his tea farm and on the untrained secondary school teacher's remittances for school fees payments.

Agesa's farm is 0.9 hectares. It is registered but he has not collected the land title deed yet. He says he cannot collect his title deed because it is very expensive. Part of this land (about 0.2 ha) had been sold to somebody else to raise money for school fees but the transactions for the transfer hit a snag as he did not have a title deed on which the transactions could be pegged. Also he claimed that the man who bought the land did not complete the full payment for the land and he was now considering raising money through borrowing to refund the purchaser. He cultivates both cash crops and food crops, keeps livestock and cultivates trees. Recently (July 1993) he sold off three local zebus and bought one Grade heifer at Kshs 15,000.

Commercial Crop Farming Sub-System

Tea Sub-system
Tea is the main cash crop grown and an important source of income. The area of land under tea

cultivation is 0.3 hectares, and this is planted with over 3,500 tea bushes established in 1968. This farmer has organized a schedule of work for both tea and food crop farm management. Four days per week are spent on the tea plot, mainly picking tea. He works from about 8.00 in the morning until about 4.00 p.m in the afternoon on plucking tea and transporting leaf to the tea buying center. The tea farm is weeded well and has, establishing a closed canopy, as recommended by the KTDA. Eight bags of fertilizer are used every year to maintain the good growth of the crop. The number of bags of fertilizer used is within the recommended limits although he could use up to 12 bags of fertilizer to increase the yield further. The average monthly tea yield in 1993 was 225 kgs and the total tea production for that year was 2,703 kg. Total tea production for 1991, 1992 and 1993 was 2,228, 2,119 and 2,703 kgs respectively, giving a three year annual average of 2,350 kgs. This production is quite good as he produced above the minimum expected for such a plot which is 1,950 kg (see Section 4.6). In other words, he reached the potential expected for this area.

In 1993 this farmer earned an average monthly income from tea-growing of 590 shillings and the end of year payment (bonus) was Kshs 21,852. His total monthly tea income for 1993 was 7,314 shillings, therefore, the total annual tea income for this household would be Kshs 29,166. Tea earnings fluctuate annually because of the differences in price on the World market. The 'end of year' tea payments for this farmer for 1991, 1992, and 1993 were Ksh 5,125, 8,565 and 21,852 respectively, resulting in a three year annual average of Kshs 11,847. He is categorized as being amongst the best farmers and his tea production and earnings for 1993 are close to the national expectations. However, compared to other smallholder tea-growing areas in Kenya such as Kirinyaga, Murang'a, Kiambu and Embu his production is still low. Nevertheless for Vihiga District he is one of the best farmers. Tea income is mainly used to pay school fees and to buy food. The main farm investment carried out by this farmer is the purchase of fertilizer from KTDA which is done on credit. This farmer experiences labour constraints but because of limited funds he does not employ wage labour to assist him. He does most of the farm work on his own. Agesa is a hard-working man and his farm management practices are above average and considered very good by the villagers and agricultural extension staff.

French Beans
A second source of income for Agesa is the cultivation of French beans. The company responsible deals with women farmers, who are also eligible for credit. However, land for their cultivation has to be approved and allocated by the male head in many households. Agesa asks his wife to get the farm inputs but he does the actual management (preparation of the land, ridging, weeding and picking) of the crop. He does not allocate his wife a piece of land for cultivating this crop and he does not want her to have any income of her own. Details for managing this crop are followed closely. The right type and measure of fertilizer and CAN top dressing to provide vegetation cover to prevent wind destruction, weeding and picking at the right time, and correct size. The wife delivers the harvest to the buying center, receives the money and gives it to her husband. French beans are cultivated on tiny plots and not much money is made from them. However, for women it can be an important source of income. French beans earn Agesa about Kshs 3,250 per annum. This illustrates the importance of land ownership and the marginalization of women as they are not allowed to use land freely as we can see in this case. It also illustrates the importance of capital within rural households and its control.

Food Crop Farming Sub-system

The food crops cultivated are maize, beans, sorghum, millet, bananas, sweet potatoes, cassava, sugar cane and some vegetables, especially sukuma-wiki and cabbages for sale. However, the main food crop is maize. The management of these crops was good as Agesa follows the methods recommended. He prepares the land well, planting on time and in rows. He weeds well, uses fertilizer and plants hybrid maize seeds during the long rains season.

The fertilizer used when planting the food crops is taken from the tea fertilizer and the CAN used for top dressing is taken from the French beans. In addition, a little compost and manure prepared at the homestead is used. This shows the inter-relationships between the farming sub-systems and how some farmers devise methods for solving the problems of soil degradation and of buying fertilizers for food crops. Sometimes Agesa receives inputs such as fertilizers and seeds, especially for vegetables, from the Ministry of Agriculture, Livestock Development and Marketing because he is a contact farmer.

The land under maize cultivation is about 0.1 hectare. From this he harvests 3 bags (270 kgs). He confirmed that green plucking contributed to a much lower total harvest. Green plucking takes place during the month before the harvest. During such a month, green maize is the main source of food in all meals. An individual needs about 200 to 250 kg of cereals in a year, especially in rural areas where there is a limited variety of food types. In fact, cereals (maize) form the main source of food for all three daily meals. This household harvested approximately 170 kgs through green plucking. It is possible that the total maize harvest of 270 kg plus green plucking would come close to 440 kgs which would be the potential for such a small parcel of land. The maize crop is inter-cropped with beans, and sorghum. Sorghum is harvested twice a year, and produces about 2 bags (180 kg). Agesa harvested about half a bag (45 kg) of beans for the two seasons (total production from maize, beans and sorghum 495 kg). This production is good, especially considering the small size of the plot. He was one of the first farmers in the area to adopt new varieties of bananas, sorghum, and sweet potatoes. These innovations were introduced on his farm by the agricultural extension agent because Agesa is a 'contact farmer'. Fruit trees are planted also in the homestead.

Livestock Production Sub-system

For a long time this farmer has kept three zebu cattle mainly for milk production. As has become the practice in this area, cattle are tethered and stable-fed. Through lengthy discussions with the livestock officers, he realized that his local zebu cattle occupied a lot of space and consumed a lot but had a low milk production. The 5 litres of milk from the three cows produced per day was all sold to raise money. He first established sufficient napier grass and in so doing, in fact reduced the farm area being cultivated under the food crop sub-system. In mid 1993, he bought a grade cow which was in-calf. However, he has not yet constructed a zero-grazing unit for recommended stable-feeding. Agesa's time is now divided mainly between the management of tea and the grade cow. The cow is provided with adequate feeds (napier grass and all green roughages and water), it is disinfected and cleaned once a week. It is hoped that its milk production will increase and bring in more farm income.

The cow-dung from the livestock sub-system is kept in heaps to manure the food crop and

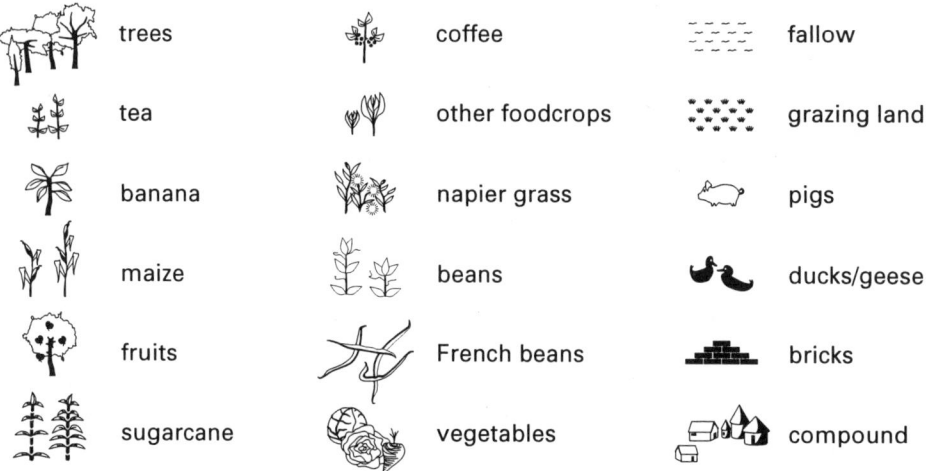

Figure 6.1
Symbols used in Farm-sketches

vegetable plots. Maintenance of soil fertility and the prevention of soil loss through erosion are keenly monitored. In addition to the deep drains established, napier grass is planted on prepared ridges or benches to prevent soil erosion.

Tree Production Sub-system
A piece of land that Agesa considers agriculturally unproductive and which lies down the valley next to the stream has been put woodlots, mainly of eucalyptus trees, (see Figures 6.1 and 6.2). The trees are weeded only when still young, otherwise they are left to grow naturally. Trees are mainly for sale, for firing bricks, fencing, construction work in the home, and the twigs are used for firewood. There is a shortage of firewood for cooking in this household because the wife only receives twigs when the husband harvests the trees for sale or firing bricks. Agesa is able to raise approximately Kshs 4,000 from the sale of trees every year.

Agricultural Extension
Agesa worked as a tea plucker for many years before retiring to his home and making the decision to become a full-time farmer. He gained a lot of experience in the management of tea and general farm planning while working on tea estates. He used these skills on his farm and his enterprise has become a model farm in his village. Agesa is one of the 'contact farmers' in his area. It was his good management, adopted from the tea farms in Nandi where he worked as a tea labourer, that led to his selection as a contact farmer and the use of his farm as a demonstration plot for the farmers in his neighbourhood. Agesa has planned his farm well and is one of the two farmers that keep a record of what they are doing on their farms. He says that farming is his only job at the moment and that is why he invests all his time in it. He has accepted farming as a job and as a commercial enterprise.

The frontline extension agent uses this farm to demonstrate methods of farming to other farmers around. Sometimes the farmer receives a few inputs, for instance seeds or fertilizers from the

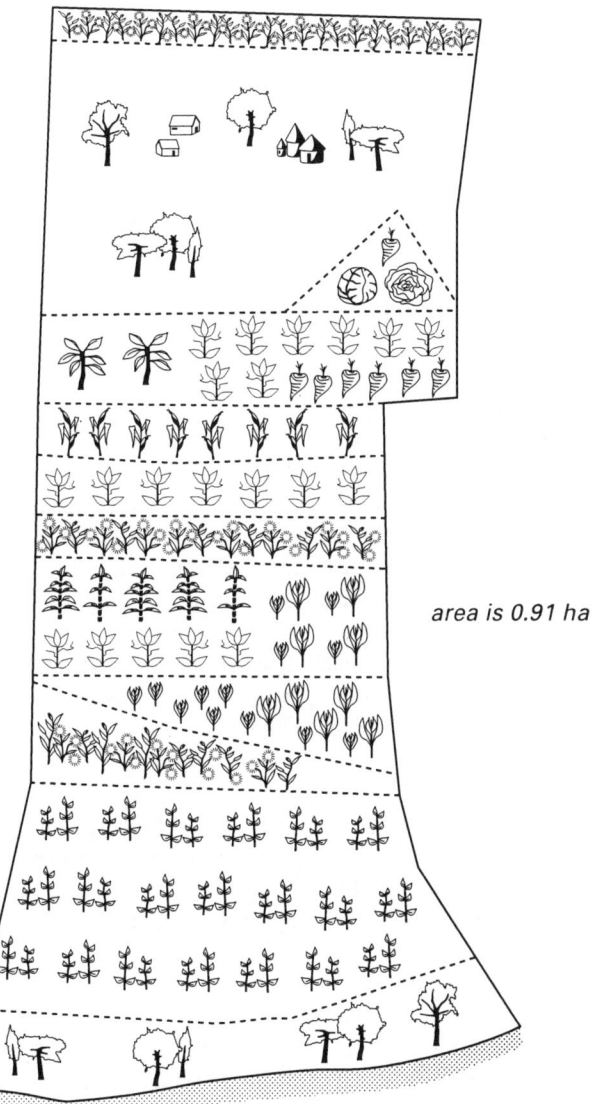

area is 0.91 ha

Figure 6.2
Agesa Farm (source: Study Results).

extension staff of the Ministry of Agriculture to enable him to prepare farm demonstrations. One can question why someone who is already experienced, good in farming and who does not need to learn about farm management and the different agronomic practices ? This could be looked at as taking a 'ready-made' farm instead of taking the trouble to train somebody whose farm is very poorly managed. More farmers could appreciate if a farmer with limited knowledge, poor management practices and low production was selected and through extension improves and achieves good farm management and high agricultural production.

It appears that extension agents prefer to work with farmers who are progressive and who already have already better education, good farm management practices and sometimes more capital and other resources than the rest of the farmers around them. Agesa is an example. He is also a member of a local soil and water catchment committee. This group is concerned with soil and water conservation and it is supposed to create an awareness in the local community about the importance of conserving and protecting the environment. Agesa receives extension advice in general agricultural production, livestock production, tea and soil and water conservation, whereas other farmers in this area have never been visited by any extension agent. Definitely this creates some of the differences that were observed in the management and production in the farming sub-systems. Agesa does not have time to diffuse these different messages from different experts in agriculture and forestry.

Other Income Earning Strategies
Other coping strategies are brick-making and the selling of trees. The brick-making business earns Agesa close to Kshs 3,000 per annum. Many parts in this area have open excavations (depressions) filled with rain water because of brick-making. The stagnant water in many parts of the location partly contribute to mosquito breeding and to the high incidence of malaria, especially among children. Many trees are felled to fire bricks and the business of brick-making is environmentally degrading, especially in Maragoli where farms are small.

Agesa's household has a total annual income of approximately Kshs 36,000. More than 75% of the total household income is spent on school fees. Much of the money is received piecemeal except for the end of year tea payment, and proper budgeting is impossible.

Farming Constraints
The major problem that hinders Agesa's further progress is bad interpersonal relationships with the members of his household and his neighbours. First, the farmer has a bad temper, and second, he drinks a lot. Heavy drinking makes him quarrel and fight with his wife frequently. His wife and children have rebelled against him and therefore, they do not assist him with farm work. The children are so afraid of him that they avoid him most of the time. He does much of the farm work all by himself. He uses most of the money left after paying school fees for drinking alcohol and there is no money left for hiring casual labourers.

His wife is not aware of the amount of money her husband earns from farming. Sometimes she has to sell her labour to rise money to purchase food for the household. The wife does not have a single plot for her own use. The husband utilizes the whole farm right up to the house. Something that is quite common with farmers is a wild expenditure patterns and it is noticeable in many households. It takes twelve months to get the end of year tea payment (bonus) and which should be budgeted to last approximately one year until the next lump-sum is earned. Yet in many households this money is spent within a month and then for the next eleven months, the household hardly has any money.

The neighbours in the community do not like Agesa because of his aggressive nature and therefore, although he is a 'contact' farmer, only a handful of farmers go to his farm for instructions. In fact, on two occasions I found the extension agent alone with this farmer giving him instructions on farm management. This can be called information blockage since the messages do not reach other farmers. For this method to succeed, the 'contact' farmer must have good relations with his family and his neighbours.

Agesa considers he is short of land and capital. He was not able to plan and manage the different crops as recommended because of the small size of the farm. He also claimed that the soil was degraded because of over-utilization and inadequate fertilizer use. Money shortage was pinpointed as the main cause of inadequate management by this farmer.

Generally, Agesa is a determined and hard-working farmer. His management is good, but he needs to improve his relationships with his wife by sharing part of the farm income, and with his children to be able to get extra labour from them. This is an example of the workings of conflict within the households and shows how such conflicts can influence decisions on labour allocation, investments and use of money, for example.

Baba Eboso

Background History

Eboso was born in 1956 in Gaigedi Sub-Location of North Maragoli location, Sabatia Division, Vihiga District. He is 39 years old and therefore belongs to the category of young household heads below the age of 45 years. Eboso is the third born in a family of ten. He started schooling in 1963 at Gaigedi primary school. He repeated a number of classes due to regular absenteeism from school. He joined Busali Union secondary school in 1974. In 1977 he sat for his `Ordinary' level examinations and passed with division four which is considered a minimal pass.

After completing his studies, he went to Kiambu to live with relatives, and to look for a job. He was employed as a 'shambaman' (gardener). He became dissatisfied with this job because it was far below his qualifications since 'one does not have to go to school to become a gardener.' This conflict between status and qualification led him to quit this job after three months. In 1979, he was employed by the Kenya Farmers Association. He worked as a recording clerk while at the same time he was getting on-the-job training. He also sat for private clerical examinations to enable him to get promotion in this company. In 1985, the co-operative union was renamed Kenya Grain Growers Co-operative Union Limited (KGGCU). He was promoted within the ranks to become a cashier. He was sacked in 1988 because of the loss of goods and money due corruption. When he lost his job, he migrated from the urban area back to the village to become a small-scale farmer in 1989.

He got married in 1981 to Libesa and they have five children, three boys and two girls. This is a young household with children in primary school and others still too young to go to school. Eboso's wife Libesa has ten years of education (secondary form two). She is a trained nursery school teacher and effectively, she is the only person in this household who is in non-agricultural employment. This household offers a glimpse of the changing roles between men and women whereby (a few) women are in non-agricultural employment while their husbands have no formal jobs but have to work in small-scale farming to produce food for the family and to raise agricultural incomes. This is something not fully accepted in many households. Men are still considered the breadwinners and therefore the right people to be in non-agricultural employment while women are still seen as housewives in many parts of Western Province. A man like Eboso who allows his wife to work outside agriculture while he is in agriculture is labelled a 'fool' and the wife is labelled 'spoiled' by the neighbours. But if both husband and wife are in non-agricultural employment and the husband has a better job and salary than the wife then this is considered 'normal'. However for Libesa, the salary in nursery school teaching is small (Kshs 200 per month) and too irregular to be of much impact on the household budget.

Commercial Crop Farming Sub-system
Eboso is managing his father's farm which is 1.35 ha. and all the farming sub-systems are found on this farm. He cultivates tea as a cash crop and maize, beans, bananas, sorghum, cassava and sweet potatoes as food crops in addition to livestock keeping and tree production. The tea plot covers an area of 0.5 hectares which is among the biggest in this area. He has an establishment of 5,000 tea bushes.

In 1993, he harvested a total of 3,371 kg resulting in a monthly average of 281 kgs. This farmer produced 0.7 kg of green leaf per bush which is fairly close to the national average of 1 kg of green leaf per bush per year. In 1990, 1991, 1992 and 1993, the production on this farm was 7,320 kg, 7,172 kg, 2,100 kg and 3,371 kg respectively producing an annual average of 4,991 kgs averaging 1 kg of green leaf per bush per year which is the national average. This is a good rate of production and an indication of proper management. From 1992 production appears low but this is mainly because pruning was carried out during the year and it takes two years for the yield to re-stabilize.

Baba Eboso is one of the best tea farmers in North Maragoli and in Vihiga District as a whole. In a year when all his plots were plucked (1990), he reached a total of 7,320 kg of green leaf which is just as good as the production by farmers in the eastern parts of the Rift Valley. This confirms the fact that this area can be as productive as the other tea-growing areas in Kenya given the right management and if there is institutional support in the form of infrastructure, the improvement in the quality of tea through proper husbandry, and high prices of harvested green leaf.

In 1993, Eboso earned an average monthly income of Ksh 693 and his total income for the twelve months was Kshs 8,314. The end of year tea payment (bonus) was Kshs 28,216. Therefore, the total tea income in 1993 was Kshs 36,530. For the period between 1990 and 1993 he earned a net income from tea of Kshs 20,565, Kshs 20,000, Kshs 8,050 and Kshs 28,216 respectively. These earnings are much lower compared to what his colleagues earned in other tea-growing zones especially east of the Rift Valley (Section 4.5). The main explanation for this is the low price for tea produced in this zone, because of its persistent poor quality. This household's farm management is good because the farmer follows the recommended agronomic practices. He uses between seven to ten bags of fertilizer a year. Though this household may be producing and in fact produces good quality tea, it does not enjoy the benefits of good management because of the zonal grading system. In a way, this system acts as a disincentive to the few farmers that follow recommended husbandry practices but who are paid the same price as the many farmers who do not practice proper farm management.

Eboso and Agesa are the only two farmers in our sample who keep farm records. The former is always ready to listen to advice in regard to farm improvements. Coupled with his experience at

Inter- and Intra-household Farm Management Situations

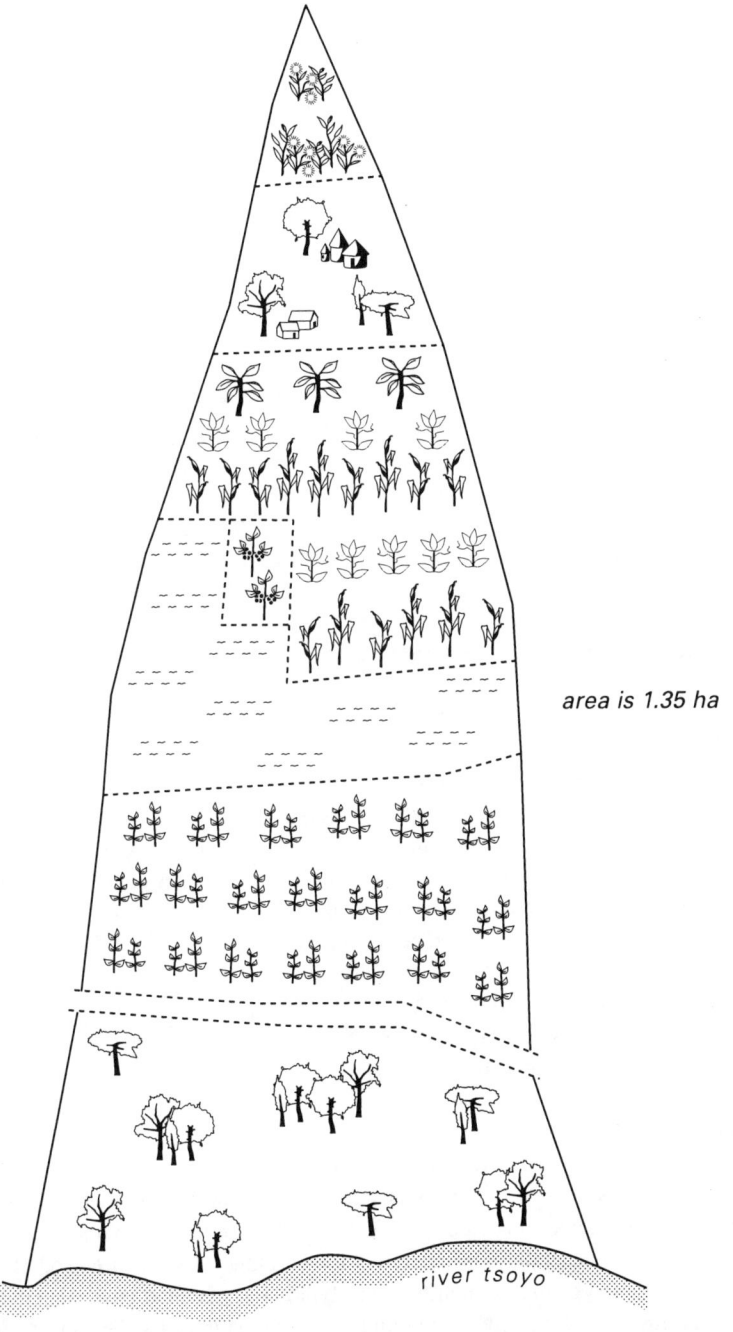

Figure 6.3
Eboso Farm (source: Study Results).

the Kenya Green Growers Co-operative Union and the fact that he lost his job and now has to rely on agricultural production for his family's and parent's livelihood gives him energy and incentive to work harder. He has accepted farming as an employment and as his main source of income. He blames the poor quality of tea and poor production in this area to low tea prices, low use of inputs, the absence of a tea processing factory, and lack of good use of technical advice which is so crucial for quality tea. As already pointed out, farmers in this area have been earning Ksh 3.10 per kg in 1991, Kshs 4.10 per in 1992 and Kshs 9.10 per kg in 1993 whereas in other smallholder tea areas farmers earned Kshs 15.50 per kg of green leaf delivered. In fact, the earnings for 1993 are the highest this area has ever recorded and this was thanks to an increase in world market prices.

Food Crop Farming Sub-system
Eboso concentrates all his efforts on the tea crop and in fact food crop parcels are neglected. He inter-crops maize with beans, sorghum and peas with very little spacing (very high density) so that the crops are not able to grow well at all. He does not plant in rows but prefers to broadcast because it is faster and cheaper. In 1993 he did not use hybrid seeds because of the high price for them and thus used the local seeds. The average maize yield from his plot of 0.2 ha was 90 kgs (one bag) instead of the potential of approximately 1,200 kg. This farmer need a minimum of 1,800 kgs (20 bags) to feed his household and the extended family of his parents (6 more dependents). He could produce half of this amount on his own farm if he used the standard agronomic practices for maize (see Figure 6.3, the farm layout.

Livestock Production Sub-system
Eboso has three cows. They produce 3.5 litres of milk a day. Of this 2.1 litres are for sale which brings him an income of Kshs 700 per month and approximately Kshs 4,500 per annum due to the short lactation period of the local zebu cows. Three cows should produce about 9 litres of milk in a day if well managed. These animals receive inadequate grass and other green foliage as he has not established napier grass on his farm. He does not buy feeds/concentrates for them and they stay tethered to one place the whole day and receives an inadequate amount of water. The animals were only dipped four times during 1993. Little attention is given to the cattle because of the shortage of labour. It is Eboso's aspiration to save enough money to enable him to change to zero-grazing as he knows that although zero-grazing is more intensive and labour demanding, financially it is more rewarding if well managed.

Tree Production Sub-system
The cultivation of trees is considered to be a coping strategy by this farmer. Several coping strategies have been devised by this household to provide sustenance to the family. He estimated that in a year he is able to earn between Kshs 5,000 and Kshs 7,000 from the sale of eucalyptus trees. The trees also provided the household with materials for building, fencing, firewood and shade in the home. This farmer was of the opinion that eucalyptus trees created environmental problems. He observed that the part of his land that is under eucalyptus trees has become drier and is no longer suitable for any other farming activity (see Figure 6.2). The reduction of the water flow in River Ezava to this farmer could be explained by the planting of eucalyptus trees along its course by all farmers in this area. He asserts that the eucalyptus trees use much water

which makes the soil to get drier. Eucalyptus trees were introduced to Kenyans during the colonial period and since then many farmers have planted them because they mature faster than the indigenous trees. Indigenous trees have been lost through continued cutting to expand agricultural activities and now the eucalyptus trees are more widespread.

Labour Use and Its Constraints
Tea is a labour intensive crop and during the long rains when tea leaves grow very fast, farmers need adequate labour to harvest the ready crop. The money Eboso earns from tea is not sufficient for him to contract labourers permanently, yet he cannot manage to pluck the tea all alone. This has contributed greatly to lower yields because sometimes his tea remains unplucked due to lack of labour. He is among the few farmers who use hired labour quite regularly. He confirmed that most of his monthly earnings are invested in contracting hired labour while the end of year payment is used for paying school fees. The only other labour is from his wife who works on the farm in the afternoons after her school duties. This household experiences continuous shortage of labour throughout the year.

Coping Strategies
This household is not able to produce adequate food from the family farm because it considers the plot reserved for food production as too small. As mentioned above the farmer only produced 90 kgs in 1993 from the family maize plot. Instead, Eboso leases land measuring 0.4 hectares in Nandi District, specifically for maize and beans production. He purchases inputs such as hybrid seeds, fertilizers and uses hired labour under his supervision for all farming operations on this rented plot. He aims at maximum benefit and this is only possible through buying the right inputs, optimal farm management and good supervision of labour. He pays Ksh. 1,000 for renting this plot of land for one year. Nandi is located in a different ecological zone and it has only one maize crop and a second crop of beans during the short rains. He harvests between 15 to 20 bags (1,350 to 1,800 kg) of maize and 3 to 5 bags (270 to 450 kg) of beans in a year from this land. This production of maize and beans is sufficient for the family's food consumption for more than three-quarters of a year. This maize and beans production is equivalent to approximately Kshs 21,000 to Kshs 30,000 using 1992/93 prices. He would have needed to have this amount of money to enable him buy food for his family had he not invested in renting land outside North Maragoli. This is a clear demonstration that conscientious decision-making in a household, investments in inputs, supervision and good farm management can contribute tremendously towards successful farming.
Eboso also has a business of buying and selling chicken. He buys chickens in Nandi District and sells them at a small profit on the local market. Such money is used on daily household needs like sugar, cooking fat, paraffin etc, and school requirements such as books, and uniforms. In total this household has an approximate annual average household cash income of Kshs 47,000.

What has led to the success of this household's farming? Eboso and his wife Libesa are very close and they make their decisions together. The cooperation and good working relationship between the husband and wife has assisted in the good management of their different farming sub-systems. Household incomes and expenditures are managed together and each one is aware of what goes on within the farm and the household. Eboso is knowledgeable about tea and food crop management because of the experience he gained with the Kenya Grain Growers Union.

His good education (12 years of schooling) and knowledge of farming gives him an advantage over illiterate farmers. He is able to read about new innovations in farming from pamphlets and papers. He also learns more about farming by listening to the agricultural programmes on the radio. He attends the 'Chief's Baraza' where agricultural information are sometimes communicated to farmers. He is one of those few farmers who like to experiment with new ideas and who takes risks. This explains his neglect of food production on the family farm and his investment in a different area for food production. He has accepted farming as a full-time job and he has the right attitude towards his work which a number of men lack. His major constraint is a shortage of money. He is responsible for his ailing parents and he spends a lot of money on their medication. He also has a responsibility to finance the education and feeding of his brothers and sisters because of his parents' poor health. He has conflicts in deciding what are priority areas when it comes to investment given the limited funds available for farming, purchase of food, medication and education.

Another limitation to farm management is the conflict between Eboso and his brother over incomes from tea. This brother, who works in Nairobi, and still has a small family of three and is expected to send remittances to the rural home especially for his parents' upkeep does not help. The brother is un-cooperative and uncompromising about the plight of the rural family. He demands that he should always get half of the earnings from tea but he contributes nothing towards its management. In 1993, Eboso took, on credit, 10 bags of fertilizer. His brother took 8 out of the 10 bags and sold them to compensate for his unpaid share of the tea income. The implication is that the tea farm will get an inadequate supply of fertilizer the following year, evidently reducing the yield and income. Inter- and intra-household conflicts play an important role in small-scale farming.
Eboso's farm is rarely approached by the extension agents because Eboso is too inquisitive and sometimes will challenge and correct wrong ideas conveyed by extension agents.

Mzee Manase

Background History
Mzee Manase was born in 1938 in Vokoli Sub-location of North Maragoli. He is 56 years old placing him in category three of household heads: 55 years and above. He was the second born in a family of nine. He was enrolled at Vokoli primary school between 1951 and 1954. He only got a fragmentary education up to class three (3 years of schooling). He cannot follow a conversation in English or Kiswahili. He is practically illiterate. He says that he was not bright and therefore it would have been a waste of money for his parents to keep him in school. At the same time he was too big and too old to be in the lower primary school, because by the time he enrolled in standard one he was already above 13 years of age and found learning uninteresting. He got married in 1964. He has had two marriages. In his first marriage he had 6 children, 2 died and 4 are still alive. These children have never gone to school. The first wife died in 1977 and he married a second wife in 1979. The second wife is disabled, can only use one hand and she is mentally retarded. Manase also appears to be mentally unstable. He has three malnourished children in this second marriage.
In the early years of his life, Manase worked in his home area as a farm labourer. In 1978, he migrated to Limuru near Nairobi where he was employed as a tea plucker. He worked in Limuru

for one year after which he left to work in Nairobi as a watchman. He lived in Nairobi for two years after which he left for Muhoroni sugar company to work as a cane cutter. He worked in Muhoroni for one year and in early 1983 he returned home, having achieved nothing and having remitted no money to his wives. The type of jobs held by this farmer shows his lack of education and special skills. The frequency with which he changes jobs may be an indication of instability, restlessness, indecision and lack of commitment.

At the moment, Manase works as a farm labourer on a neighbour's farm in Maragoli. Effectively he sells all his time to others. He is employed by a primary school teacher as a gardener, managing and feeding cattle, plucking tea and transporting it to the buying/collecting center. He earns Kshs 200 a month as a farm labourer. At night he works as a primary school watchman where he earns Kshs 338 per month. His total monthly earnings amount to Kshs 538. Most of this money is spent on drinking local alcohol.

Commercial Crop Farming Sub-system
This household is one of the most impoverished male-headed households we came across in the study. Manase has a farm of 1.2 hectares. This farm is still registered in his late father's name. The farm is supposed to be shared between Manase and a brother who works as a 'houseboy' in Nairobi. Since Manase lives in the home he should effectively manage this farm or part of it to feed his household. Tea covers an area of 0.2 hectares with 2,500 bushes established in 1964 by his late father. He produces on average 6 kg of tea monthly and an average of 72 kg per annum. This meager production earns this household an average of Kshs 15 a month and Kshs 180 in a year. The extremely poor production and low payments are related to the poor farm management.

The tea crop has been neglected for the last five years and Manase puts the blame for the dismal performance on his brother. He would prefer that the tea parcel were subdivided between him and his brother so that he can work on his section of the farm, yet he takes no action to legalize the subdivision of the farm. The brother in Nairobi also appears to be little concerned. He is not interested in farming and he has no wife to stay on the farm. Manase uses this tactic as an excuse for neglecting the management of his farm. He has no time for his own farm as he is too busy working elsewhere. No weeding, no application of fertilizer and no pruning had been carried out on this farm for the last five years. His wife cannot work long hours because of her disability. Lack of investment, commitment, limited education and indecisiveness on the part of the head of the household has led to poor production and limited income.

Food Crop Farming Sub-system
The remaining part of the farm which measures about 1 ha is rented out to neighbours at about Kshs 300 per annum or is laying fallow. The production of maize is zero bags (00 kg). The household has a few bananas. At the beginning of this survey the farmer had two cows which had been left by his late father. These cattle were sold in mid- 1993 by his brother who then gave Manase Kshs 200 from the total of Kshs 10,000 that the animals fetched.

This household had the worst management of all the different types of households that were visited in the survey area. The members of this household were on the brink of starvation and they relied on the kindness of the neighbours who, once in a while provided them with food. The

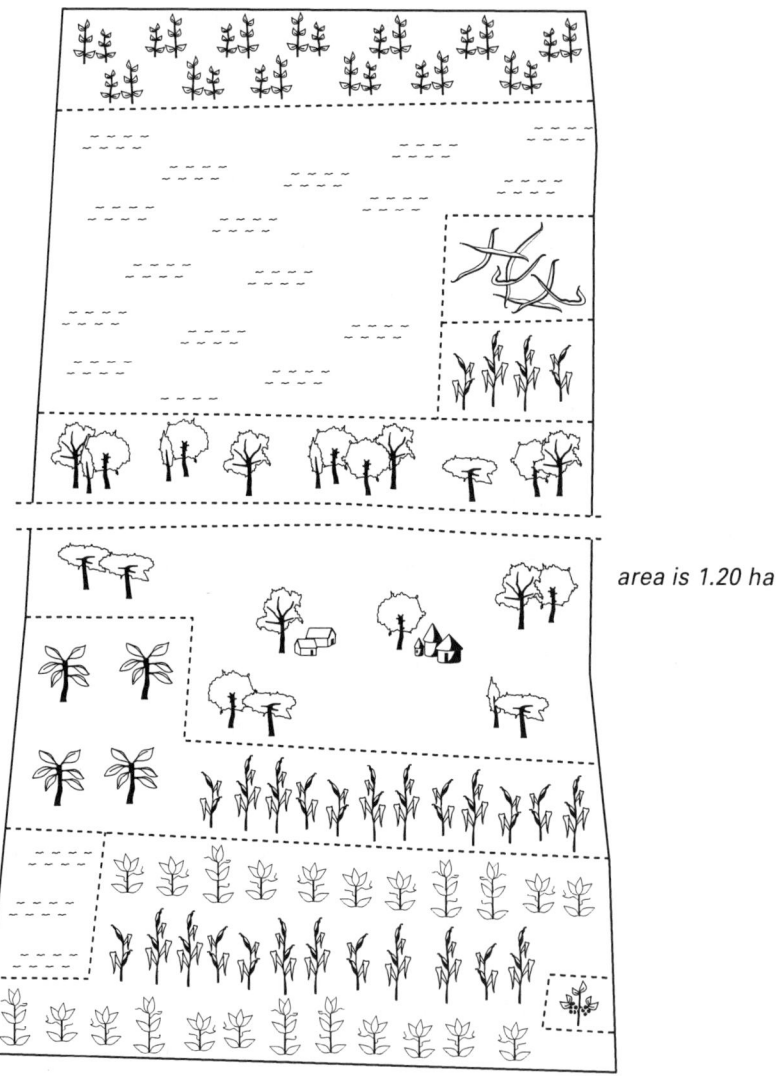

area is 1.20 ha

Figure 6.4
Manase Farm (source: Study Results).

wife's disability renders her incapable of working on the farm. The older children have all moved away from the home. The girls are married and the boys are working as gardeners or herdsmen for other farmers in the area. These working children, though earning limited amounts of money, do not remit any of their earnings to their father as they are well aware of his character. Manase's wife believes that the husband mistreats her because she is disabled and therefore is a liability to him. She said that she could not get separated or a divorce from her husband and return to her parents' home because it was considered a blessing that had found a man to marry her in her disabled condition and this had given her parents a great relief. Therefore she would not be welcomed in her parents' home under any circumstances. It is a belief (custom) among the Maragoli people and the larger Luhya community that once a woman is married, it is not right to divorce a husband. Society judges the woman harshly and she is seen as a social misfit. Manase's relation with his wife is confrontational and they are always quarrelling, regardless of who is present. Manase avoids his home in most cases. Every time I visited his home it took almost an hour to trace him and once he and his wife were together he would be exchange bitter words with her or arguing for no apparent reason. His wife had abrogated her role as the household's food provider. Her disability led her to assume that her husband should be able to provide the family with the food required food.

Coping Strategies
This household had no coping strategy. A computation of Manase's earnings gives him approximately Kshs 7,000 per annum. He earns small bits and pieces which he spends on his own leisure items. Manase is not serious about farm management. He is by nature a lazy person and just seems to be passing through this world. He is not the type of man worthy of being assisted either financially or technically. He had not devised any coping strategies for his household. No extension agents has ever visited this farm for as long as he could remember.

This example illustrates that the success of the farm household system depends on decision-making, set priorities, education, permanent sources of income, and harmony between the members of the household as well as favourable external factors. Inter- and intra-household conflicts, low level of investment in farm labour, and inadequate agronomic knowledge amongst household members partly explain poor farm management and subsequently low agricultural production. Poverty contributes greatly to poor farm management as it has been shown by this household.

Mzee Musasa

Background History
Musasa was born in 1949. He was born into a polygamous home of four wives and he remembers with consternation the inter-and intra-household conflicts over resource ownership and control, and school fees. His mother was the second wife. He was his mother's first born child. His parents passed away while he was still young and so he was brought up by a step-brother who was a successful businessman. He entered primary school in 1957 and went through the school system quite successfully. He passed his secondary school certificate of education, form four (12 years of schooling) examination with a second division which would have enabled him to proceed to 'A Levels' and eventually to University but, because of lack of school fees, he stopped after secondary school in 1968. In 1969, he went to Mombasa to look for a job. In 1970 he was employed by the Ministry of Works as a clerk based in Mombasa. He worked for this Ministry for five years after which he resigned and joined the teaching profession as an untrained primary school teacher. He was admitted to a primary teachers college in 1980 and graduated in 1982 as a trained primary school teacher. Since then he has continued teaching in the same school and he is now a primary school headmaster. He is also the Kenya National Union of Teachers' representative of his branch. He is a well-to-do man because he earns a good salary. He has built houses which he rents and this brings him a monthly income.

He married in 1974. His wife Jesica is also a primary school teacher. They have nine children, five boys and four girls. Two of these children have completed secondary education and are training, one is in a medical nursing school and another is studying at a technical training college, four are in secondary school and three are still in primary school.

This household belongs to one of the young households where farms are expected to be small, mostly less than 0.8 hectares. Musasa and his wife are literate and are in government employment. This is one of the few households that can be categorized as having a common production unit and a common economic pool. Musasa and his wife have a very good relationship and understanding in the sharing of household and economic responsibilities. They have one bank account and they are both signatories to that account. The payment of school fees for the children is not a problem as they plan everything, especially their household incomes, together.

This household has a farm of 1.2 hectares, 0.7 hectares was inherited from his parents and 0.5 hectares was bought with a loan from the Teachers Savings, Credit and Co-operative Society. Musasa and his wife have taken several loans from the Teachers Co-operative Society and two loans from a commercial bank, which have assisted them to invest in land purchase, buying plots, building of rental houses and the building of their permanent house in the village.

Commercial Crop Farming Sub-system
They have invested in all the farming sub-systems found in North Maragoli: cultivating cash crops, food crops, livestock and tree production. Musasa established coffee in the early 1970s. However, due to the problems associated with coffee (a) low and delayed payments, (b) lack of inputs, (c) inefficiency and corruption in the co-operative society, (d) poor pricing and high taxation, (e) the labour intensive character and (f) no incentives to the growers, he decided to uproot the crop even though it is against government policy. He is of the opinion that since farms are small in North Maragoli, farmers must grow the most rewarding crops. He further says that "There is no point in keeping land under a crop that does not benefiting the household especially when the household has no land for food production". The portion of land that was under coffee has been planted with tea to increase the hectarage under tea. Tea is on a 0.2 hectares plot established in 1987 with approximately 3,000 bushes. The tea crop is still young and so production is still low. The canopy had not yet developed fully at the time of the survey and it had a number of open spaces. Musasa and his wife were in the process of getting seedlings to fill up the open spaces. The right quantity of fertilizer is used. This is the only household where the tea was being sprayed to strengthen and improve the quality of the tea leaves. This household relies on hired labour since both the husband and the wife are in formal employment. Musasa normally uses school drop-outs, especially boys, to work on his farm. These are drop-outs from his school, either because there were problems with school fees or because they were unable to proceed to secondary school.

He is the only farmer in the whole study area who insisted on paying farm labourers according to the number of kilogrammes of tea they plucked. He pays Ksh 1 per kilogramme and those labourers that do not agree are free to go elsewhere. This gives the labourers an incentive to pluck more kilogrammes and earn more money. He has succeeded in using this method because he has money and he is quite powerful in this area. Some villagers approach him for assistance, especially for loans and to solve personal problems. The other poor farmers in the study area could not bargain with farm labourers because they often lack money and hire labour on credit. Without cash money to pay on the spot, their power to bargain and dictate to the workers is minimal. Musasa's children help with farm work on Saturdays and during holidays. Sunday is not a working day in this area but a day of worship.

Tea is registered in the wife's name but all the earnings from the crop are deposited in the joint bank account. In 1993, the total tea production was 1,552 kg, an average monthly tea production of 129 kg (0.5 kg of green leaf per bush). The total monthly cash income from the crop was Kshs 4,834 and the end of year payment (bonus) was Kshs 13,744. The total annual income was Kshs 18,577 in 1993. They are confident of being able to get higher yields and earnings in the years to come because of the investments made in tea. They have increased the area under tea, use adequate fertilizer, spray and weed the crop regularly.

Food Crop Farming Sub-system
The food crops planted by this family are maize, cassava, sweet potatoes, bananas and sugar cane. The size of the maize plot is 0.6 hectares. They purchase hybrid maize seeds which is planted during the long rainy season, and local maize seeds are planted during the short rainy season because they mature quickly. They use fertilizer during planting and CAN for top dressing later. Weeding is carried out twice. Hired labour is used on food crops. Twenty-five bags (2,250 kg) of maize was produced in 1993 (total for both long and short season rains in 1993). This was the highest production of maize in all the households in this survey. In addition to the green plucking, this production was close to the expected production potential for the area (see Section 4.4 for details). An additional two bags of beans was produced. This example shows that local farmers can produce maize within the expected range if they practice good crop husbandry and have the money to buy the necessary inputs and hire labour so that the farming activities are not delayed.

Livestock Production Sub-system
Musasa has three local zebu cows. However, he hopes to sell them and to buy one grade cow after he has established a zero-grazing unit and planted enough napier grass. Livestock husbandry in this household is poor because the members of this family are away from home for most of the day. The cattle do not receive adequate water and grass. Also the area of napier grass is too small to feed three cows. Milk production is therefore quite low. An average of six bottles (4.2 litres) is produced per day and milk is for home consumption. Cattle are not taken for dipping but once in a while when he has time, Musasa disinfects the cattle at home. Cattle was not taken for dipping because he suspects there was no drug in the dip and he has lost confidence in the effectiveness of cattle dipping just like many other farmers in the area.

Tree Production Sub-system
Some 0.1 hectares of land is under eucalyptus trees. The trees plot is well weeded and doing very well. The trees provide timber and firewood for household needs. No income is earned from the trees. This farmer also expressed the fear that eucalyptus trees were dangerous for soil water retention.

Agricultural Extension

Musasa is one of the 'contact farmers' in the area. He is obviously a very busy person and is rarely at home during the day to meet with the frontline extension agent. He confided to us that he did not know why he had been chosen as a 'contact farmer' because he has no time to meet with agricultural extension agents and has no time to go out to disseminate information to other farmers in the study area. This household has access to agricultural information through a number of channels. Musasa stresses the fact he can read agricultural information for himself from pamphlets, news papers, he can listen to agricultural programmes over the radio, he can watch the agricultural programmes on television and is able to learn also during the agricultural shows. He is of the opinion that he and his wife are more knowledgeable in agricultural matters than the young frontline extension worker who has no experience and confidence in practical farming. He thinks that the poor illiterate women in the study area need more extension than men who are literate and are able to get information from many sources. He states that, in fact, if he has an agricultural problem, he can ring any agricultural officer in the district and discuss it over a drink after working hours. He did not meet the frontline extension worker in 1993 although the extension, agent came to Musasa's farm several times whilst he was away. It would again appear that when it comes to choosing 'contact farmers' extension officers prefer to select and work with progressive or rich farmers in the village instead of dealing with poor illiterate farmers.

Constraints to Farming

The main problem experienced by this household, which also affects other households where both wife and husband are in non-agricultural employment is inadequate farm labour supervision. When both wife and husband are working outside the home - which is the normal daily pattern- and all children are away in school, the home is left under the care of a domestic servant who is not assertive enough to supervise farm labour. As a result, the hired labourers sometimes work very badly especially when weeding of food crops. At other times labourers steal farm inputs such as maize and bean seeds and fertilizers.

The members of this household first pointed out that problems of pests and diseases - especially in beans and bananas - were increasing. Second, Musasa claimed that the theft of green maize, sweet potatoes, cassava, bananas, and sugar cane from his farm during the day when they are away at work and at night was part of the bigger and growing problem in the area. Many unemployed and disgruntled youths and other poor landless people get food mainly by stealing from farmers. Third, it was observed that one of the problems with tea is that it is destroyed by hailstones. The nature of rains in this area is such that it falls with many storms and hailstones. Musasa was disappointed with the KTDA's method of grading and pricing of tea. He says that farmers like himself invest a great deal in tea, trying to produce a higher grade. However they end up getting the price of tea of a lower grade because other farmers in the same zone produce low quality tea. He also complained that farmers waste too much time at tea buying centres, because they must wait for tea to be weighed and for transportation. He suggested that these are some of the problems that will only be solved when Western Province has its own tea factory. Improving the overall quality of tea requires a considerable amount of enforcement by tea extension officers who must supervise the management of the farms. Close proximity and quick transportation to the factory and overall KTDA policy are also factors affecting tea quality and attempts to improve it.

Other Income Earning Strategies

Musasa's family has many sources of income. Two permanent sources of income are the salaries of the husband and wife. There is also the allowance for being a representative of the Kenya National Union of Teachers, the monthly earnings from renting out houses and from tea. This household had a total monthly household income of about Kshs 16,000 and an approximate annual household income of about Kshs 192,000 which is the highest among all the farm households in the survey. However, most of Musasa's income comes from outside agriculture - an indication of the importance of non-agricultural employment and the potential benefits of a good education. Many families can no longer rely on agriculture alone to provide incomes for development and household needs. Agriculture in this area is at times considered an unreliable business, earnings are low and payments are delayed in such a way that farmers seldom have money when they need it most. Musasa carries on food farming just to produce food for the basic needs of the household and to fulfil a social and cultural function as everybody else does in this area. The future of agriculture as he sees it lies in labour intensive, high income earning activities such as horticulture, zero-grazing and tea production.

Mr Musasa is comfortable and has a very good permanent house. He is among the few farmers in small-scale farming who has a telephone, electricity, water, television, a radio and gas for cooking. This is because of the large proportion of non-farm income, which is 90% of the total household income. He enjoys good relation with his wife and this in a way has contributed to good household and farm management. Understanding one another's needs and sharing resources and responsibilities within the household can very much enhance farm production and other developments within the household. Households that experience conflicts such as Manase's had inferior farm management, low agricultural production, and experienced more impoverishment.

6.3 The Example of an Independent Female in a Male-Headed Household

Mama Iminza

Background History
Iminza was born in 1943 and got married in 1960 to a primary school teacher. She is 52 years old and belongs to the middle age group. The husband's basic salary is Kshs 5,000 per month which amounts to Kshs 60,000 per annum. He teaches in a local school and he lives in his home, not on the school's compound. This is a male-headed household and the couple have six children, two girls and four boys. The total farm size is 1.5 hectares (see Figure 6.5, farm layout).

Commercial Crop Farming Sub-system
Tea was established in 1965 on a parcel of land measuring 0.3 hectares with approximately 2,000 bushes. The tea is registered in the husband's name. The woman has continued to manage the tea farm for many years whereas tea payments come through the man's bank account and the husband uses the tea income. The tea brings a monthly income of between Kshs 800 to Kshs 1,500 to the man's account. The end of year payment (bonus) normally brings between Kshs 15,000 to Kshs 20,000 to the husband's account.

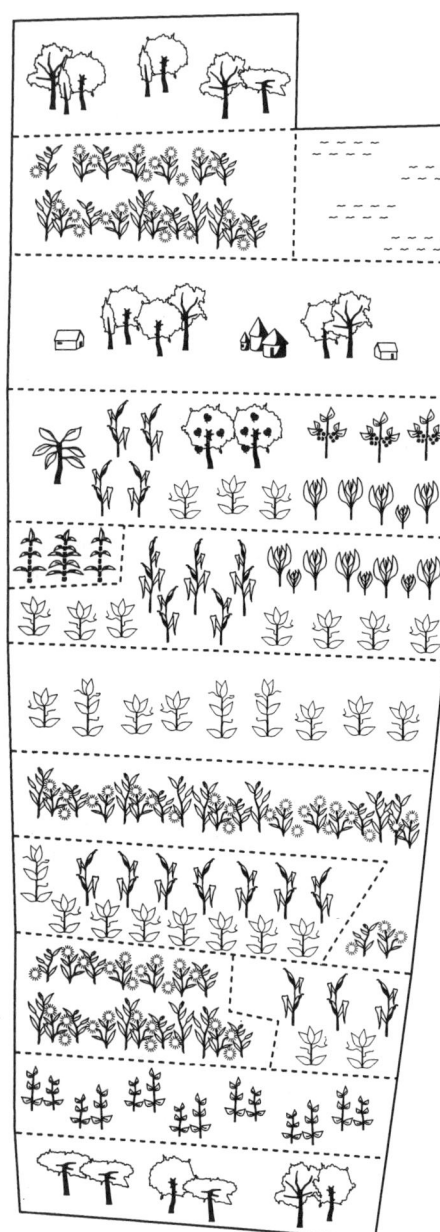

area is 1.50 ha

Figure 6.5
Iminza Farm (source: Study Results).

For many years, Iminza's husband has earned and used the money from the tea although his contribution towards labour and investment has been almost nil. The couple have bitter quarrels and fights whenever the woman asks to be given part of the tea earnings so that she can invest the money in the tea farm and also to buy essentials for the household. On many occasions Iminza has been beaten very severely by her husband, a number of times she had to be hospitalized with broken limbs and a black eye. The husband always reminds her of the fact that farms belong to the men and a woman is married to work on the farm.

Eventually, early 1992, Iminza abandoned working on the tea plot. She stopped weeding, plucking and pruning tea. There are no more tea earnings and interestingly, the husband does not seem disturbed by his wife's decision. Now the wife is no longer beaten by the husband as she no longer asks to be given money from tea. She even stopped asking for any financial support from her husband for the maintenance of the household.

Although Iminza's husband is a primary school teacher with a salary and house allowance totalling about Kshs 5,000 per month, he does not support his family financially. He alleges that since he is paying his children's school fees, the wife should provide other essentials like food, clothing and work on the farm. Yet, the husband knows very well that his wife has no income of her own to manage the farm-household system without financial support. Iminza's husband, after paying school fee and buying other school necessities, spends the rest of his earnings on alcoholic drinks such as 'chang'aa and Busaa', and smoking. The man goes straight to the bar or to local brewing joints after school and he only arrives home very late in the night to demand for food and to sleep.

The management of the whole farm is sub-standard because of the poor intra-household relationship between husband and wife. Excessive use of alcohol often leads to social animosity, irresponsible behaviour and neglect of the needs of the household members.

Food Crop Farming Sub-system

Maize husbandry was little different to the management of tea. The plot under maize is 0.4 hectares. The effort being extended to maize production was inadequate. Local maize intercropped with beans and sorghum are planted both during the long rains and during the short rainy seasons. None of the required inputs for higher maize production are applied. In 1993, 3 bags (270 kg) maize was produced on a plot of 0.4 hectares. This is far below the potential of 2,400 kgs per annum in two harvests.

Coping Strategies

A number of coping strategies have been devised by Iminza to help maintain and manage the household. In 1980, she borrowed Kshs 1,500 from a woman's group savings and credit fund (ROSCA or revolving fund or Merry go-round) of which she is a member. This money was used to rent 0.4 hectares in the Trans Nzoia District through relatives who are settled there. This land was put under maize with intensive crop husbandry using mainly hired labour and all the required inputs such as fertilizers and hybrid maize seeds. The crop husbandry and farm labour was supervised by a male relative who lives in Trans Nzoia. Iminza was able to harvest 20 bags (1800 kg). Iminza sold the total harvest and was able to raise Kshs 7,000. Part of this money was re-invested in renting land (at 1980 prices, renting an acre of land or 0.4 hectares cost approximately Kshs 400) to plant maize for the following year and this is what she does every year. Part of the maize production is transported to Maragoli for home consumption while the rest is sold. Many of these activities are carried out without her husband's knowledge. If he knew knowledge about such a project it would create a lot of conflict between him and Iminza. The husband would demand money and if he were not given it, he would forbid her to go to Trans Nzoia.

In 1989 Iminza was able to purchase a cross-breed cow for Kshs 4,000. This can be considered her second coping strategy. The cross-breed cow receives good care. It is provided with sufficient water because there is running water in the home, enough grass and supplementary feeds. Consultation with livestock officers about its veterinary check-ups is a continuing process. One of Iminza's sons is responsible for taking care of the cow. The cow produces 12 bottles (8.4 litres) of milk a day. Ten bottles (7 litres) are sold while 2 bottles (1.4 litres) are for home consumption. She earns about Kshs 3,000 a month from the sale of milk. This money is used to buy essential foods for her family and sometimes to contract labour to work on the farm. Her third coping strategy is petty business. She travels to Migori and sometimes to Kisii where she buys dried maize, beans, millet, dried cassava, groundnuts and sardines. These foodstuffs are sold at her farm gate. A daily income of approximately Kshs 50 is earned from this petty trade which is used to satisfy daily needs such as buying sugar, cooking fat and paraffin for example. Iminza's fourth coping strategy is her involvement in the firing of bricks (brick-making) on a small scale. Together with her son she fires a number of bricks each year and this brings the household about Kshs 1,500 a year. Iminza makes an average annual income of Kshs 27,000. Her husband as we have seen earns approximately Kshs 80,000 in a year.

Iminza claimed that she has never been visited by an agricultural extension agent. She goes to Vihiga to talk to the livestock officer when her cross-breed animal has problems. Sometimes she has to buy fuel for the government vehicle and pay the livestock officers for their time and services. She is very disappointed with the artificial insemination services (A.I) because she paid for this service three times in 1992 and her cow still did not conceive. In 1993 she had to pay for the same service two times before her animal eventually conceived. It was disappointing and made her lose confidence in the AI service.

She stresses that rural women need considerable financial and technological support. She feels that there should be ways and means to assist rural women to get small supervised loans to improve their living standards either carrying out petty businesses or by going into intensive farming like zero-grazing and horticulture. Women need a lot of information about farming and business. Rural women also need to be trained to enable them to function optimally in different areas of intensive farming and business. This is particularly so because in most rural areas in Kenya women are responsible for the up-bringing of children, and the provision of food for the household. Iminza believes that for the smooth running of the home and good farm management, there must be good a understanding and working relationship between the husband and the wife. Most men who take to illicit drink and beer have serious problems with their families. Many of their homes become impoverished unless the woman herself is innovative and hard working.

6.4 Female-Managed Households

Mama Chagenda

Background History
Chagenda was born in 1943 in Mudete Sub-location, North Maragoli location of Vihiga District. She is 51 years old and belongs to those household heads in the middle age category. She was the second born in a family of ten children. She started school in 1950 at Mudete primary school. In 1962 she sat for her class eight examinations (8 years of schooling). She did not pass high enough to continue with secondary education. Many parents still considered more befitting for a woman to get married than to go on to higher education. The local community says that because a woman's responsibility is to prepare food for the family, education can never make her a better cook and so one should not waste money on someone who specializes in cooking in her marriage.

Chagenda got married in 1963. Her husband was working with the Power and Lighting Company under the Ministry of Energy. In 1964, he left that Ministry and joined the Ministry of Education. The same year he joined the Primary Teachers College to train as a P1 (Primary One) teacher. Immediately after she got married, Chagenda started managing the farm because her the husband was not living at home. The husband completed his training in 1966 and since then he has taught in many primary schools. He retired from teaching in 1991.
When the husband was paid his gratuity after his retirement, he bought another piece of land in Nandi District, migrated and married a second wife. His main reason for marrying a second wife was because Chagenda had only given birth to two children. She was unable to have more because of medical problems. The second wife was married to bear children. The Maragoli farm has been left under the jurisdiction of Chagenda.

The household has a farm of 1.8 hectares and is considered a big farm in North Maragoli. Chagenda claims that her husband neglected her because she could not give birth to more children. Despite the fact that her husband had received a good salary and is now getting a pension, she has had to pay for the school fees and the upbringing of the children on her own. However, she is happy with the fact that her husband has constructed a permanent house for her. To prove to the husband that she could cope without his financial support, she decided to go into full-scale farming.

Commercial Crop Farming Sub-system
The size of the tea plot is 0.7 hectares with 4,000 bushes (stems) and in fact this population could be increased further. This is one of the biggest tea farms in our sample survey. Tea was established by Chagenda in 1969. She told us that she has accepted farming as a full-time job.

She takes 12 bags of fertilizer on credit every year but she asserts that she should be taking about 20 bags to obtain a good harvest. However, she does not take 20 bags because repayment could be very high. She has kept her tea farm very clean by weeding regularly every two months. She prunes in good time and she hires a group of five women (sometimes more) who work for her on a daily basis. Because she applies an almost adequate amount of fertilizers and weeds regularly, she is able to pluck three rounds a month and sometimes four rounds if the rainfall has been

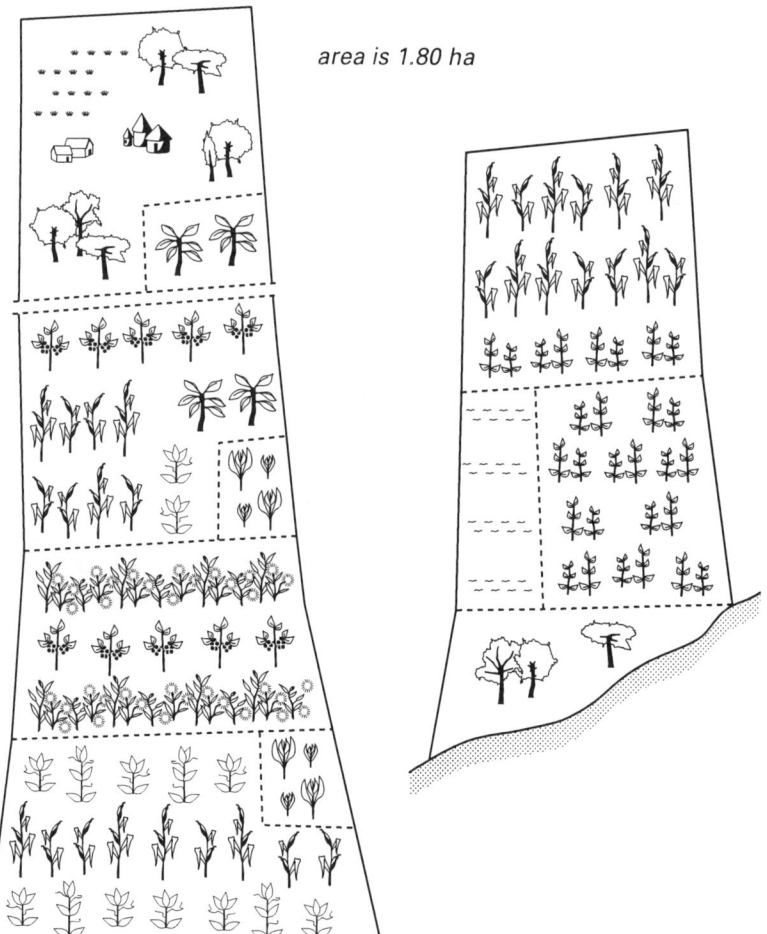

Figure 6.6
Chagenda Farm (source: Study Results).

favourable. If 20 bags of fertilizer were applied to the crop, it might be possible to pluck four rounds which is the recommended number of pluckings for well-managed tea. In 1993 she produced a total of 3,614 kg of tea at a rate of 0.9 kg of green leaf per bush. The average monthly tea production was 301 kg. When compared to other tea growing areas, a plot of 0.7 hectares should produce a minimum of 4,550 kg per annum. Chagenda is producing below the expected minimum yet by North Maragoli standards she is among the female-managed households considered to have the best management and with the highest production.

In 1993, she made an average monthly income of Kshs 842. The total annual income from the monthly tea earnings was Kshs 10,000 and the end of year payment (bonus) was Kshs 32,456. Therefore the total annual tea income in 1993 was thus Kshs 42,455 for this household. Such a plot would have rendered close to Kshs 50,000 in other smallholder tea- growing areas. The relatively good results on this farm are achieved by good investments. Chagenda uses hired labour because she has the money and her family is too small to manage all the tasks on the farm. The monthly tea income provides money to pay her labourers and buy the essentials for the household. One problem experienced on this plot however is that farm labour is poorly supervised and therefore green leaf is not plucked clean.

Chagenda has a parcel measuring 0.3 hectares with about 1,000 coffee bushes. She has neglected the coffee. She believes the coffee farmers in this area have been cheated on as far as coffee prices, coffee taxation, and the prices of coffee inputs are concerned. Coffee payments are irregular and very small and she for example, only earned Kshs 4.50 from the coffee crop for the whole of 1993. Despite this, the management of the coffee crop is very demanding in terms of inputs, particularly labour. Instead of wasting money buying labour and other inputs for such a non-paying enterprise, she has decided to uproot the coffee crop slowly and replant coffee parcel with a maize crop (see Figure 6.6, farm layout).

Food Crop Farming Sub-system
The food crops she plants are maize, beans, bananas, sorghum and sweet potatoes. Maize is planted twice a year and she adopts the recommended methods of farming. She prepares the farm well and in good time and plants hybrid seeds during the long rain season, using fertilizer and weeding twice. The maize is grown on a parcel measuring 0.4 ha and produces 20 bags (1,800 kg) of maize, and four bags (360 kg) of beans which is more than enough for the household's food consumption needs. Part of this produce is sold to earn money to buy farm inputs. This was the only household in the survey where more food was produced than required. Small family size and high production explains the surplus of food in this household.

Livestock Production Sub-system
Chagenda has three local cows and two calves. When they produce milk she sells some of it to make extra income. Enough napier grass has been established to feed these cows. The household has running water in the home and so the cows receive adequate water throughout the day.

This farmer has been successful because she devotes most of her time to farming. She also gets much support from her two children who are grown-up. The husband lives in Nandi and he does not bother her with one exception. Once a year he comes to claim his share from the 'end of year' tea payment. They share the earnings in equal halves. She has been given the mandate by her husband to manage the tea crop and to get all the earnings she can without any interference. Although the farm is registered in her husband's name it is earmarked for her son. From this case study, it would appear that where a woman has freedom to use and control resources, farm management is at a high level. However, she cannot apply for loans from banks because of lack of legal ownership and title deed.
Her main complaints (1) sharing the final tea payment with the husband who does not contribute to the management of the farm. (2) The inadequacy of agricultural extension information. She told us that ever since she became a farmer she has never been visited by any general agricultural extension officer, so she wonders who they visit. It is only the tea extension officer who comes to visit her. She has never been invited to farmers' seminars, farm demonstrations, or to agricultural workshops. The knowledge she is using on her farm was acquired from parents, neighbours or

what she sometimes reads in papers or agricultural books. (3) Lack of money. If she had capital or if she was given a loan her farm could produce much more. She would also invest in zero-grazing, but she is handicapped as the land title deed belongs to her husband and therefore she has no collateral. In her opinion there is very scant information on the operations of agricultural credit institutions. She wishes that those concerned would supply rural small-scale farmers with such information. She observed that once villagers realize that a husband does not care about his wife, some neighbours organize themselves to steal from such a woman at night because they know that she is defenseless. Some of her maize crop and bananas are always stolen from the farm at night. This demoralizes women farmers and makes them live in fear.

Theft has discouraged many small-scale farmers constructing of recommended structures for storing maize, beans and other products outside the main house because such a store would be raided and the farmer would lose all his/her harvest. Finally, she recommends that the government should review its policy regarding provision of loans, technology and information and emphasize the role of women. An important aspect is the allocation and ownership of resources, especially land in the rural areas of Kenya.

Mama Mmbone

Background History

Born in 1953, she is 42 years old and belongs to the category of young heads of households. She was second born in a family of seven. Mmbone started school in 1962 at Gilwazi primary school. She was an average student and she sat for the certificate of primary education in 1968. She passed well and went to Pangani Girls Secondary School in Nairobi in 1969. She dropped out of school after two years of secondary education (10 years of schooling) for medical reasons and because of frequent shortage of school fees. She returned home where in early 1971 she become involved in petty business, selling soap, vegetables, fruits, potatoes and sardines. According to her, this was the only business open to her. She had very little capital at that time to venture into a larger profit making business. Her petty business helped her raise money for essentials, such as clothing and soap. She married in December 1971 and her husband works in Nairobi with a company that deals in sportsware. She has five children, three boys and two girls, one is in secondary school and all the others are in primary school.

Commercial Crop Farming Sub-system

Mmbone manages the farm on behalf of her husband. The total size of the farm is just 0.3 ha. All the farming sub-systems are squeezed into this small farm. The tea parcel is approximately 0.1 hectares with 1,500 bushes and was established in 1978. This number of tea bushes is below the recommended 2,500 which the KTDA considers acceptable but still the farm is registered. She provides all labour. The children provide extra labour on Saturdays, especially with weeding tea and other crops. Plucking takes two days a week. Fertilizer is rarely used. Only two rounds of plucking a month is possible, probably due to low amount of fertilizer used. In 1993 she harvested 1,650 kg green leaf tea resulting in a monthly average of 137.5 kg. The average monthly income from tea for the same period was Kshs 420. The total annual income from the monthly income was Kshs 5,045 and the 'end of year' tea payment was Kshs 14,253. In 1993, this household earned a total of Kshs 19,298 from tea production (see Figure 6.7).

Food Crop Farming Sub-system

The food crop farm is 0.1 hectares in size and maize, beans, sweet potatoes, vegetables, and bananas and a few trees are grown on it. Food crop production is very small due to the small size of the farm. All crops are inter-cropped. This farm produces a maximum of 0.5 bag (45 kg) of maize in a year.

Farmers may inter-crop certain crops but not all food crops can be inter-cropped successfully.

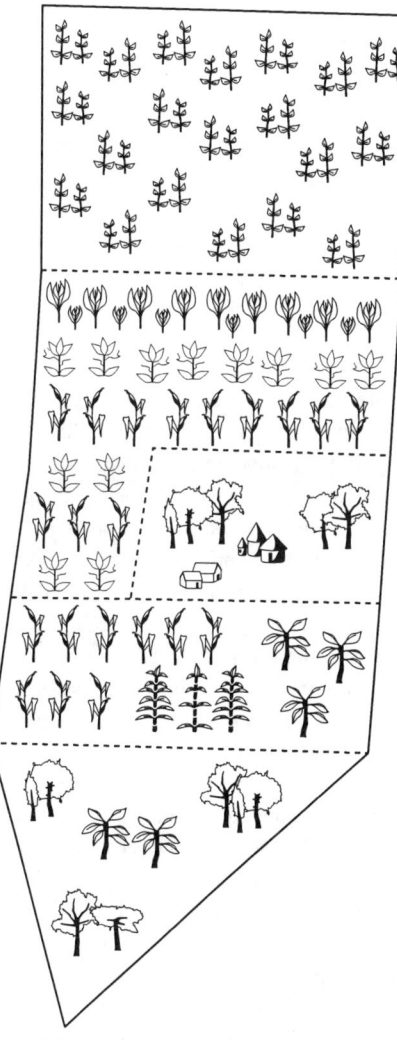

area is 0.30 ha

Figure 6.7
Mmbone Farm (source: Study Results).

Maize can be inter-cropped with beans, peas, millet or sorghum, but inter-cropping all of them at the same time drastically reduces production. It was observed that some farmers in North

Maragoli inter-crop maize with cassava, bananas and napier grass for example which led to a low maize production. Mmbone inter-crops all these crops and barely harvests more than 45 kg of maize a year. The amount of land she has limited her possibilities very much and has created many uncertainties. Most of the maize was, in fact, plucked green. Mmbone did not try to invest in maize production. Local maize was used as seed and no fertilizer was used at all. The reasoning behind the lack of investment in food production was the fact that whether or not there was paid inputs, food has to be bought for this household anyway because the farm was very small.

Livestock Production Sub-system
At the beginning of the in-depth case study, this household was in the possession of one local breed cow. Before the end of 1993, the cow was sold because of insufficient land and no place to grow napier grass. The cow was poorly looked after and buying grass for it daily was very expensive. It produced only half a bottle (0.4 litres) of milk a day.

Tree Production Sub-system
The land under tree cultivation is small but there are a number of trees grown specifically for fencing, construction and firewood. The trees are only cut when the husband is home on holidays.

Coping Strategies
Mmbone has devised her own strategies for coping with life. Knowing that the farm is too small to provide food for the family and bring in much income, she decided to get involved in petty business. Her main line of business is buying and selling fruits, groundnuts, cowpeas, bananas, avocados, eggs, paraffin, shoes made from plastic materials, Irish and sweet potatoes, cabbages, rice, beans, tomatoes and dried maize. This involves travelling to other areas such as Kisumu, Migori and parts of Nandi District to buy the wares at a slightly lower price in order to make a small profit. Much more earnings accrue from this business than from the small farm. She makes an average monthly income of Kshs 4,000 and an annual income of approximately Ksh 46,000 from her small business. An advantage for business people in Maragoli is the ready market there created by high population density and limited food production on many farms. Most of the population relies on the market for its food supplies. To assess and evaluate the profitability of her business, good records are kept indicating what was bought where and at how much. The records also show how much has been sold and the amount received. Mmbone is able to feed her household throughout the year and her household is much better off than most in North Maragoli. Funds are raised from trade to pay school fees and to buy all the necessary household materials. The husband sends her Kshs 500 every month but she reflects that sometimes she has more money than her husband. On a number of occasions she has provided him with transport to go to Nairobi. The husband is among the small percentage of men working outside their homes who send regular remittances to their wives. Their incomes and accounts are separate and they rarely discuss how much each one is making. The important issue is that the household is able to meet its needs, the children are well fed and the school fees are paid. The total household income is approximately Kshs 71,300 per annum.

She observes that some small-scale farmers are being pushed out of agriculture by reduced farm sizes and increasing landlessness. For such farmers there is little or no future in farming. She is of the opinion that women should quickly begin thinking of other ways of coping, apart from agricultural production. She is confident that if women who are interested in business are assisted financially through loans and grants, they could do very well and the living standards in the rural areas would improve tremendously. Women in the rural areas are suffering because they are not independent culturally, socially and economically. Women should be liberated from a cultural bondage that is regressive so that they can make their own decisions without having to deal with feelings of fear and uncertainty as far as their husbands are concerned. Another reason given for dismal agricultural performance are the conflicts that arise between relatives over land. In her case, the problem lies with her father-in-law who refuses to release land to his sons because he has married a second wife. The household is forced to consult the father-in-law on all aspects of farming. The conflict is with the father-in-law in this household. The relationship between the wife and husband however is good. The husband has given her the freedom to carry on her trade without his interference.

It is the contention of this female farmer that the agricultural extension staff are not working effectively. She has not received any advice about what are the best strategies or options for small farms like hers. She thinks that agricultural extension officers should be able to advice farmers with very small farms on what to grow or on the best methods of intensification.

Mama Muhonja

Background History
Muhonja was born in 1955. She is 39 years old and belongs to the category of the younger household heads. She went to school up to form two (10 years of schooling) and dropped out because of a teenage pregnancy. After baby was born, she worked as a domestic servant in Nairobi for a short time before changing to nursery school teaching. She got married in 1972 and moved back to live in the rural area. She has five children, four girls and one boy. Her husband works in a brown sugar factory where he earns Kshs 1,600 a month and he also runs a butchery on a local market. This business brings him about Kshs 5,000 a month. The husband is married to a second wife and lives with his second wife in another small market center. Muhonja is neglected by her husband. She must work on her farm to raise money to pay school fees for her children and buy food for the household. She has been involved in many physical fights with her husband over the issue of paying for school fees for the children and general household support. During this survey, the husband burned all her clothes and documents and declared that she should leave his home. She says that she cannot leave her husband's home because of her children and she has nowhere else to go as she would not be accepted on a permanent basis in her parents' home. Two of her children who were in secondary school dropped out of school during this survey because they could not pay the school fees.

Commercial Crop Farming Sub-system
The total farm size is 0.3 hectares, of which 0.1 hectares is under tea. There are 1,456 bushes. Tea management is very poor in this household. Tea weeding only occurred once in the course of 1992 and 1993. Fertilizer has not been applied for years. The neglect of the crop stems from feelings of frustration due to marriage conflicts with her husband. She says that if the tea is well managed and produces a good yield bringing a good income, her husband will demand money. Muhonja comes from a well-to-do family and all her brothers and sisters have well-paid jobs. She is the only one in

this family who did not complete school successfully. She knows a number of officials within her local tea branch. By paying a small bribe to some of these officials, the leaf officer at her buying center gives her additional tea taken from illiterate farmers during the course of weighing and recording faulty weights. She harvested a total of 1,030 kg resulting in a monthly average of 86 kg in 1993. The average monthly earnings during this same period was Kshs 303. The total annual income from this monthly income was Kshs 3,633. The bonus was Kshs 8,260 resulting in a total tea income of Kshs 11,893. This income appeared to be somewhat exaggerated given the poor management on this farm. She had actually plucked only one round a month and could barely make 50 kgs. This is a good illustration of the laxity is tolerated in some tea-buying centres.

Food Crop Farming Sub-system
About 0.1 hectares of land is under food production. Local maize is planted during the long rains and the short rainy seasons. Not much is done in terms of using inputs for food crop production except for weeding the maize crop. Only 2 bags (180 kg) of maize are harvested per year. This harvest does not provide enough food for the household. Muhonja has to buy food from the market to feed her family.

Coping Strategies
Through financial assistance from a brother, Muhonja purchased a grade cow and she is practicing zero-grazing. There is a considerable cost involved in starting of zero-grazing and once started, it calls for continuous commitment, otherwise all the money will be lost if the animal dies because of poor management. Many farmers who adopt this innovation endeavour to provide the best management and establish adequate napier grass, reliable feeding, provision of water and ensure the cow is disinfected against tick-born diseases once a week. Napier grass is supplemented by artificial feeds. Optimum management pays off through high production of milk. Seventeen bottles (12 litres) of milk are produced every day. Two litres are used for household consumption and ten litres are sold at Kshs 150 per day. A monthly income of Kshs 4,500 is earned by this farmer from the grade cow.

Another method of increasing the household income is through horticulture. Emphasis is put on the growing of vegetables, especially 'sukuma-wiki' (a kind of kale) which is in great demand. Vegetables grown by this farmer are sold to local schools and bring her an income of about Kshs 300 per month. The market potential, especially for vegetables, is large in Maragoli. This area, in fact, relies on Nandi District to supply it with most of the vegetables it requires. Vegetables are very expensive during the dry season. Muhonja's household has an average annual income of approximately Kshs 45,000 and she earns this without any support from her husband, mainly from the sale of milk, tea and vegetables. This illustrates that zero-grazing and horticulture have high potential in this area and that farmers cay make a sizeable income from these sub-systems thus enabling households to meet many of their needs. Some women who have starting capital are earning more money from their coping strategies than their husbands are getting as wages from their none-skilled or semi-skilled jobs in the urban areas.

Muhonja's situation illustrates the way tensions and conflicts between wife and husband can greatly slow down farm management and production and destabilizing the farm-household economy. Such conflicts can also totally disrupt and pose serious problems for the other members of such households, especially children. Competition and indecision as to who should earn the income from tea and who should be contributing to the welfare of the household has led to the neglect and poor management of farming sub-systems, particularly affects the commercial crop sub-system. It also appears that such conflicts lead many women to try and find suitable

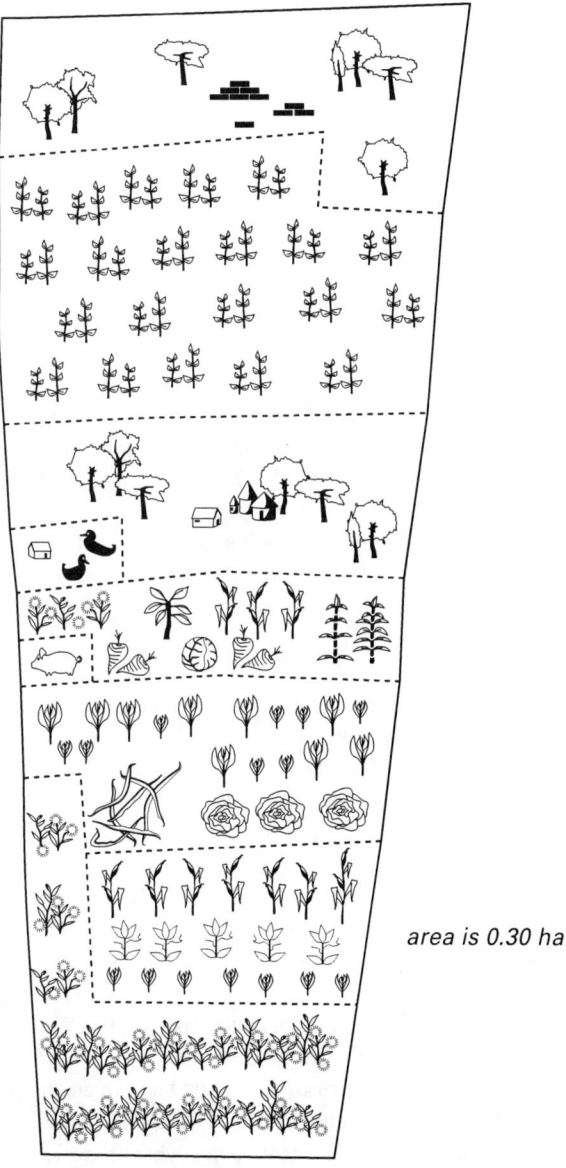

area is 0.30 ha

Figure 6.8
Muhonja Farm (source: Study Results).

coping strategies through which they can earn a sufficient income to meet household needs. Despite the fact that some women have been deserted and neglected by their husbands, their households continue to operate, though with many constraints.

Muhonja emphasized that if credit were available to women in the rural areas it would enable them to start income generating projects. For instance, her zero-grazing project brings in more money than she earns from tea. She pointed out that women need to be financially independent from men because they feed families more often than men do. She observed that she was only aware of the services of tea extension officers and not of the general agricultural extension. She consults livestock extension officers privately, buying all the inputs required for the grade cow. She invests much in the grade cow because it earns her more money than any other enterprise on her farm.

6.5 Female-Headed Households

Mama Ajema

Background History
Ajema was born in 1951 in Kisatiru Sub-location in West Maragoli location. She is 44 years old and so belongs to the younger household heads. She started schooling in 1958 at Wangulu primary school. In 1965, she sat the Kenya Primary Examination (K.P.E) which she passed and she joined one of the Secondary schools in Nandi in 1966. After two years in secondary school she sat for Kenya Junior Secondary Examination (KJSE). She did not perform very well and left school after 10 years of schooling. She got married in 1968. Her husband was then working as a secondary school teacher in Nandi District. She confides that, instead of studying she was spending a lot of time with this teacher and that is why she could not perform well in her examination. She enrolled to re-sit for K.J.S.E as a private candidate in 1969. She passed well and in 1972, joined a primary teachers college where she was trained for two years to become a primary school teacher. After training she taught in primary schools in Nandi District staying with her husband. However, the husband was such a heavy drinker that he neglected his job and this led his being reprimanded on a number of occasions. Eventually, both of them were transferred to their home district, Vihiga District, where they commuted to work from their own farm.

The husband's drinking habit became worse and he would beat her senselessly whenever he was drunk. Whenever the husband received his salary, he would spend all his earnings in the bars returning home only when the money was exhausted to demand for some from Ajema. Ajema was very frustrated and insecure because of the disrespectful behaviour of her husband. She was among the very few women in the village with a non-agricultural job and other people there started ridiculing her, especially when she was beaten badly and she still had to report for duty. All the household responsibilities were left to her. The insecurity and frustrations drew her into an attempt to commit suicide by drowsing her body in paraffin and setting light to it. She was rescued by neighbours and was hospitalized for over six months with third degree burns which have left her with permanent scars all over her body. However, this did not deter the husband from his bad habits. One morning in 1987, the husband was found murdered, his body dumped in a coffee plantation. Ajema became a widow with five children, four boys and one girl. She has continued to pay

Inter- and Intra-household Farm Management Situations 197

school fees for the children, her elder son is a lawyer, the second born is working as a clerk, the third born has just completed form four and the last two are still in school. With the realization that she no longer had a husband, she embarked on farming to supplement her salary and ease the up-bringing of her children. She inherited a farm measuring 0.5 hectares on her husband's death. Being a primary school teacher, she is a member of the Teachers Savings, Credit and Co-operative Society. She took a loan from the co-operative and bought an additional 0.4 hectares of land.

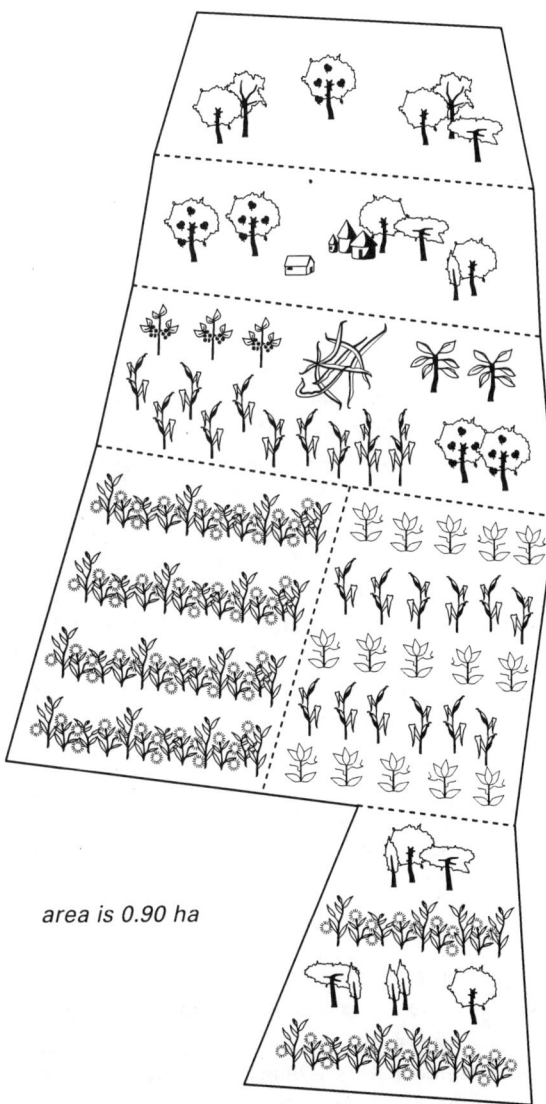

area is 0.90 ha

Figure 6.9
Ajema Farm (source: Study Results).

Commercial Crop Farming Sub-system

The total farm size of this household is 0.9 hectares and 0.4 hectares is under tea. She established 2,500 bushes in 1990 after her husband's death. The tea is still young and therefore the harvest is not very high. Production is about 42 kg every month and 500 kg a year. At the moment the average monthly income is Kshs 137. The annual income from the total monthly earnings was Kshs 1,640 and the 'end of year' payment was Kshs 3,200. The total tea income was Kshs 4,840. However, she is sure that after four years the crop will be doing well because she is following the practices recommended, applying sufficient fertilizer and hires labour for weeding and tea plucking. (see Figure 6.9, farm layout).

Food Crop Farming Sub-system

Maize and other food crops are inter-cropped on a plot measuring 0.1 hectares. Hybrid maize is planted during the long rains and local maize is planted during the short rainy season. Ajema buys fertilizer for the cultivation of maize and this is supplemented by manure from the cow shed. Three bags (270 kg) of maize is produced from this plot. Our case studies demonstrate that those with non-agricultural jobs are able to buy farm inputs because of a permanent source of income.

Other Production Sub-systems and Income Earning Strategies

Government employment and salary assures one of job and income security. The first and main source of income in this household is the monthly salary from primary school teaching. She earns a monthly salary of Kshs 2,335 and an annual total of Kshs 28,020.

The second source of income is from dairy farming (zero-grazing). Ajema, by virtue of her job, ownership of land and land title deeds, is in a position to take loans from commercial banks or from the co-operative society. She took a bank loan using her land title deed as collateral and bought two cross-bred cows. She has also invested in the construction of zero-grazing units and the establishment of napier grass on a parcel of land 0.3 hectares in size. Her cattle are very well managed. The cows receive adequate grass, other supplementary, artificial feeds are supplied to them, and there is plenty of water, because this home has piped water. The units are kept clean and disinfected. Once every two months, livestock officers are invited to examine the animals. Sometimes the livestock officers deworm the cattle, or give some medication to prevent certain diseases. An approximate income of between Kshs 1,500 to Kshs 2,000 per month is generated by milk production, though this could be higher than this (maybe an under-estimation by the farmer). A high level of investment is necessary for a zero-grazing project; this explains why many small-scale farmers without good permanent employment and high incomes are unable to adopt it. Many farmers would like to adopt zero-grazing because it is pays more than some of the crops planted at the moment.

The third source of income is from renting out a house. Ajema took several other loans from the bank and the Teachers Savings, Credit and Co-operative Society, using the land title deed and her job and salary as security. The loan money was used in the purchase of a plot to construct a permanent house that could be rented out. An income of Kshs 1,200 is earned from this project every month. The total household income from these different sources is approximately Kshs 6,000 a month and about Kshs 80,000 a year, including the end of year tea income. Ajema was in the process of securing another loan so she could build another house to rent out.

Constraints

Ajema is of the opinion that widows and women in non-agricultural employment have a number of problems that another affect farm management and agricultural production in one way or another. Her contention is that there are many social and cultural stigmas attached to a woman who is widowed. Local residents shun such a woman because they label them as 'husband killers'. A widowed woman may have very few friends as some men may tell their wives not to associate with her, especially when a husband dies under suspicious circumstances such as being murdered as happened to her husband. Other married women see a widow as a potential ' husband snatcher'. The relatives of the dead man see such a woman as the enemy, especially when inheriting property and land from the deceased is at issue. For all these reasons, it is difficult for such a woman to find farm workers easily, to consult others when she has a problem or even to socialize normally.

More than 90% of widowed women are housewives and therefore it becomes economically very difficult to cope when the husband dies, even though some of these husbands never supported wives financially when they were alive. This was quite puzzling as many women appeared more insecure after the husband's death. Men apparently are important because they provide necessary cultural and social security within the household.

A working widow relies on hired labour for the management of the farm and she lacks time to supervise this labour. Under such circumstances, hired workers sometimes do a shoddy job. Sometimes, work tools are stolen in her absence. There is more theft from the farms and households of widows because without a male head, security is reduced. Sometimes with the death of a husband, some children become problematic as the level of discipline may decrease, especially in homes where the man was an autocrat. If the children are grown-ups they start demanding their father's property and the problems then normally centre on tea, because it is the cash crop in Maragoli. Ajema is better off than many women in small- scale farming because she has a permanent job and salary. She is well aware of her legal rights and that is how she changed the land title deed from her husband's name to her own, something many other widows have not done. She has used the title deed on several occasions to take out loans which have been well utilized. She points out that widows need legal and financial support.

Mama Kaveza

Background History

Kaveza was born in 1920 at Kivagala Sub-Location in South Maragoli location. She is over 70 years of age and belongs to the mature household heads. She never went to school as in those early years there were hardly any schools and where they did exist they were for boys. She therefore was engaged in household chores and assisted her parents in farm work from a very early age as was expected of girls.

She got married in 1940 and her husband worked in Nairobi as a cook. After ten years he changed his job and become a watchman. He worked as a watchman for two years until the building he was guarding was attacked by thugs. He sustained serious injuries during this raid which led to his retirement and death.

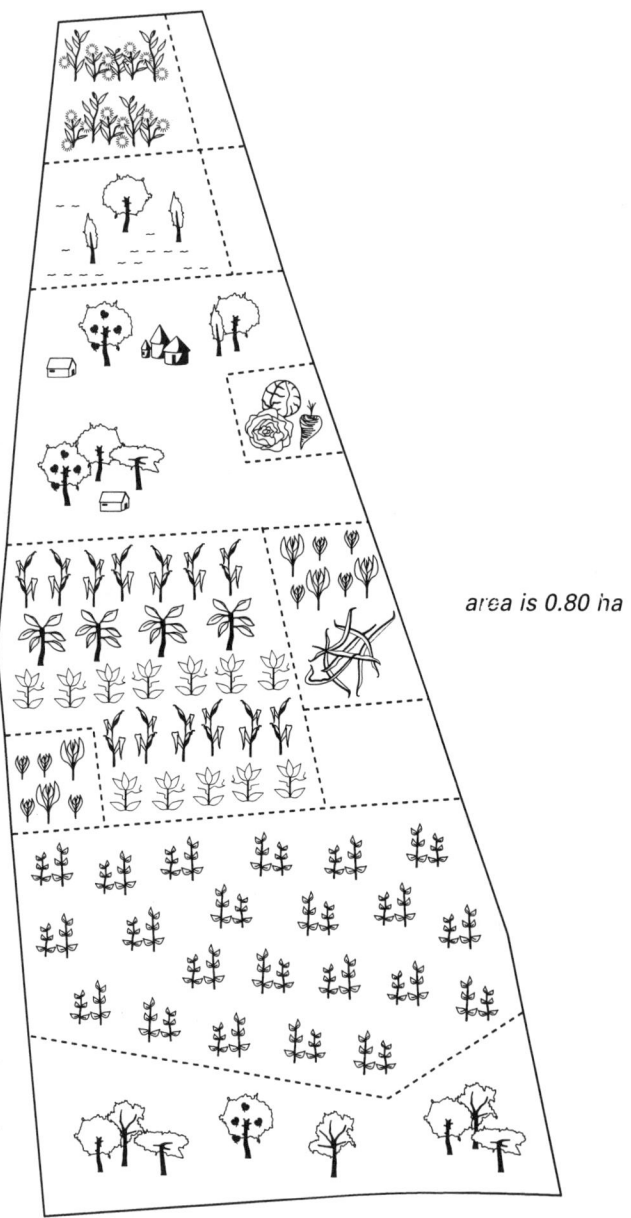

Figure 6.10
Kaveza Farm (source: Study Results).

Kaveza has nine children, three boys and six girls. Four of the girls have died and she has now only five children. Only the boys in this household were given the opportunity of studying and not the girls. The sons all have good jobs, one of the girls is married, and the other was married but because she could not have children she was chased away by her husband and now she lives in the mother's home.

The total farm is 0.8 hectares. The land is still registered in the husband's name. She claims that she is not aware that a woman can change the title deed after the husband's death to her own name. It is her opinion that land belongs to the men and therefore it is the responsibility of the sons to get the land subdivided and shared among themselves. The implication here is that quite a number of women are unaware of their rights, the legal aspects of land ownership and land use, especially, what they are entitled to when their husbands die. This appears to be more common among illiterate women or women with a few years of schooling. This ignorance explains why some women are denied ownership and access to land by the husband's kin when the man dies. This also may explain the insecurity and uncertainty that face women when a husband dies, because they are unsure of their position as far as use of the land and any property left by the husband is concerned. (see Figure 6.10, farm layout).

Commercial Crop Farming Sub-system
Tea is an important source of income on this farm. There are 3,000 bushes of tea which were established in 1961 by the husband on a parcel of land measuring 0.3 hectares. Because of her advanced age of Kaveza, as the household head, she does not do much work on the farm. She relies on her daughter and hired labour for the management of the crop. The tea is fairly well managed, is weeded frequently and is pruned every three years. The supervision of hired labour appeared to be one of the problems on this farm. Kaveza takes only two bags of fertilizer on credit from KTDA which is inadequate for 3,000 bushes. Tea production is low because of the small amount of fertilizer used. In 1993, 2,502 kg of tea were harvested achieving a monthly average of 209 kg. The average monthly income was Kshs 686. The annual income derived from this total monthly income was Kshs 8,233 and the 'end of year' tea payment was Kshs 22,308. Thus this household earned a total of Kshs 30,540 from tea in 1993. This household could well earn more from tea by using adequate fertilizer.

Food Crop Farming Sub-system
Apart from tea, Kaveza also concentrates on food crop production. Maize and beans are grown on a parcel of land measuring about 0.1 hectares. This plot has been under these crops for many years with little improvement being made to soil fertility and therefore the land is no longer productive. Local maize is planted in both seasons of the year and very little fertilizer is used. Kaveza states that she is not sure about the right type of fertilizer to use and how it should be used because no agricultural extension staff have ever visited her. One bag (90 kg) of maize is produced in a year.

Other Production Sub-systems and Income Earning Strategies
Some bananas are grown for home consumption and there are a large number of trees for fencing, construction work in the home, for firewood and for sale. Kaveza has two traditional cows. The milk produced is for home consumption. Kaveza says that she has no problem with her children because she thinks they were well brought up. Their father trained them to respect their mother. She says that these days children do not respect their mothers because they always see fathers abuse or beat their wives and so they grow up believing that women should be

abused. She has a very close relationship with her children and they give her both social, cultural and economic support. Every month she gets remittances of about Kshs 1,000 from her sons. This support has assisted Kaveza to eat well and therefore she is not sickly like other widows in the village. This household makes on average makes a total annual household income of approximately Kshs 42,540.

Mama Keyonzo

Background History

Keyonzo was born in 1940 in Vokoli Sub-Location, North Maragoli Location. She is 54 years old. Her parents died when she was still young and she was brought up by an uncle and his wife. She never attended school and the uncle and aunt used her as their domestic servant and gardener while their own children went to school. Because to frustrations and the mistreatment she received at her uncle's home, she was forced to look for a husband and she got married in 1958. Her husband used to work in Nairobi as a gardener. She has seven children, four girls and three boys. Her husband tried to send most of them to school. He died in 1985. Two of her children are working, three have no jobs and two are still in school.

Commercial Crop Farming Sub-system

Keyonzo has a farm of 1.5 hectares. This farm is very badly managed. She has a tea plot of 0.3 hectares with 2,500 tea bushes established by her husband in 1977. Since her husband's death, Keyonzo lost control of her life. As a young girl, she grew up surrounded by intimidation and fear, relying on decisions that were made for her by others. She was always ordered around. When she got married the husband made all the decisions for her and she worked according to the instructions she received from him. The children also listened to the instructions they received from their father as she did not have any instructions for them. Once her husband died, there was nobody to make decisions and give instructions. All the operations on the farm ground to a halt. The tea farm is very poorly managed. It has not been weeded for the last four years and there has been no pruning for more than six years. She is not aware of the procedure she should follow to make an application for fertilizer and therefore since her husband died she has never used fertilizer on this farm. The average yield from this tea farm is approximately 1.7 kg a month, 20 kg annually. The average earnings are Ksh 8 a month, and Ksh 168 annually. The support linking a woman with her children and society at large is often disconnected with a man's death and this is demonstrated by this household. Low incomes and lack of education are some of the important variables explaining the low performance of female-headed households. (see Figure 6.11, farm layout).

Food Crop Farming Sub-system

The maize parcel is about 0.5 hectares and the most Keyonzo can harvest is one 90 kg bag. Yet, if the maize plot was well managed it could produce about 33 bags (3,000 kg) of maize a year. The maize plot is poorly prepared, is planted late with local maize and no fertilizer or compost manure is used. Only a small section is weeded and the rest is not weeded at all. The woman works all alone and therefore it is not possible for her to complete all the different farm operations that require her attention. The children do not provide labour and she does not have money to use hired labour. The children sit at home most of the time while Keyonzo is busy working alone on the farm.

Inter- and Intra-household Farm Management Situations

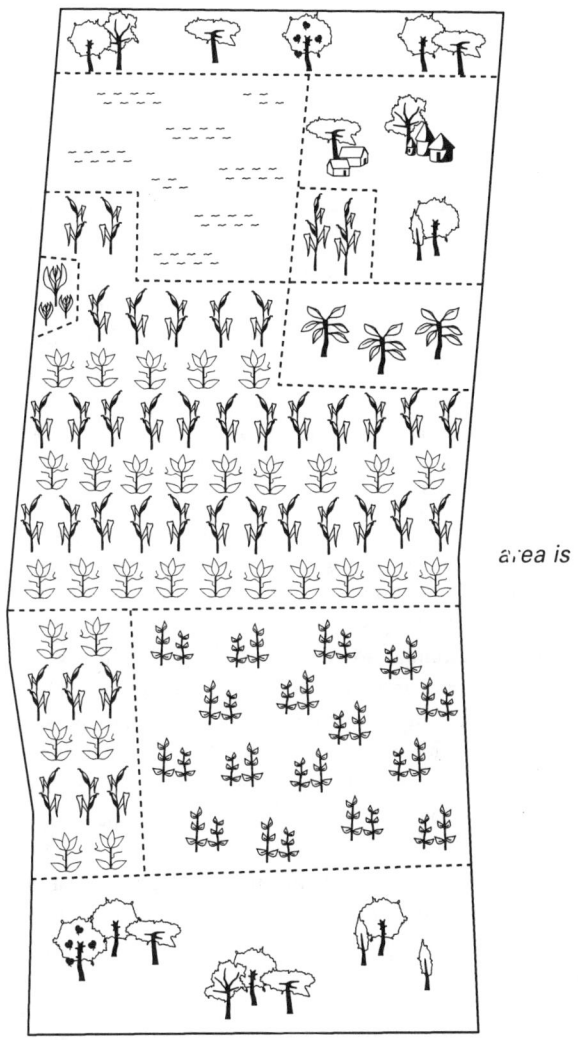

area is 1.50 ha

Figure 6.11
Keyonzo Farm (source: Study Results).

The other crops on her farm are beans, sweet potatoes, bananas, and cassava which are badly managed. She has no source of cash income. Because Keyonzo overworks, she looks much older than her real age. She has health problems.

She has two local cows which are not well fed and the highest quantity of milk that she gets is 0.5 litres a day. She says she is not aware that there are agricultural and livestock officers who give advice to farmers. This household has not devised any coping strategies (see Tables 6.1, 6.2 and 6.3 in the Appendix which contain the summary information on production and incomes of the cases described).

6.6 Conclusion

This chapter has illustrated a number of points about the various household types and their effect on the overall functioning of the farming system (for a summary of major crop statistics of the case studies see Appendix 4). It shows that each household type has specific needs and constraints and these should be understood in order to explain the differences in farm management situations and agricultural production. If such needs and constraints are understood then it may be possible to draw policies and programmes that specifically address specific household types and narrow the gap between male-headed households, female-managed households and female-headed households in terms of accessibility to factors of production and lead to an improvement in the livelihood of all households. Such an analysis would also address the inter- and intra-household differences that were discussed above.

A clear difference appears between households that are male-headed, female-managed and female-headed. Many male-headed households have better access to factors of production such as land, labour and capital. Heads and members of this household type appear to have more education, and therefore the types of non-agricultural jobs held by some of them are well paying. Male-headed households also seem to develop more rewarding coping strategies and, overall, they earn more money from outside agricultural production than any of the other two types of households. High incomes in some of the individual houses of this household type make it possible to make investments in the different farming sub-systems. More male farmers are selected as 'contact farmers' and have access to agricultural information. They use other methods than female farmers. Thus the level of crop and livestock husbandry was of a higher standard than in male-headed households than in the two other types of households. Men also have better chances of getting loans than female farmers and this helps to explain differences in farm investments, farm management, agricultural production and farm and household incomes.

Female-managed households have demonstrated that they are better in farm management and achieve higher agricultural production than female-headed households. Some female-managed households receive financial support (remittances) from their spouses and they are enable to use some of it to hire casual labour. Female-managed households are younger and can work longer hours than female-headed households. The female-managed households also have young children who provide some labour during school-free days. In fact, female-managed households have more education which seems to give them a better chance when it comes to managing of the exotic crops than female-headed households where the level of illiteracy is the highest for all three types of households.

By examining the members of each household in this way we have shown that women and men have different roles, females being mainly concerned with the smooth functioning of farming. This involves such tasks as the provision of labour in all the farming sub-systems and within the homestead; the production of food for consumption within the household; the generation of income for the household or for the male-head of the household by managing the commercial crop sub-system and the management of livestock and milk production, despite the fact that such produce in many households still usually benefit the men. Incomes accruing to many households belong to men; many women have no right to such incomes. Farming is affected by inter- and intra-household conflicts. These lead to a slowing down of the system, poor farm management, low agricultural production and low incomes. Farmers pointed out that such conflicts are increasing because of decreasing agricultural farm size and population increases, harsh economic realities, recession and the inflation of the Kenyan shilling. All too often women seem to bear the burden of these conflicts because of their lack of ownership and in some households, lack of access to land and the fundamental lack of incentives. Women suffer more because of their low level of education and lack of money. These problems are made worse by a weak agro-support system in which only a few farmers are targeted while the rest of the farmers, especially women are neglected. Men's role, however, is distinctively that of head of household, and who often only gives instructions and is responsible for making certain farming decisions. They rarely provide the required labour. Moreover the man is often a migrant labourer and provider of remittances. However, these remittances are irregular, sometimes insignificant and frequently earmarked for specific functions. They are, therefore, negligible as far as agricultural development is concerned. It is also quite clear that the sustenance of rural households depends on women and men are not the bread-winners as proclaimed.

Finally, it has been shown that agriculture as a production sector can barely provide jobs, food and incomes for the growing population in small-scale farming in the North Maragoli area. Many families will no longer be able to depend on agriculture to satisfy household needs unless there is a change in emphasis on the types of farming sub-systems. Because of this, many households have devised coping strategies to help meet those household needs not met from agricultural production. Indeed, it has been found that 94% of the households cannot feed themselves from the food crop sub-system on their own farms. Food production is therefore loosing out and many households have to rely on food purchases with the implication that they must always have money.

Notes

1 All names used for our case studies have been changed.

2 Figure 6.1 shows the symbols used in all farm sketches. Refer to this figure when viewing the other farm sketches.

Female Farmer Tending to her Milk Cow

7

The Impact of Selected Agricultural Policies and Programmes on Farm Management and Agricultural Production in Relation to Gender in North Maragoli

In this Chapter, we will investigate selected agricultural policies and programmes which have direct relevance for farm management, agricultural production and the living conditions of small-scale farmers. Attention will be paid to policies and programmes with respect to land tenure, agricultural extension, credit, cash crops, food security, and livestock development, as well as to recent changes and prospects for the success of the various policies and programmes. The purpose of this Chapter is to show the differential effects of selected agricultural policies and programmes on the farm management and agricultural production of both male and female farmers.

7.1 Introduction

Kenya's agricultural policy started off as a dual policy, catering primarily for the European farmers in the former White Highlands, and to a smaller degree for the areas occupied by African farmers. Up to the end of World War II there was virtually no interest in African agriculture because colonial agricultural policy was primarily geared towards safe-guarding the privileged position of the settlers (Hinderink and Sterkenburg 1987). During and immediately after the Second World War the emphasis was no longer only concentrated on settler farmers but now attention was also given to the plantations and ranches where British capital had been invested. At the same time more attention was given to the requirements of the urban population. From 1947 onwards the development of African agriculture became a matter of government concern since many of these areas were experiencing dense population, soil degradation, low food production and famine. An expansion of output in the 'African Reserves' was stimulated by improving roads, the development of marketing and processing facilities, and the provision of various types of agro-support services. Initially, food crops received priority but later export crops such as coffee and tea were also promoted (Hinderink & Sterkenburg 1987).

Between 1947 and 1955 the foundations for the development of the small-scale agricultural sector were laid down. This culminated in the adoption of the Swynnerton Plan for the intensification of African agriculture in the country, which **de facto** became the most important colonial policy shift since the creation of the colony. The decisions taken following the inauguration of the Swynnerton Plan have influenced agricultural development in the country up to the present day. Kenya was categorized into areas of 'high potential', 'medium potential', and 'low potential' agricultural land for purposes of planning agricultural and livestock development.

For areas of high potential agricultural land such as North Maragoli, the Plan stressed the planting of cash crops such as coffee and tea which were previously restricted to the European farmers, as well as the introduction of improved dairy cattle.

Until Independence virtually all the progress made in African agriculture could be attributed to the Swynnerton Plan. However, this progress was still marginal in terms of output and in terms of the number of farmers and the acreage involved (Hinderink & Sterkenburg 1987). Since then, new policies have been formulated with respect to specific topics such as extension and education, rural credit, the intensification of livestock production through the adoption of 'zero grazing', and the improvements of land tenure. A fundamental change in agricultural development policy started in 1972. This was brought about through the well-known ILO mission which visited Kenya to study the problems of employment and unemployment, and which paid considerable attention to the agricultural sector and to agricultural policy. The mission concluded that increasing landlessness, unemployment, and income inequalities called for a drastic re-organization and change of policy. It recommended an integral approach to rural development with the redistribution of land from large owners to landless households, the seizure and subdivision of large farms if their owners were in arrears with loan repayments, a progressive land tax in relation to the size of the holdings (Hinderink & Sterkenburg 1987, Livingstone 1986, Hunt 1985 and ILO 1972). However, until the early 1990s the issue of gender was not addressed in any agricultural and land policy implementation procedure.

It is only recently that the government has realized the importance of gender and now some of the policies formulated have objectives relating to gender. In Sessional Paper No. 1 of 1986 on Economic Management for Renewed Growth and in the Development Plan 1994-96, it is stated that the government is very much aware of the importance of gender in food and overall agricultural production. It is clearly stated in the Development Plan that one of the long-standing objectives of Kenya's agriculture and food policy is to ensure both national and household food security. It is further stated that an important pre-requisite in addressing the underlying and basic factors behind food insecurity in many households in Kenya is a knowledge of the country's gender structures. It points out that age and gender have not been adequately addressed in the government's attempts to improve food production, distribution and consumption. The government claims that issues such as gender equity, dependency ratios and education participation rates will no longer be ignored or only partially considered in policy. Women's participation in agricultural production it is claimed will be promoted through: (a) encouraging and ensuring that women farmers have access to agricultural information, particularly in female-headed households; (b) improving women's access to and control of resources, especially to land which is the main factor of production in agriculture. Customarily, the inheritance of property including real estate is biased against women, husbands being the legally acknowledged title-holders. However, with the recent changes in succession laws, women should be able to buy and sell land under their own names even if they are married; (c) targeting women through women's groups, co-operatives, schools, etc. in terms of agricultural education, extension and the dissemination of information on improved technologies (appropriate inputs, production and storage); (d) training and employing larger numbers of women as extension workers; (e) establishing special credit schemes for women in women's groups, and (f) translating extension messages into vernacular languages since many rural women may not know Kiswahili or English (Kenya 1994:128).

However, more important than policy formulation is its implementation and the government's commitment to the stated policies. The institutional machinery for implementing general and particularly gender-sensitive policy appears to be lacking in many government departments, and policy matters with respect to gender are approached in a lukewarm manner.

7.2 Land Tenure Policy

Land is one of the factors of production and includes everything attached to it by nature, such as water, soils, forests and mineral resources. All activities, whether economic, social or physical, must be located and take place on land. Land is fixed in size and therefore requires proper planning for optimal utilization (Ellis 1992). Among the Luhya people land is viewed with much emotional, cultural, psychological and economic attachment. The livelihood of people in Vihiga District and Western Province whether they work in the urban areas or not is very much tied to the land in the rural areas.

> "Land is not merely a means of subsistence; it represents a person or persons. It can be thought of as ancestors, mother, bride or unborn descendants. Land is the source of life, the mother of human kind...It contains the ashes of the ancestors and hence the roots of the present... If a person has no land or very little land, if he is young, he has to go to town to get money to buy his own land" (Andreasen, 1990:163-164).

Therefore, a land tenure system, whatever its nature, should not merely be seen as an isolated aspect of the economy of a society. A land tenure system should define the amount of access which members of the society may have and the degree of control that may be exercised over land resources and, consequently, the manner in which land may be used and the manner in which benefits accruing there from may be distributed. Land tenure reform must therefore be evaluated as part of an attempt to restructure society as a whole and not just its management and agriculture (Okoth-Ogendo 1976). But is there any clear policy on land tenure reform in Kenya? The answer must be negative because land tenure policy is in fact rather general and vague.

As already described in Chapter two, there are two types of land tenure systems which existed side by side during the larger part of the colonial period. The Africans lived in what was known as the 'African Reserves or Non-scheduled areas' where each group applied their traditional form of land ownership, a combination of inheritance systems and communal lands. On the other hand, the settlers who occupied the larger parts of the highland and rangeland areas used the individual or company freehold land ownership system. However, in the late 1940s it was felt that every encouragement must be given to the African to farm his land well and to participate in general agricultural planning and co-ordinated development of both his local community and of the specialized agricultural industries in which he may be concerned. Special attention was given to the land tenure reform. According to the Swynnerton Plan "Sound agricultural development is dependent upon a system of land tenure which will make available to the African farmer a unit of land and a system of farming whose production will support his family at a level, taking into account perquisites derived from the farm, comparable with other occupations. He must be provided with such security of tenure through an indefeasible title as will encourage him to invest his labour and profits into the development of his farm, and as will enable him to offer it as security against such financial credits as he may wish to secure from such sources as may be open to him. The commitment that land and chattels will be mortgaged as security against loans and that he will be 'sold up' if defaults must be fully accepted by the farmer in applying for loans and by the government in preparing any legislation covering land tenure and agricultural credit." (Swynnerton:1955:9). This indeed was a complete reversal of the previous government policy when little attention was paid to the African farmer. At the same time, this new land tenure policy completely neglected the existence of women on the farms as individuals who may require land of their own.

It was observed then that African lands were suffering from low standards of cultivation and income. In particular, and as a result of African customary land tenure and inheritance, they suffered from fragmentation whereby any one family could possess several fields scattered at wide intervals. This meant that land could not be developed economically either to the system of farming best suited to the area or to the inclinations of the farmer himself. It was perceived that it was impossible under such circumstances to develop crop rotations, to apply manure, to establish and manage grass, to improve the management and feeding of livestock or to tend cash crops in a satisfactory manner. It was therefore thought that if, by suitable reforms of the system of land tenure and inheritance, fragmented land could be combined into economic farming units which could be exploited by proper farming methods, these lands would yield the greatest return to the economy of Kenya, given suitable injection of staff and finance (Swynnerton 1955:9).

It was believed that once the land was registered, farmers would be able to buy and sell land amongst themselves and to mortgage titles to land against loans from the government and other approved agencies. It was also envisaged then that with the acceptance and implementation of this policy, able, energetic or rich Africans would be able to acquire more land and bad or poor farmers less, thus creating a landed class and a landless one which would provide cheap labour. This was indeed perceived as a normal step in the evolution of a country (Swynnerton 1955:10). The principal aspects of land policy during this period were land consolidation, registration and titlement, and individual ownership. However, this policy has greatly contributed to the inequalities in resource (land) ownership that are prevalent in Kenya today.

On the attainment of Independence, the Kenya government continued this policy of land reform: consolidation, registration and titlement in the high and medium potential areas of Kenya. It has been pointed out that there has not been a major review of land policy since Independence: "The existing situation combines colonial land tenure laws with recent practice in a complex pattern that makes it difficult to operate a land use policy" (Kenya 1986:90). In 1986 the government appointed a commission to consider the following elements of land tenure policy in Kenya:
(a) taxation and other measures that provide an incentive to use land more productively;
(b) regulations limiting the extent of subdivision to ensure that farmland can produce adequate income for the family unit, including potential criteria governing subdivision, which must vary by agro-ecological zone;
(c) laws that could encourage and protect holders of large tracts who lease their land to those able to farm it more intensively;
(d) means by which authorities in urban centres of all sizes can obtain land expeditiously for

needed expansion, especially to accommodate small-scale manufacturing and service industries;
(e) the appropriate infrastructure to utilize land allocated for public facilities to promote a rural-urban balance (Kenya 1986:90).

The process of land registration commenced in many parts of Western Province, including North Maragoli in the early 1960s. By the early 1970s the whole of the former Kakamega District had been adjudicated and registered, and this made it possible to introduce such cash crops as coffee, and later tea to many farmers (men). Actually, in the late 1950s a few farmers in this area were already cultivating coffee (Lavrijsen 1984). The impact of the policy relating to land tenure is very visible in North Maragoli, with all farms registered and many farm-households cultivating cash crops (see Table 7.1).

As to land use the government emphasizes that private owners of land have a social obligation to put their land to its best use. The sanctity of private land ownership is respected in Kenya, but only if private land is used in socially responsible and productive ways. The government insists that two mis-uses of land must be prevented if the strategy of agricultural and economic growth is to be realized. First, there must be limits to the subdivision of small farms, despite growing population pressure on land. However, no indication is given as to how the excess population on such farms is to be dealt with. Subdivision should be prevented beyond the point where total returns to land begin to diminish. Here the policy is not clear as to the size limit and the limit of total returns. Second, land should not be allowed to lie underutilized in large landholdings and so the government suggests that steps must be taken to induce landowners to put underutilized land to more productive use. It is also not clear how the government can implement such a policy. Moreover, nowhere is there a clear definition of the gender issue although the government claims that it recognizes the importance of gender in agricultural production. The question of how women are expected to play an important role in agricultural development without the necessary resource, therefore, remains unanswered.

Whereas the policy on registration and ownership of land title deeds has made it possible to introduce cash crops in North Maragoli, the introduction of an individualized land tenure system has led to the displacement of many people such that landlessness and near landlessness is a major problem today in the study area. This can be seen by the figures showing reduced farm sizes (see Section 3.3.1). Some of those who were left landless have gone to swell the numbers unemployed in the urban areas. Others have been fortunate enough to find land in the settlement schemes, or to purchase land in some other districts of Kenya (Livingstone, 1986) or to have found work on labourers in large-scale agricultural estates. However, the vast majority still clings to the little land owned by their parents, such that the land formerly registered in one name has now been so subdivided that many families own very little land (usually less than 0.8 hectares, 44% of the sample survey). Thus efforts to introduce more intensive methods of farming are severely hampered by the absence of adequate land, and this is felt keenly at the individual farm level. Table 7.1 below shows the status of land ownership based on the sample survey and the case studies.

In this study 45% of the participants in the sample survey population and 67% of those in the in-depth case studies have collected title deeds. Land is predominantly registered and titles written

Table 7.1
Land Tenure System and Ownership in North Maragoli (source: North Maragoli Sample Survey, 1992, and In-depth Case Studies, 1993).

	Land registered		Title collected		Name on Title of those collected	
Type of hh	Yes	No	Yes	No	Male	Female
Sample survey						
Male-hhh	60	00	36	24	36	00
Female-mhh	60	00	27	33	27	00
Female-hhh	60	00	18	42	17	01
Total	180	00	81	99	80	01
Case studies						
Male-hhh	13	00	11	02	11	00
Female-mhh	11	00	06	05	06	00
Female-hhh	9	00	05	04	04	01
Total	33	00	22	11	21	01

in men's names as illustrated in the table above. Only one farm is registered and the title written in a woman's name. Gender differentiation is still an important factor in land ownership in this area as it is in many other small-scale farming areas of Western Province. Despite proclaimed policy that women are recognized as equal partners in land ownership, this policy has not been implemented and land is still predominantly held by males as the legal owners.

The land tenure policy was justified as a means to remove fragmentation (Kenya 1994, De Wilde 1967, Swynnerton 1955) and to prevent subdivision beyond the point where total returns to land begin to diminish. However, more than thirty years later one is back to square one since fragmentation is once again rampant and the development of agriculture and the improvement of individual family incomes is once more at risk. The situation in North Maragoli is such that the subdivision of land remains a continuing process. Indeed, some of the parcels of land are as small as (0.1 to 0.3 hectares). Twenty-four percent of the family farm holdings were found to have farm sizes of less than 0.3 hectares in the actual farm measurements (see 5.4). There are many of these types of farms in North Maragoli and Vihiga District, especially if accurate farm measurements are done. As has been mentioned before, it is not clear from government policy when subdivision should be stopped and how such a policy can be implemented. It is even more unclear what should happen to people on mini-farms in such densely populated areas as North Maragoli, and many other parts of Western Province.

Yet, land registration is still regarded as a pivotal policy in Kenya's agricultural development in spite of the problems outlined above. So far no alternative policy is being envisioned, apart from efforts to introduce more intensive systems of land use such as zero grazing or even mixed cropping which is being revisited after having been condemned as unscientific for years, especially at the time of the introduction of hybrid maize seeds (Kenya 1965, 1981).

The economic justification for revolutionizing the land tenure system in Kenya was that once land titles were individually held, market forces would henceforth take over. Thus armed with freehold land titles, small-scale farmers (men) would begin to obtain bank loans for agricultural development using their titles as collateral for the sums borrowed. This was done with the understanding that since many ethnic groups in Kenya followed the patriarchal system, granting land title deeds to men would benefit all the members of the household. However, as it has been shown in a number of the chapters of this study, privatization of land and granting land titles deeds to men has only undermined and marginalized women in Maragoli, in Vihiga District and in many parts of Kenya in the following ways:-
(a) women have no control over ownership of land and cannot make decisions about it; (b) women have limited access to land use and major decisions on land use are made by men. Land is subdivided for different uses by the men and many women work on these farms as unpaid labourers; (c) on most farms women's role is to provide labour for both food and cash crops, but men were found to earn most of the farm income in many households: 92% in male-headed households, and 70% in female-managed households; in female-headed households 33% of the farm income is earned by sons; (d) women have no access to credit because they lack collateral as land is the major collateral for rural households; (e) land is a stock of capital, and therefore with privatization and registration of land into men's names it has acquired a market value and it can be sold. Instances of males either mortgaging land by taking big loans which they fail to repay or selling off the land quietly without informing members of their household have occurred, and women and children have found themselves being evicted and becoming landless; (f) women experience many uncertainties and insecurity as far as their continued use of land is concerned, especially when husbands enter into polygamy. In this study, these uncertainties and insecurity were often found to affect the management of the farming system negatively; (g) conflicts between husbands and wives were observed over cash crop ownership because women have no land of their own where they can cultivate their own cash crops. Such conflicts in some households led to women being domestically abused by their husbands. This sometimes leads to some women boycotting farm work and refusing to supply labour for the commercial farming sub-system. In approximately 35% of the households tea has been neglected because of inter- and intra-household conflicts over land ownership or tea incomes.

A number of issues can be addressed from our field findings with regard to the land ownership question and the land tenure system as related to Government policy. Due to the changing socio-economic parameters in Kenya, cash income is becoming an increasingly important resource for the subsistence of all people whether residing in urban or rural areas and whether they are women or men. This implies that individuals must endeavor to earn money from their daily activities in order to purchase and satisfy their daily household needs both in reproduction and in production functions. In the North Maragoli setting of small-scale farmers, this implies that one of the main sources of income comes from agricultural production, particularly from the commercial farming sub-system. Farmers work better when assured of the income to they can expect from their labour. This was noticed during the course of this study. It was observed that in a good number of households, women did not have any rights to farm income because of the question of land ownership. The farmers confirmed that there was no collective pool of incomes and resources in the household. Men and women had their own incomes and each of them had priorities for the use of their respective incomes. Women use theirs mainly for consumptive goods. Expenditure patterns varied from one individual to another and from one type of household to another. The different priorities of men and women influence the use of money within the household. Many men would prefer to invest their money in the purchase of additional land, building a good permanent house, and buying additional cattle for example, while women think of buying food, clothing and for medical treatment for the household. This implies that the two operate at completely different levels, and men view women's needs as trivial while women think that their needs are important for the sustenance of the household. This indeed make some women conclude that it is the ownership of land that give men power over them. Agricultural production may improve when the issue of land is solved and when the government realizes that households are not made up of single units but form multiple production units whose interests should be taken into account when formulating and implementing policies.

The issue of land ownership is a delicate and sensitive one and it will take a long time before a solution is found. It has been stated by Livingstone (1986), Odingo (1985), Hunt (1985) and the ILO mission report (1972) that no clear policy has been found on the question of land redistribution and land ownership. If it seems difficult to resolve the land problem among men themselves, it may be even more arduous to consider the issue of women. Most of the men in the study area have refused to accept that women are marginalized and they resist trying to understand the gender question. The approach to gender in this area and in many other parts of Kenya is very sexist. Women do not, in fact, want to take away the land from men but want to have equal access and decision-making power as fa as use, earnings and expenditure are concerned. The laws are made by men and it needs much courage to find men who are prepared to stand up against other men for the needs of women in order to harmonize agricultural production.

However, given present-day economic realities, and the fact that now we have an increasing number of single-parent managed households and female-headed households in particular, there is a need to revisit the land issue to satisfy the interests of both men and women in the small-scale farming sector. A mechanism for sharing household responsibilities and for the equal distribution of resources should be put in place and enforced through legislation and policy instruments. Farm incomes should go directly to those managing farms irrespective of ownership. Ownership of land should be bestowed on those managing the land and earning their living from the land. A re-examination is needed of the local traditions to make them fit into present day political and socio-economic reality. The problems and solutions to farm production and economic efficiency are greatly influenced by the land tenure system and the ownership of and access to land.

Liberalization of land ownership and the granting of joint land title deeds in the names of both husband and wife could be one solution. In a way this would make a wife an equal partner to the land and so make her feel protected. At the same time the wife may use the title to obtain credit where necessary. This would greatly reduce the problem of men mortgaging or selling land without the knowledge of their wives and families. Farming investment would be carried out together. However, this can only be successful if there is some degree of mutual understanding, friendship, harmony and sharing within the household especially between the husband and wife. Understanding and sharing may only be cultivated where partners respect one another as equals. This is lacking in many households in North Maragoli as we have seen. Some men claimed that because they pay a dowry to marry wives, it is in fact a way of buying them and women are therefore bonded through dowry to work and obey men. Within the Abaluhya tradition, the dowry was seen as a gift of appreciation to the parents of the girl, but in present day Kenya, marriage has become commercial. Some parents demand far too much money as a dowry from their perspective sons in-law so that some men feel that they are buying wives since some have to take bank loans to pay this dowry.

7.3 Agricultural Extension Policy

The traditional concept of agricultural extension is information delivery to farmers, a function related to the transfer of officially sponsored technologies being promoted by a commercial firm or a Government Ministry (Moris, 1991). This is also the way agricultural extension is considered by the government of Kenya. According to the Swynnerton Plan (1955), the policy on extension was that African farmers were to be given technical assistance (extension) to develop agricultural land along sound lines, and with regard to ecological conditions. Extension mainly concentrated on the big progressive farmers and on specific cash crops, the improvement of livestock, management and conservation matters. At that time, there was one agricultural extension officer available for 83,000 people and the Government's objective to train more extension personnel (Swynnerton, 1955:12). The Government has continued this policy of providing agricultural extension to as many farmers as it is possible. More approaches have been devised to try to reach more farmers especially the small-scale farmers.

> "To improve the flow of information to farmers, the extension services of the Ministry of Agriculture, Livestock Development and Marketing will be strengthened through the expansion of programmes for staff-training and through dissemination of research station findings to officers and more effective demonstration of approved management systems. Greater attention will be given to more effective and widespread group extension techniques involving demonstration farms and field days. Emphasis will also be given to on the spot training of farmers in the basic principles of crop husbandry, the use of fertilizer and other inputs, crop rotation, on-farm storage for subsistence crops, record keeping and financial management." (Kenya 1981:19).

The main objectives of Kenya's agricultural extension policy are:- (a) education of farmers, mainly the spatial diffusion of new technologies and information in agriculture by the simplification of reporting procedures and reaching many farmers; (b) development of linkages between farmers and agricultural research stations, researchers, NGOs and agricultural staff; (c) provision of additional transport facilities to agricultural staff to enable them to reach many farmers; (d) improvement in the training of extension staff; (e) development of more demonstration plots to illustrate the effectiveness and profitability of improved production techniques; (f) improvement in the dissemination of farming guides and other materials to farmers and producer co-operatives and (g) the creation of employment in agriculture. The value of agricultural extension is therefore mainly to pass on technology to farmers in the rural areas on how to become efficient in the production of food crops, cash crops and livestock. Newly recommended practices (from research) are supposed to be passed on to the farmers (Kenya 1986 & 1994).

Although extension is by far the most widely used technique in Kenya for disseminating new scientific knowledge and skills to enhance production and improve farm management, from its inception the policy on extension has not been clear on gender differentiation and on the definition of a farmer. During the colonial period, agricultural extension was provided only to the White farmers. The emphasis was on a single crop approach, and the improvement of livestock and ranching particularly in the Kenya Highlands or the settlement areas. With the introduction of cash crops in the African areas (see Swynnerton Plan, 1955), extension was provided to the big progressive farmers and to the African male farmers as cultivators of commercial crops. Since Independence the government has followed the same approach with very little modification to serve women farmers. As already pointed out in Chapter 2 the extension programme suffers from an over-emphasis on individual farm extension methods and an inability to cope with large numbers. Our research shows that the general objective of educating a large number of farmers is not being met in North Maragoli.

A number of authors have correctly criticized agricultural extension for the linear, unidirectional flow of information between extension agents and farmers. Information flows have not been multi-directional, and not much importance has been attached to giving feedback to researchers on how farmers respond to new technology (Farrington 1994, Ellis 1994, Due 1988, Dey 1981, 1984, 1985, Jones 1986 & Moock 1986). A great problem is the extreme fragmentation of extension over the various Ministries and, as far as the Ministry of Agriculture is concerned, over the different departments. The crops department is divided into different sections for specific crops. The livestock department is divided into production and veterinary sections for matters of extension. The Ministry of Environment and Natural Resources and the Ministry of Energy both have departments of forestry to provide extension on forestry, agro-forestry, and conservation matters, and finally the Ministry of Tourism and Wildlife is responsible for fisheries extension. Many of these ministries and departments do not work together, and sometimes they are not aware of what the others are doing. This appeared to confuse the farmers and to overload a few contact farmers with many, sometimes conflicting messages because the same contact farmers are approached by many agents of these ministries and departments. Moreover organizational changes at ministerial and departmental level create overlaps and sometimes even resistance to co-operation in the field. Such multiplication and duplication of tasks clearly demonstrates lack of co-ordination and weak organization.

The research station in Kakamega with many field demonstration plots all over the district was found to have very important information, both practically and theoretically for agricultural staff and farmers. Yet, there was very limited if any connection between farmers and the research station. We also found that, in fact, frontline extension workers and other junior agricultural workers who deal directly with farmers did not have any opportunity or access to research station staff and work. Farmers did not know what went on in the research station. The Provincial and

District agricultural staff consult with agricultural research station staff, but this linkage appeared to be very weak. Given these prevailing circumstances, the linkages that the government describes in its extension framework does not exist. Unless this situation is righted agriculture extension will continue to suffer from the lack of information available to farmers. The small-scale farmers in North Maragoli complained that frontline extension workers did not have any meaningful new information and were repetitive (see 4.9). These poor linkages created complications in implementation of new technologies and the flow of information.

In public it is professed that women farmers need assistance but privately and in policy formulation this is looked on as a political and ridiculous demand (Moris 1991). Our field findings show that very limited advice is given to women in small-scale farming. As far back as the 1960s, it was difficult in Eastern Kenya to locate men to interview on farms which were proclaimed to be under male control. One mainly encountered women in the fields, carrying water, selling beans and harvesting maize. The men were away in urban areas or at the market centers doing business (non-agricultural activities), chatting with friends or drinking in bars (Moris 1991). Today, this is also the situation in many households in North Maragoli. Women in this area constitute the largest single category of active farmers in crop husbandry and animal husbandry. They are in charge of both food and cash crop production, livestock production and the care of the household. Yet, they have a very limited interaction with the extension services despite the Government's promise to encourage awareness of and ensure access to agricultural information by women farmers, particularly in female-headed households. In the in-depth case studies approximately 15% (3 out of 20) of the female-managed and female-headed households were 'contact farmers' and about 20% in the general survey had been in contact with extension staff in the period 1992-1993. The women expressed the opinion that the extension agents avoided them and only visited the homes of male-headed households. In a few households, women are not even allowed by their husbands to talk to strangers in his absence regardless of the sex of the visitor. There would be a problem in disseminating extension to women in such households even if extension services were easily available. Women said they are not selective about the type of extension officer who visited them so long as they receive the information they required. However, the specific needs and problems of women farmers are not addressed in the design and implementation of agricultural support services. The Government has yet to fulfill its policy objective of targeting women through employing more female extension officers and using women groups. None of the frontline extension workers in North Maragoli were females at the time of this study. It was therefore not possible to evaluate the impact of extension if women were to be used as frontline extension personnel in this area.

As early as 1981 the government pledged to provide transportation and increase the number of agricultural extension personnel in the field. However, during this study not much had changed. The agricultural section of Sabatia Division served approximately 19,000 farm-households and has only one vehicle. Often the vehicle was non-operational because of mechanical problems or a lack of petrol. Many of the officers remained desk-bound, rightly claiming lack of transport and operational capital. Two officers however did use motor cycles.

Farming demonstrations and field days are some of the methods used to provide extension information in North Maragoli. These two are the group methods the Government hopes that extension staff can use to reach many farmers. In Sabatia Division one field day was held in 1992 and one in 1993 instead of the 40 field days which should have been organized. Not many farmers had advance information about the field days and so both were poorly attended. Shortage of finance was listed as one of the reasons for their poor organization of field days (see 5.12). Farming demonstrations are held on 'contact farmers' plots. The demonstrations are organized and supervised by the frontline extension worker using the training and visit (T&V) approach. This approach, approved and partly sponsored by the World Bank, was thought to be the best one to improve extension and turn a cadre of poorly supervised, unmotivated and poorly trained field personnel into effective extension agents (Kenya 1994). Unfortunately, the T&V approach has not had a great impact in the study area for various reasons. First, the T&V approach has led to a considerable increase in government staff, but this increase has taken place amongst office staff at the headquarters rather than field staff. Yet, what is urgently needed in this area is field staff to interact with farmers.

Second, there is a lack of operational funds for travel, training, demonstrations, field days, and farmers' workshops which severely limits extension's ability to be responsive to the client's needs. This lack of funds is all the more serious as T&V requires field staff to interact intensively with the farmers. Under these circumstances extension is unable to establish a political base sufficient to ensure long-term financial support above that required to meet salary needs. Third, supervision of the extension agents by district and divisional agricultural officers is minimal which subsequently leads to a high frequency of absenteeism on the part of the extension agents. Fourth, the T&V approach appears to further institutionalize the hierarchical tendencies that already existing supporting top-down, centralized management, despite clear aims to the contrary. The bureaucrats appreciate the T&V approach because it is a new way of holding staff at different levels accountable. However, in the end no-one is held responsible. The top-down flow of information which stems from national planning objectives, does not reflect the objectives, conditions and constraints of the farmer. Extension is, therefore, unable to deal effectively with the site-specific needs of farmer problems and opportunities as our North Maragoli case shows.

Apart from these operational and organizational problems, a more fundamental reason for the minor impact of the T&V approach has had in North Maragoli is the fact that under prevailing conditions of increasing population and decreasing farm sizes, the extension agents are not able to advice farmers adequately on the most efficient systems for such small farms and how further subdivision should be halted. The T&V approach appears too narrow a model in many situations and therefore of limited utility, particularly under circumstances of considerable household-level and farm-level heterogeneity as is the case in North Maragoli. Such a narrow focus consequently leads to a concentration on a few contact farmers who are mostly men. This does not only limit the trickle-down effect to other farm-households, it also hampers the dissemination of knowledge within the household. Indeed, cases of poor husbandry practices such as the under-utilization of fertilizers or their wrong applications, inadequate use of hybrid seeds, wrong tea plucking and bad pruning methods are cases in point that call for the dissemination of extension to all household members.

Despite the fact that the officers in the field know that women are heavily involved in small-scale farming, they still prefer to deal with men most of the time. The officers are quick to point out that women are normally too busy to attend agricultural courses and that men are relatively free, flexible and easy to deal with. Women felt that they were neglected and many showed interest in attending agricultural training. One weakness of this agricultural training is the fact that the officers do not use the farming systems approach, but insist on a single commodity approach, disaggregating the farm into different units, whereas farmers have an aggregated approach to management. It would appear that a cheaper and more effective approach would be to offer the courses as one package using the farming systems approach. Another problem observed is the poor rate of evaluation/follow-up because of the shortage of finance and transport.

The chief's 'Baraza', demonstrations, and field days are considered by the Ministry as the group approach techniques. The chief's Baraza (meeting organized by the chief and his assistant chiefs) takes place once a week. During these meetings the chief discusses the development projects in his location and social problems (land problems and domestic or family conflicts). Extension officers are encouraged to use such forums to meet farmers. Our assessment of the chief's barazas show that the larger percentage of those in attendance are men (at least 80%). Here again extension is mainly communicated to men. Many of the people who turn up for the chief's meetings either have a problem for the chief to solve or appeared to be idle.

According to the 'divisional office agricultural calendar', there should be four field days at the divisional level, eight field-days at the locational level, and 40 field-days (on contact farmers' farms) at the sub-locational level every year. This creates a total of 52 field or teaching days in a year. Observations in the field shows that only one field day was organized at the divisional level. The 40 field-days with the contact farmers are left to the discretion of the frontline officers.

Demonstrations have the same problems as field days. One of the extension officers pointed out that field days and demonstrations are expensive to hold and on many occasions the Ministry's budgetary allocation for this purpose was too scant. In fact, very few farmers are reached. It also turned out that more than 80% of those in attendance are men and not women. Information as to when these meetings would take place do not reach many farmers and sometimes the meetings are held too far away and therefore are expensive. Extension staff have problems with transport when they have to visit farmers in the interior, and they lack demonstration materials.

Other media like the radio, television, agricultural shows and newspapers do not appeal to most small-scale farmers especially to women. Households with a good income have a radio. The radio is the man's property and often its use is limited to a few members of the household. The time agricultural programmes are on the air, are times when women are very busy on the farm or involved in household chores. There may be conflicts as to which radio station/channel to listen to. Some members of the household may want to listen to an agricultural programme while other members may want to listen to something else. Also language remains a barrier especially when technical words are used. Television sets are very expensive, and in our sample population only two farmers had a TV set. Newspapers are also expensive for the small-scale farmers.

Agricultural extension has traditionally been regarded as the most logical, scientific, and systematic method of disseminating new knowledge and skills to farmers to help them in successfully adopting innovations and making a more efficient use of their land and associated resources. Extension is seen as the heart of the trickle-down and diffusion theory of communication. However, agricultural extension which once seemed so promising for improving productivity and efficiency by encouraging a greater adoption of new practices, appears, nevertheless, to have failed to make a significant positive impact on agricultural production in this area.

The foregoing discussion illustrates that the specific needs and problems of small-scale farmers and women in particular, are not addressed in the design and implementation of agricultural extension services in North Maragoli (and many small-scale farming areas of Kenya). A reorientation of extension seems necessary to improve effectivity of the technical messages and communication strategies and to adapt these to the reality of small-scale agriculture, viz, the reality that many small-scale farmers are women. Special approaches are needed for several reasons: many male policy-makers, and extension and research staff lack understanding and knowledge of women's roles and constraints; agricultural technical messages concentrate on the resources, commodities, tasks and activities which are considered more relevant to men, whilst extension to women has traditionally concentrated on home economics subjects; women's attendance at extension activities is constrained by their lack of time and mobility and this is a result of cultural norms, their domestic responsibilities, and their heavy workloads. Women have also a more limited access to factors of production including inputs, equipment, land and credit whilst their low levels of education and literacy hinder their understanding of extension messages and active participation in extension activities. (see also Saito et al 1994, Moris 1991, Egsmose 1990).

Perhaps the most important change would be a re-orientation of research and extension to what the clients need and ask for or, in the words of Robert Chambers 'to put the farmer first' (Sims and Leonard 1990). These words could be rephrased 'to put the women first' for small-scale farming in rural Kenya. A number of scholars have suggested ways in which the extension service can re-orient its field activities to meet the special needs of female farmers. These include recognizing women as heads of households in procedural and legal matters, so that they are more free to make farm investments and to get farm incomes; design programme content with an eye to low-resource farmers who need adequate family nutrition but lack land; give extra attention to improving household energy and the efficiency of time-consuming processing and water conveyance tasks; devote more research and extension to those enterprises that are crucial to women especially to the traditional crops that have been neglected since the introduction of new cereals and cash crops. Information on some of these crops such as sweet potatoes, yams, millet, bananas and horticultural crops should be disseminated to women farmers and not only to the male contact farmers as is the case at present; extension packages for landless farmers who have only the space around their houses for farming and those with very small farms as in North Maragoli and Vihiga District should be devised, and these should include single-parent families, especially female-headed households, as a demographic category for planning rural services (Antholt (1994, Moris 1991).

Suggestions by Farrington (1994) may also be applicable in this context. He emphasizes that approaches based on farmers' extension should include participation in diagnosis, testing and dissemination. This should be organized with groups of farmers, especially women groups, rather than individuals. The success of this approach depends on the possibility to capture the complexity, diversity and risk-facing low-income farmers, again especially women. These farmers' knowledge and household strategies/management are very important and farmers themselves are the best placed to interpret how relevant new technologies might be. Farrington further stresses extension through non-governmental organizations. The usual difficulties of maintaining a public sector field-based extension service on limited and often diminishing budgets and the bureaucratic and disinterested attitude of many agricultural officers should encourage efforts to link extension with field-based organizations (NGOs) whose mandate includes the provision of technology to small-scale farmers.

7.4 Agricultural Credit Policy

Credit as defined by Ellis (1992) is a sum of money in favour of the person to whom control over it is transferred. Credit involves a lender and a borrower. Credit involves a price for the transfer of control over money, which is the interest rate charged by the lender to the borrower. The lender may be an individual or an institution. Agricultural credit in Kenya is normally from institutions (Agricultural Finance Corporation, AFC), commercial banks or cooperatives.

The Swynnerton Plan stipulated clearly in its policy that more credit be made available to African farmers for the development of cash cropping and animal husbandry. "Access to sources of agricultural credit big enough to meet the requirements of very large small farmers, administered preferably through district land development boards and co-operative societies will be made available" (Swynnerton 1955:8). If Africans were to develop their lands to their full potential and near the standards of European farmers it was recognized that they required greater access to finance and this was linked to the achievement of land title deeds. There were basically two types of loans to African areas (a) loans to African farmers or groups of farmers to develop more intensive farming systems and (b) loans to groups of farmers, district cash crop or irrigation boards or African District Councils or to departments to develop large schemes for improvement, settlement, ranching, irrigation, processing plants for crops or stock, afforestation, etc.

In 1948 the body appointed to work on the policy, administration and distribution of loans to African farmers was the African Land Utilization and Settlement Board later renamed 'The African Land Development Board' (ALDEV). Despite this policy statement, the rate at which loans were extended outside the settlement schemes and European farming areas was minimal and very slow (Heyer and Waweru, 1976). In 1963, the lending activities of the two boards of Agriculture (for Africans and Europeans) were combined under the Agricultural Finance Corporation whose main policy objective was to promote and intensify agriculture, especially in the neglected small-scale farming areas of Kenya through the provision of loans without discrimination (Gerhart 1975).

Thus Kenya government's policy and objectives on credit from 1963 remained pegged to providing credit for farmers so they could develop their land, intensify agricultural production and improve livestock, purchase the required inputs, and machinery and adopt new innovations to boast, foster and intensify agricultural development in the small-scale farming sector (Kenya Development Plan, 1994-1996, Sessional Paper No.1 of 1986, African Socialism and its

Application to Planning in Kenya, Development Plan 1970-74 and Sessional paper No. 10 1965). The prevalent poverty and the scarcity of money in the small-scale areas prompted the government to develop a policy on farm credit (see Section 2.2.3).

Since the 1970s, the various agricultural credit schemes established in Kenya for small-scale farming areas can be divided into those aimed at expanding commercial and cash crop production and those aimed at assisting poorer farmers, principally, the Integrated Agricultural Development Programme (IADP). The objectives of the IADP programme were stated as:- (a) to reduce smallholder farming production constraints through improved agricultural extension, farmer and staff training, input supply, marketing, credit and price control; (b) to help build infrastructure and institutions for smooth output and input delivery systems in the small farm sector; (c) to promote a whole farm approach (farming systems approach) through the establishment of viable technological packages and support systems; (d) to develop a project planning and implementation system at all levels and (e) to increase national production (see Section 2.3.3).

Despite these rather broad objectives, the focus has been very much on agricultural credit, directed particularly towards poorer farmers not catered for under other schemes (Livingstone, 1986). The policy again was not clear on gender. It is common knowledge in Kenya that when one talks about a farmer, one is referring to a man in agricultural production. Therefore, this scheme was designed for poorer male farmers. Our study area was one of the regions where it was implemented and a limited number of male farmers benefitted. Other schemes organized to provide credit for the purchase of farm inputs such as seedlings, and chemicals for example, and which have also only benefitted a few men in this area are the Smallholder Production Service and Credit Project (SPSCP), which was a precursor to the IADP and started in 1975, and the Co-operative Production Credit Service (CPCS) which began in 1972. These are examples of some of the credit schemes organized by the government with international funding. However, in general they recognized men as the farmers and the poor in the small-scale farming sector. This might be one of the reasons why many male-headed households perform better in farm management, produce more and earn higher incomes than female-managed and female-headed households. To some extent this illustrates the level of marginalization of women in the rural agricultural sector and how policies play a leading role in subordinating women farmers and negatively affecting the agricultural sector. At any rate, not all male farmers benefit from credit as we observed due to a number of conditions that also excluded the very poor male farmers.

A policy document stated that "if small-scale farmers are to make effective use of the improved supply of agricultural and livestock inputs, they should be provided with adequate financial resources. These resources are provided through the expansion of seasonal and long-term credit programmes. Particular emphasis is placed on the timely disbursement of seasonal credit for land preparation and for the purchase of seeds, fertilizer, and other inputs. Where possible, seasonal credit is paid in kind (given in material form, for instance, as fertilizers) rather than in cash. In the longer term, the government hoped that policies would be developed to strengthen the institutional framework with the aim of mobilizing rural savings and providing increased agricultural finance. The policy of the government was to move towards a decentralized agricultural finance system." (Kenya 1981:18).

That socio-economic progress and agricultural production increase at a faster rate than population growth is a basic condition in Kenya. Therefore, the introduction of suitable and more efficient new production techniques to increase productivity and production capacity, and the quick adoption of technical innovations by many producers, is seen as being of decisive importance for social change and economic development (Kenya 1986 and 1994). To achieve these objectives investments in labour, fertilizers, seeds, capital or land resources are needed which implies a considerable demand for credit. However, producer credits to implement agricultural innovations tend to create and strengthen social status inequalities in rural communities (Livingstone 1986, Leonard 1982, Heyer 1976 and Hansel 1974). Large and wealthy farmers, mainly men, are the first to benefit from credit and innovation programmes, whereas the poorer peasants' social position deteriorates, as evidenced in North Maragoli (see 4.9.1 and 5.12).

Institutional credit for small-scale farmers in Kenya and in North Maragoli in particular, is limited for this reason despite the official aim of providing agricultural credit to large numbers of farmers (Livingstone 1986 and Heyer 1976). This shortage of funds for small-scale farmers is all the more serious under the present conditions of structural adjustment programmes and the liberalization of the market in Kenya. These conditions have led to a high rate of inflation and the high prices of goods on the market, and these include agricultural inputs. Small-scale farmers in North Maragoli have been adversely affected because the prices of such farm inputs as fertilizers, herbicides, hybrid seeds and even labour have increased by almost 300% from what they were in 1990. This las left many farmers unable to purchase farm inputs without credit. In the same period the retail market price of a 90 kg bag of dry maize has increased from Ksh 450 to approximately Kshs 900 (200% increase) at the Maize and Produce Board. This increase of dry maize does not, however, benefit the small-scale farmers in North Maragoli since they are basically consumers.

The Government hoped that the Agricultural Finance Corporation (AFC) and Commercial banks would become important sources of credit once titles to land had been issued making it possible to secure loans by mortgages on the land. Commercial banks have indeed extended a considerable volume of credit to farmers especially in the large-scale farming sector, but they have now shown a tendency to reduce their lending. Mortgages in small-scale agriculture have not proved a very effective security, because procedures associated with such security are time consuming and costly, and in case of default it is difficult if not impossible to sell land when a court judgement has been obtained. The bad experiences which banks in Kenya have had in security loan repayments have made them very cautious about extending additional credit and have led them to reduce the maturity deadline on their credits to extremely short periods. The small size of farms in North Maragoli has been an obstacle that has prevented many small-scale farmers from qualifying for credit. Commercial banks have tended to confine their lending to people who have a regular source of non-agricultural income and who can pledge the income they regularly receive through, for instance, a bank. Small-scale farmers without regular non-agricultural employment are disadvantaged. Many farmers in North Maragoli cannot qualify for these loans. However, if a comparison is made been men and women, then more men than women qualify for this type of loans because of their considerable involvement in formal employment. Commercial banks are not particularly well equipped to assess the credit needs of a large number of small-scale farmers, or to control the disbursement of loans in such a way as to ensure that expenditure is really productive. Despite the Government's policy that it will assist the small-scale farmer by availing credit to him/her, the major programmes of the AFC and commercial banks have been and continue to be directed towards medium and large-scale farmers and companies.

A number of credit schemes have been tried in the study area, and a great proportion of them benefit men rather than women. The main body providing credit to farmers is the Agricultural Finance Corporation (AFC). The study area qualifies for both seasonal crop credit, industrial crop credit and improved livestock or zero-grazing credit. Our findings demonstrate that few farmers have benefitted from AFC loans in this area. Only a limited percentage of the total population of North Maragoli has used AFC loans at one time or another. Information from the general survey shows that only five farmers out of 180 had benefitted from AFC loans. All these farmers were men and were well-to do and in formal employment. In 1993, only one farmer in a male-headed household in the in-depth case study was found to receive credit from AFC. This farmer was a teacher (see Sections 4.9 and 5.12). Many farmers complained that the AFC, lacks information about the small-scale farmers' situations, problems, needs and aspirations in its jurisdiction. A visit to the AFC. headquarters for this region revealed that the officers had no records about the number of farmers they were dealing with.

This institution has set up very stringent conditions that have to be fulfilled before loans are disbursed to farmers. Some of these conditions are used to exclude a large number of small-scale farmers from applying and acquiring loans. Some of the conditions are: the land should be suitable for the crop or activity the loan is applied for; if possible the land should be owned by the applicant; if leased, then the lease should be through the Land Control Board for a period not less than three years; the size of the land should not be less than 5 acres (2 hectares); the applicant should have tangible security, for example land title deed, machinery log book, fixed deposit in a bank, a life insurance policy which has almost matured and which is one and a half times more than the loan being applied for, and a letter of legal undertaking accompanied by a past repayment record for those who are not new applicants. On the basis of such loan conditions, most women in the study area are not able to qualify for these credits because they lack the required collateral and they may not fulfil the conditions outlined above. The procedure is too long and too complicated for the financially poor and illiterate women who dominate the small-scale farming sector. It was observed that there was a general lack of information on credit systems since they do not fall under the portfolio of the extension service. The limit of 2 hectares for farm size disqualified many farmers in North Maragoli and many parts of Vihiga District where the average size of a farm was 0.8 hectares.

Formal credit schemes in North Maragoli seem to be typically gender-oblivious, and they are in practice gender-biased towards men since it is the male head of the household who is approached and registered for the provision of institutional credit. It is the male head of the household who is in formal employment and it is the male head of household who holds a land title deed in his name that can be used as collateral. Women in the small-scale farming sector fail to fulfil the conditions set by the lending institutions .
Some farmers have benefitted from savings, credit and co-operative societies. These co-operatives are organized by the workers themselves who contribute money monthly. Such co-

operative societies are formally organized in the formal sector and so they are more applicable to people in non-agricultural employment. Loans from these sources are only accessible to farmers who are in formal employment such as civil servants, teachers, and those working with parastatal organizations or banks. As already examined in Chapters 4 and 5, men have better education and so more men are in formal employment than women. More than 90% of the women in this area are involved in small-scale farming with no linkage to non-agricultural employment. Two females, one in a female-headed household and one in a male-headed household had benefitted from this loaning scheme because both are teachers. In our 1992 general survey, 5 out of the 60 males (8%) had benefitted from this scheme and in the case studies survey 2 men (15%) benefitted from savings, credit and co-operative societies' loans (SCCS). Informal sector co-operative associations that are favourable for women groups have very limited funds. Members usually borrow money for school fees and for consumptive goods for the household from such revolving funds (Chapter 8). Throughout the study area, approximately 70% of the women were involved in these welfare organizations. However, despite promises, the government has so far not organized any definite loan scheme for rural women and for women groups.

The other method of getting loans either in money form or 'kind' is by joining farmers co-operative societies. It is obligatory for all farmers cultivating cash crops to join co-operatives in order to get loans and also to be able to sell their produce. During the early years when cash crops such as coffee and tea had just been introduced, it was mandatory that the those who joined such co-operatives were landholders. Membership of co-operatives had been mainly open and limited to male farmers. Coffee co-operative societies are the major co-operatives in this area. Coffee as a crop is associated with many disincentives as discussed in Section 4.5.2. Many of the coffee co-operative societies have collapsed, but when they were still active, most of the members were men. In 1991, 7% of the farmers benefitted from such loans, all of them men.

The KTDA has a different approach to assisting its farmers. All farmers cultivating tea get fertilizer on credit from the KTDA annually regardless of gender as long as they are registered by the institution. However, many men in this area still insist on being registered as the tea farmers, so that they can earn the income from the crop. The KTDA has been flexible and monthly incomes are paid to women so long as the husband gives his approval. Women in female-headed households who have registered upon their husband's death may apply for fertilizers just like men, and even in female-managed households with a written notification from spouses women may apply for inputs on behalf of their husbands. Some husbands are reluctant to grant such notification being afraid that the wife may be granted the income leading to a failure to request for fertilizer in the absence of the migrant husband. A problem noticed in the field was discrimination, especially when the fertilizer delivered from Nairobi was inadequate for all the farmers. Sometimes priority was given to male farmers and especially the progressive and well-to-do farmers.

To conclude, the general narrow resource base of small-scale farmers in general limits their ability to bear the risk of innovation and credit and this contributes to the disadvantage of not being able to make full use of credit facilities. Under these conditions, the small-scale farmer, unlike his well-off colleague, finds that credit and innovation involve a greater amount of risk because he has a narrow basis of existence and this could fail. Because of this some male farmers fear becoming involved with credit.

7.5 Policies Affecting Cash Crops, Inputs and Seeds

As has been pointed out earlier, the cultivation of cash crops was introduced into many small-scale areas only after the Swynnerton Plan (1955) had gone into effect. Its main policy objective was to improve the economy of Kenya and, at the same time, bring development to the rural areas and improve households' incomes. It was envisaged that by developing cash crop farming and industries, a very wide cross-section of the African community would benefit. The growers would augment not only their own wealth but that of the district, the coffers of African District Councils and the income of the Colony from exports. The policy further stressed the need to work up the interest and enthusiasm of Africans to participate in the schemes, not only in planting the cash crops but in appreciating the needs of high standards of cultivation (farm and crop management), crop preparation and the development of co-operative marketing (Swynnerton 1955:14). The Kenya government has continued to improve on the same policy since the country continues to rely on cash crops for foreign exchange earnings and for employment. The policy lays emphasis on continued expansion and intensification of the area under cash crop farming. Indeed, it a criminal act for farmers to destroy or uproot a cash crop once it has been established (Kenya, 1963, The Agricultural Act).

> "The Government will promote primarily for export, the production of coffee, tea, pyrethrum, sisal and cut-flowers. The Government will create an enabling environment in the form of various policies (e.g credit, extension, inputs, regular payments etc) that will motivate farmers to higher production in all areas of Kenya." (Kenya 1994:115, see also Kenya 1965, 1970, 1974 and 1979).

Coffee

From 1955 - 1965, cash crops were being introduced slowly into the study area because of lack of financial resources. From 1965 - 1976, there was a very significant increase in the promotion of cash crop development, especially coffee and tea. Cash crop expansion in North Maragoli and in Kenya in general was also encouraged by the good prices paid. This was especially the case where coffee was concerned and led to the 'Coffee Boom' in 1976. Cash earnings were high, and were used for coffee expansion. More farmers started growing coffee on their farms.

Unfortunately, five years later coffee prices throughout the country began to fall after another five years (from about 1980) and the trend has continued to up to the present time, except in a few good years such as 1994/95 when there was frost problems in Brazil. Coffee payments have become unpredictable and very small, especially on tiny holdings such as those found in our study area. In North Maragoli and Vihiga District at large, the early excitement has disappeared and most coffee growers are secretly uprooting the crop in opposition to the stipulated government policy. Many coffee farmers have neglected the crop's management, and coffee plots are mixed with other crops - especially bananas.

Actually, coffee has not performed well in the area from the beginning. It was plagued with the coffee berry disease because the study area is one of the wettest regions in Kenya. After the coffee boom of 1976 other problems such as lack of inputs, poor agro-support services, and co-operative society inefficiencies as discussed under section 4.5.2, have also contributed much to the collapse of coffee growing in the area.

From the late 1980s to the present day, the Government and the World Bank have introduced five programmes related to coffee policy in order to give a new impetus to coffee growing in the country. These programmes are not being successfully implemented in North Maragoli and Vihiga District although they are known to the agricultural personnel and to the farmers. The programme can be described as follows: First, extension efforts are focused on existing coffee areas and on acreage expansion and replanting, using the new Ruiru 11 variety and other varieties under development. These varieties are estimated to double the yield produced compared to the present existing varieties, they do not require spraying against coffee berry disease or leaf rust and consequently the unit cost of production can be reduced by about half. Farmers are to be encouraged to use better crop management and gradually to replant the existing coffee area with new varieties by the year 2000. However, the trend in North Maragoli and Vihiga District is more towards reducing the farm area under coffee, quite contrary to the government programme (see 4.5.2). Second, there has been an increase in the extension-related activities of the Coffee Research Foundation, supplemented by the Ministry of Agriculture's extension service. However, extension efforts towards coffee expansion in North Maragoli appeared to be both very limited and practically non-existent. Farmers asserted that they are not ready to accept the mere promises of future benefits from coffee. They want to be assured that they will receive a subsidy in years of crop failure and low world prices. On the other hand, the extension staff seem indifferent and helpless in the face of the coffee situation in the study area. Third, there has been an expansion of the Coffee Research Foundation's Seed programme for Ruiru 11 and other high yielding varieties to encourage farmers to renew coffee plants. However, our findings show that none of the farmers in the survey are renewing coffee plants or expanding plot size. All of the farmers appeared ready to uproot the crop. Fourth, an integrated programme for the planting, production and marketing of Robusta varieties in Western Kenya and the Coast, has been developed and this involves a special research, extension and credit programme. However, most of the credit has gone into the construction of coffee processing facilities and not into agronomic coffee field improvements. Fifth, a programme has been established to reduce the delay in payments to growers with the assistance of aid donors. This is in conformance with the KTDA payment policy, where farmers receive a monthly payment and a lump sum annually. However, payments for coffee to farmers in North Maragoli and Vihiga District in general continues to be delayed and farmers receive payments only once a year despite the government promises to regularize payments.

Coffee management and production in the area have slowed down progressively and by late 1993 the main coffee processing facility in North Maragoli was no longer functioning any more. This points to the impractical, unsustainable and dysfunctional policy objectives that rely heavily on donor money. Areas with strong and rich co-operatives such as Central and Eastern Provinces, have been successful in coffee production but areas with poor co-operatives have not been able to sustain their coffee farmers during difficult years. Many of the coffee societies and co-operatives in the study area are inefficiently run and mismanaged. Most of the money for coffee expansion in Vihiga District was used for the construction of coffee processing plants with very little credit being extended to farmers to expand and renew coffee plants. Some of the coffee processing plants are 'white elephants' such as the one in North Maragoli, that had been operating at a quarter of its capacity since 1991 and eventually came to a halt in late 1993 because insufficient coffee beans were being delivered by farmers for processing and the staff who had not received salaries for more than six months were very demotivated.

Tea

According to the Swynnerton Plan "Tea will be a much more difficult crop to develop because of the high cost of factories and the difficulty of collecting green leaf from numerous scattered peasants. Nevertheless, every effort must be devoted to developing sound and practical peasant tea-growing schemes in suitable areas... Tea growing by Africans must be approached with utmost caution and I have only allowed for the development of 12,000 acres (4724 ha) in the next 15 years in selected areas" (Swynnerton 1955:18-19). Emphasis was also placed on the provision of extension and credit to enable African peasants to follow the right tea crop management in order to promote high production both for export and local consumption.

Tea growing had a slower start than coffee because of the high capital amounts involved in introducing it. There were many barriers to its introduction. Early colonial tea managers had insisted that small-scale tea was to be cultivated only on farms east of the Rift Valley (Leonard, 1991). After much political pressure tea was also introduced into some parts west of the Rift Valley. North Maragoli was one of the areas included in the western tea growing zone. The Kenya Tea Development Authority (KTDA) was given powers to expand tea production by assisting farmers with planting materials, fertilizers, tea roads, the transportation of green leaf, and the establishment of tea factories. Some farmers in the area were already growing tea by the 1960s, but tea was introduced to most farmers in the 1970s and early 1980s, and today North Maragoli is one of Kenya's established small-scale tea producing areas. Tea is a valuable cash crop here as appears from data from the general survey and the in-depth case studies. It is also an important source of income for many households. Making it possible for women to earn from tea growing is a process that has been slow to develop in many households, though women are the main suppliers of labour to this crop.

Tea prices have been rising steadily over the years, and coupled with the good organization of the KTDA many small-scale farmers are increasing the hectarage under the crop. All over North Maragoli and Vihiga District farmers are introducing tea onto their farms or are expanding their planted acreage by increasing the number of tea stems. It is a Government policy that agricultural commodities undergo primary processing prior to consumption or further processing, and therefore increased production of coffee, tea and milk for example call for expansion in primary processing capacities. Yet, it has already taken more than five years to put up the promised tea processing plant in Western Province so that tea production in this zone can be increased and improved. Tea from Western Province still has to be processed at Chebut Factory in nearby Nandi District, about 40 km away. This may exemplify the differentiation or gap that exists between policy formulation and its implementation.

Government's objective and policy is to try and triple tea exports by the year 2000 through production programmes for smallholders, and the establishment of commercial tea estates and new government-owned tea estates (Nyayo Tea Zones) on the forest fringes. The KTDA is conducting an infilling programme which it anticipates will greatly boost the yields of small-scale farmers (Kenya 1994). In the Sessional Paper No. 4 of 1986 on Economic Management for Renewed Growth and in the National Development Plan of 1994-96 it was pointed out that a set of measures were expected to increase the average smallholder yields by 1.5% a year on existing farms, and that there was considerable scope for expanding the area planted with smallholder tea on present tea farms, on other farms in tea areas, and in potential new tea areas. The aim was to increase the tea area planted by smallholders by 50% by the year 2000. The combined effects of the targeted area expansion, infilling programme and yield increase was to raise smallholder tea production in Kenya to a projected 128 million kg by the year 2000, compared to a record smallholder production in 1984 of 53 million kg. However, this target was already surpassed by end of June 1993 when a harvest of 483 million kg was produced by smallholders (see 3.4.1).

Why is it that tea growing and tea policy has been so much more successful than coffee?
It is said that organizational efficiency, strategic managerial control and effective accountability and incentive systems are the keys to KTDA's institutional success (Lamb & Muller 1982). These values have been achieved in a number of ways:- First, by the substantial autonomy permitted KTDA to carry out its development task in terms of policy translation and implementation. It controls the resources necessary to do its job, it is not subjected to substantial operational interferences, and it does not have to tax heavily, subsidize or deal with other burdens placed on it which could cripple its financial and operational performance. This organizational autonomy is supported by the external incentive of a high world market price at the moment. Second, by the quality control at the factory. The factories have the power to buy or refuse poor quality leaf. This forces the field staff and farmers to aim at producing high quality tea. Third, by relating the returns of tea growers directly to world market prices and by a regular payment system to the farmers (monthly and annual/ bonus payments) which have positively motivated many small-scale farmers. Fourth, by the close supervision of field staff by KTDA superiors; their performance is also subject to the influence of growers through representation on district tea committees and on KTDA's board. The field staff are well motivated because they are better paid than their counterparts in the field. Fifth, by the KTDA's concentration of institutional resources on activities which are important to the whole cycle of tea production and which have a technical nature, making them susceptible to efficient intervention by the Authority. Sixth, by a simple technical package which KTDA administers in the field and factory, and by the smooth coordination of the managerial and financial departments. Seventh, the authority like most rural development institutions is dealing with a large clientele of potential peasant growers, as well as its own personnel, and although responsible only for a single crop it is involved in the integrated provision of most components of the productive system line for example credit, extension, nurseries, transport, processing and marketing. A small-scale tea farmer therefore only has to deal with one organization for everything that concerns his/her crop. This results in better communication, management and production. Finally, and perhaps most important virtue that KTDA management has utilized to boost faster tea development lies in less corruption, transparency, and qualified and committed staff in all areas (see also KTDA 1971, Lamb and Muller 1982, and Leonard 1991).

Since the late 1980s three major policy measures by Government have been proposed to expand tea production in smallholder tea-growing areas supervised by KTDA. First, the Government formulated a plan to expand the smallholder tea areas. The KTDA has implemented this policy and tea hectarage is on the increase all over the country, including North Maragoli. Second, funds will be mobilized with the assistance of donors to increase the size of the first payment to smallholders. The KTDA has implemented this policy and farmers now earn more in monthly payments. Third, the domestic price of tea will gradually be increased towards export parity to encourage exports. Many of these objectives are being implemented and tea production by smallholders is on the increase.

Horticulture

Recently the Government began to give considerable encouragement to the production of horticultural crops in densely populated areas such as North Maragoli and Vihiga District. The Government has stipulated that an expansion of domestic market potential and an increase in marketing margins in favour of growers should be achieved by an improved market infrastructure - the construction of wholesale marketing centers in urban areas to relieve constraints to the expansion of urban markets and improved market organization - including a more effective utilization of co-operatives. French beans are the main horticultural crop grown in North Maragoli. This crop is privately managed. It appears to be no plan for the organization of growing and marketing vegetables and fruits (avocadoes, mangoes, quavas) through co-operatives or other government agencies as suggested in the policy guideline. In fact, home economics staff are more concerned with household food nutrition rather than the selling of horticultural products. At the local level there is no implementing officer.

To reduce costs and minimize price fluctuations, the Government has pledged to provide improved access roads, introduce new varieties of vegetables, and fruits, encourage co-operatives to promote more intensive husbandry methods, and establish an improved network of price information. These promises are still awaiting implementation because the officers on the ground do not seem sure of the Government's approach in this matter.

Government's plan included: the strengthening of the Horticultural Crops Development Authority and redefinition of its role to include the provision of technical advice to growers and to the government. Improving the collection and dissemination of information on market demand and prices. Specialized inputs to farmers to be provided and to assistance given in marketing produce through the operation of collection, storage and transport facilities. These activities were to be directed towards small-scale farmers in particular.

These blanket policy objectives on horticulture for small-scale farmers have lacked direction from the start. The few horticultural officers at the district level are not able to serve the whole district. No funds have been raised for small-scale farmers in North Maragoli. Many farmers in the area appeared to lack the right information and capital for investing in horticultural farming. Since farms are becoming smaller, more intensive and well-paying crops would be more beneficial to the farmers in this area. Women are important in horticultural production as illustrated in the case of French beans. But, so far the Government has not clearly demonstrated that it is willing to assist women who want adopt intensive adopt horticultural farming. This has been left to women groups who are themselves short of money and therefore have little to invest in the technology appropriate for horticultural farming (see Chapter 8).

Agricultural Inputs

The central objective of the Government's agricultural inputs policy is to ensure that adequate inputs are made available at the lowest possible prices at the farm-gate and that, as far as possible, they are used at the right time and in the correct quantities. This input policy is closely linked to fertilizer and hybrid maize seeds for our North Maragoli study area.

The Government stated that fertilizer policy would be focused on achieving efficient and timely importation and distribution. It is further claimed that subsidies would be used when Government financial resources permits the maintenance of profitable input-output price ratios, thereby encouraging wider usage. This would result in the distribution of fertilizer being in good time and at the lowest possible cost to the farmer and the nation's foreign exchange reserves (Kenya 1981:17). The Government's plan to sell fertilizers and other inputs to farmers at the lowest prices and at the right time would be very beneficial if implemented. Inputs as a whole have become too expensive and beyond the reach of many small-scale farmers as was observed before in our discussion of agricultural credit (see also Chapters 4 and 5). Fertilizer deliveries to distributors come late or sometimes do not arrive at all, forcing farmers to travel to Kisumu, Kakamega or Eldoret to buy them. Little effort has been made to show farmers in North Maragoli how to use fertilizers correctly.

The Seed
The main aim of the seed policy (especially hybrid maize seed) is to ensure a steady increase in the supply of improved varieties and to keep their price to the farmer at a minimum. The Kenya Seed Company (KSC) is the main supplier, while the Ministry of Agriculture participates closely in the forward planning and pricing decisions of this company. The KSC has been directed from the early 1970s to ensure that adequate supplies of seed, (particularly of maize), are available at the beginning of each crop season. To encourage farmers to use improved seeds, the Ministry of Agriculture has to ensure that their price is kept at the lowest possible level.

"The Agricultural Development Corporation (AFC) has urged Kenyans not to panic over the current hybrid maize seed shortage in the country. The personnel said that no amount of blame on those 'in authority' will bring about a solution if the issue affecting the causes of production shortfalls are not addressed. However, the personnel was unable to answer as to what would be done if 'those in authority' were part of the causes when farmers complained about the pricing systems, sub-standard, shortage and late delivery of seed, and other issues..." (Daily Nation, Tuesday, March 15, 1994).

The use of hybrid seeds is modest in North Maragoli, contributing partly to the low yields found in the area. Shortage of seeds or the late arrival of seeds and the production of poor quality seed are an annual annoyance to farmers. Seeds are very expensive and many farmers in North Maragoli are un able to buy adequate quantities for planting. The Government policy is actually in contradiction with the reality on the small-scale farms (see Sections 4.4.1 and 5.5).

7.6 Food Policy

In North Maragoli more than 50% of land and 50% of labour is devoted to food production. Even where cash cropping has been introduced on a considerable scale, farmers still tend to insist on food self-sufficiency. The principal determinant of land cropped is the size of land available to the family, the food requirements of the household, the availability of manpower and the farming implements. In all the households, women were responsible for food production. In the past, inability to produce sufficient food for one's own family and to meet eventual obligations to help ones kin was regarded as a source of shame in many homes. The aim is to produce, if at all possible, somewhat more than is needed so that, taking the bad years with the good, basic requirements can be met at all times. However, many households are not able to meet these objectives anymore as seen.

The Kenya Government understands the importance of providing for the subsistence needs of both the rural and urban populations and has made this the basis of its own national food policy. In the first two decades after Independence, there was more emphasis on cash crops than on food crops. Since then, however, there has been a deliberate effort to encourage self-sufficiency in food production within the individual families, and to improve access to basic foods where these cannot be produced within the family. Accompanying this has been the decision to promote more research (by the Kenya Agricultural Research Institute, KARI) on basic food crops. The national food policy consists of a series of guidelines for decision-making on all issues related to food production, processing, marketing, distribution and consumption in the country. The government emphasizes food security to guard against the 'bad years' characterized by food shortages (cf. 1960-61, 1965, 1979, 1984 and 1992/3). As pointed out in the Kenya Government Food Policy White Paper of 1981 and emphasized in the National Development Plan of 1994-96, 'a sound food policy is one which encourages increased production without making the plight of the poor worse'.

> "The central objective of the national food policy is to ensure that an adequate supply of nutritionally balanced foods is available in all parts of the country at all times. Given current constraints, the immediate aim of food security policy will be to obtain a calculated degree of security at the lowest cost. This will be achieved through:- (i) increasing food production in all areas of the country, (ii) emphasizing drought-resistant crops such as sorghum and millet in the dryland areas, (iii) improved monitoring and forecasting of weather conditions in the main agricultural zones, and wider dissemination of information on expected weather trends, (iv) regulation of food exports to main domestic supplies and importation of food as necessary to meet nutritional requirements; and (v) accumulation of a multi-commodity strategic food reserve from domestic surpluses and grain supplied on concessional terms to be used during periods of crop failure or other emergency situations. (vi) Availing to farmers the necessary extension, credit and inputs to enhance production." (Kenya:1981:2).

It is further stated that the government will maintain a position of broad self-sufficiency in the main food stuffs in order to enable the nation to feed itself without using scarce foreign exchange to import food. Secondly, the aim was to achieve a calculated degree of security in food supply for each area of the country. Thirdly, these foodstuffs should be distributed in such a manner that every member of the Kenyan population has a nutritionally adequate diet.
Therefore, the Government committed itself, through its national food policy, to expanding food production. This was primarily based on plans to increase yields by increased intercropping, increased multiple cropping, increases in the volume and efficiency of fertilizer, seeds and other inputs and improvement in cultural practices. Also a reduction in storage and handling losses had to be encouraged by improved extension advice and investment in on-farm storage facilities.

The national food policy lays less emphasis on the on-going food and farming systems research of the Kenya Agricultural Research Institute (KARI). The research findings from KARI are conveyed to the agricultural staff who have the responsibility to disseminate the findings to the farmers. However, information is only partially disseminated or is poorly conveyed with little or no follow-up. The poverty of small-scale farmers in North Maragoli does not allow them to grow crops under the same conditions as KARI because they lack the capital to purchase the farm inputs. The inputs are not available at the required place and time for those farmers who have money. Improved management in small-scale farming depends on proper government support to individual farmers which is not the case in the study area. This support should also be disaggregated on gender lines. It is unfortunate that most of the up-to-date information and innovations in agriculture are displayed during the Kenya agricultural shows by the Ministry of Agriculture, Livestock Development and Marketing because most small-scale farmers cannot afford the money and time to attend these shows. The agricultural shows are motivated by competition and money-making rather than by educational considerations and so the atmosphere is not conducive to learning. More than 90% of the sample farmers in North Maragoli had not attended the Kakamega Agricultural Show.

The national food policy is not geared to exploring all possible means of supporting and stimulating relevant research. Likewise, the policy does not project (predict) either future trends in food production or the resources to be committed to the production of foodstuffs. Most important, the policy ignores and does not clarify how individual farmers, especially women who are the main producers of food should be assisted to increase food production. On the issue of price policy, the policy fails to indicate ways of getting consumer prices to levels which cover the domestic prices of production, processing and distribution.

Large numbers of small-scale producers in North Maragoli and Vihiga District do not produce sufficient food to feed their own families for the whole year and have therefore become consumers relying on the market, an indication that the food policy has not helped them to meet their food production needs. On the other hand, there is a great shortage of food stuffs on the local markets at certain times of the year, an implication that the government distribution system is not so effective. Indeed, if the government is unable to distribute food countrywide then the private sector should be encouraged to do so. Officially in Kenya, the private sector is free to move food stuffs around but in the actual practice all kinds of impediments (for example, permits requirements, movement bans and roadblocks) discourage traders in food products.

"The Government is spending at least Kshs 2 billion to import 2.5 million bags of maize according to the Minister of Agriculture. A bag of maize is imported at Kshs 900 while farmers used to be paid Kshs 300 before the Ministry raised it to Kshs 375. The demand for maize increases by at least 4.7% annually. The Government is trying to boast the declining morale among small and large-scale farmers by eliminating policies that restrict internal trade while favouring a few middlemen or organizations. Up to 40 bags of maize can be transported anywhere in the country without permits. According to the Minister, the country can easily double its maize production per hectare because researchers have developed hybrids for various national climatic regions. However, better maize storage facilities, techniques of farming and incentives are needed both at the village and national levels." (Otula Owuor in the Daily Nation, Thursday, July 9, 1992).

The national food policy also pays little attention to social and cultural constraints that negatively affect agricultural productivity. In North Maragoli and Vihiga District, the food situation of poor families has become very problematic. While much else has changed, agricultural production remains gender-driven with crop labour being, to a very large extent, female. However, the policy guidelines remain very quiet on the issue of gender and agricultural production. Inter- and intra-household relationships in different communities should be taken into consideration, including how these determine the labour availability to sustain agricultural production. Formulating food policies is a complicated issue. The policy guidelines are too general and not specific about the needs of different areas. At the district level not much has been done to try and translate and implement policy guidelines. It therefore becomes quite difficult to measure the impact and results of this policy.

Not much has happened in North Maragoli and Vihiga District since the national food policy was formulated in 1981. Scarcity of food continues to be an accepted annual event in the area. Most farmers have not received the support promised in the policy documents (inputs at affordable prices, credit and extension) showing the inadequacy in its implementation. The policy envisaged to accumulate food reserves to guard against famine. Yet, since its formulation Kenya has experienced food shortages in 1984, 1992 and 1993 with many areas such as North Maragoli having very inadequate food supplies. The country has relied heavily on donor and relief food and not on the planned food reserves which may be an indication that no reserves were kept. No impetus for increased food production seemed evident in the study area as exemplified by shortage or low use of hybrid seeds, sub-standard hybrid seeds, inadequate and incorrect use of fertilizers and poor management on many farms. Food production is made worse by the labour constraints experienced by many farm households.

7.7 Livestock Development Policy

The national livestock development policy which came into effect earlier in 1980 states that : "The main objective of our future livestock development policy is to help the nation avoid any shortfalls in livestock production by the production of sufficient animal proteins to ensure adequate nutrition for our people; production of the necessary raw materials for our agro-industries; intensification in use of high potential land to ensure higher land and other resource productivity; full development of our extensive rangelands." (Kenya 1980:2, The National Livestock Development Policy). Therefore, the aim is the expansion of livestock production in order to meet the growing demand of the increasing population for livestock produce. The livestock policy is meant to contribute to three national development objectives: the reduction of poverty, the increase of foreign exchange, and the conservation of the country's natural resources (Sterkenburg 1990; Kenya 1988).
The Government's livestock development policy for mixed farming areas such as North Maragoli has been specified especially with regard to dairy production, through the adoption of zero-grazing or near zero-grazing policy. Measures were inspired by the fear of continued reduction in the area available for livestock as a consequence of the increasing competition with arable agriculture for land. In order to curb this reduction, there should be an intensified use of land in livestock husbandry, herds should be expanded, and output per animal raised. Concrete measures comprised research to identify new technologies for intensive livestock-keeping on small farms, and the setting up of an effective extension service to disseminate these new

technologies and to improve animal husbandry at the farm level (Sterkenburg 1990 and National Livestock Development Policy Paper, 1980). In addition, the cultivation of fodder on land unsuitable for food crops, the control of cattle diseases, including an obligatory dipping programme, and cheap domestic production of vaccines and medicines were all part of the approach. Policy measures also comprised a more effective functioning of artificial insemination (A.I) services and the creation of a national livestock credit programme (National Development 1994-96, Sterkenburg 1990, National Livestock Development Policy Paper 1980 and Luning mimeo, no date).

Supporting services to livestock are provided by the Department of Livestock which at one time was a separate Government Ministry as mentioned earlier. As already pointed out the Department renders two broad types of services. The veterinary division is responsible for the execution of vaccination programmes, the extension and information on livestock diseases, the treatment of sick animals and the promotion of dipping and spraying against tick-born diseases. The animal production division is responsible for the promotion of the economic importance of livestock by upgrading herds, extension services and ensuring the expansion of fodder crops (see 4.6 & 5.7).

Farmers in the North Maragoli area hardly benefit from or have very limited access to these mentioned services. The livestock policy is not specific on gender. As already discussed under Sections 4.6 and 5.7, livestock management is poor and milk production is low in many of the households. The optimum amount of 5 litres from the zebu, 12 to 18 litres from the cross-breed and 30 litres from the grade cow per day that are attained on many farms in other small-scale farming areas such as Kiambu, Kirinyaga, Kisii, Nyeri, Murang'a for example are a dream in this area (Kenya 1992). A number of explanations can be advanced for the meager output. The main explanation lies in the fact that policy neglects women and there is a lack of commitment in policy.

The following reasons were given by farmers and household heads as part of the explanation for the dismal performance of livestock production:- (a) many farm households are now managed by women, who provide all the labour required in the various sub-systems. Women are therefore constrained by lack of time and are therefore unable to manage livestock optimally; (b) many households headed and managed by women are impoverished and so there is little investment in the improvement of livestock husbandry; (c) the neglect of livestock in many farm households is influenced by the problem of ownership and incomes from livestock sale. Livestock are men's property and despite women managing this sub-system the income that accrues from the sale of livestock belongs to men; (d) the Department of Livestock insists that cattle should be dipped once a week. Field analyses show that this service is rarely utilized by many farm households because (i) there is no provision for the enforcement of this policy and therefore many farmers neglect it, (ii) zebu cattle are considered to be resistant to tick-born diseases, (iii) farmers are discouraged by the distances they have to walk from their farms to reach the dips, (iv) the rigid schedule of one dipping per week is disliked by many farmers. Some claimed that the amount charged which varied from 1.50 to 3.00 shillings per animal per dipping per week, four times a month and forty-eight times a year was too high; (v) farmers lack confidence in the effectiveness of dipping and believed the quality of the dips was affected by shortages of drugs or underdosing; and (vi) since many households are run by women, they simply ignored taking cattle for dipping because the community regards it to be a man's responsibility. (e) the artificial insemination (A.I) services which were supposed to improve and expand cattle have collapsed due to cut-backs in government budgetary measures and the freezing of international funding in this sector. Before its total collapse, many farmers had already given up using the AI service. They had frequently brought their cattle when in heat to the meeting points, only to find that artificial insemination personnel sometimes did not show up or the farmers had to wait for long hours. Other farmers claimed that insemination had to be repeated many times before conception occurred, which reduced the farmers' confidence in this service (see Chapter 4 and 5); (f) The extension service has not been effective. Using the T&V approach only a few male farmers had been reached.

Until recently, livestock improvement in North Maragoli and Vihiga District has not been given much attention because the high population densities of the area left little land for grazing. However, in the last decade improvement in livestock farming, especially the adoption of zero-grazing which is considered an adequate way of intensifying land use in North Maragoli and other small-scale farming areas. The general approach has been to introduce zero-grazing on individual holdings. Zero-grazing is becoming increasingly popular because it provides an important additional source of cash to the farm families, and especially for the women. From a farm management point of view, it is attractive for the woman to look after the dairy animals on the farm because they supply an important source of milk for home consumption as well as for sale. About 10% of the farmers, including women group members in the sample population, were found to have adopted zero-grazing. If the right group is targeted policies may succeed, as is demonstrated by the successful zero-grazing projects of women groups targeted by NGOs.

7.8 Recent Changes in Agricultural Policy

In early 1993, the Government decontrolled the prices of maize and wheat seeds, fertilizers and other chemicals and further liberalized grain marketing in a major shift in food policy. This occurred immediately after negotiations with the World Bank and the International Monetary Fund (IMF) to implement the Structural Adjustment Programmes which Kenya had been resisting to since the mid 1980s. The implementation of the Structural Adjustment Programmes led to prices of inputs sky-rocketing and costs of living becoming very high, contrary to stated government policy guidelines.

In addition, inflation - both domestically generated through fiscal and monetary policy and imported through the exchange rate - has dramatically changed the cost of agricultural production. A glance at the prices of some of the inputs clarifies this further. The price of a 10 kg bag of seed maize was sold for Kshs 120 in 1992 and Kshs 500 in 1993, an increase of about 400%. The same bag was sold for about Kshs 50 in the 1980s. As for hybrid seeds the prices per 25 kg bag were respectively Kshs 480 in 1992 and Kshs 1,150 in 1993, an increase of 240%. The same bag was sold at Kshs 120 in the 1980s. Fertilizer increased by 257% in price from Kshs 350 per 50 kg bag in 1992 to Kshs 1,250 in 1993 (KGGCU and KNTC price lists 1994). The prices of inputs have continued to increase regardless of the depreciation or the appreciation of the Kenya shilling. Small-scale farmers, especially poor women in North Maragoli and other rural areas are not in a position to purchase inputs at such high prices and this explains the continuing poor management and low production in the area.

It is observed that in some markets liberalization creates competition which sometimes leads to stabilization and, to some extent, a reduction in price of a few items. The Kenyan market is different because it lacks enough competition and businesses are in the hands of a small number of people who therefore determine the fate of the larger population. Prices thus increase always. Indeed the high prices of farm inputs are contrary to policy guidelines that promise that inputs will be sold to farmers at the most minimal and affordable price. This shows the disharmony between the stated policies and the reality, and how difficult it is to implement stated policies when relying on the interactions and influences of the global market.

The Agricultural Finance Corporation (AFC) is currently just as badly off as the Kenya Grain Growers Co-operative Union (KGGCU) as the financiers of the agricultural sector and distributors of inputs. These two organizations are riddled with corruption and are inefficient in the provision of services to farmers. The AFC's reserves have run down and its lending ability has been seriously cut back by non-repaid loans. Its decentralization policy of constructing new branches at district level to move closer to the farmers and its increase in its personnel have not been successful strategies. Some branches have some money to lend to farmers while other branches have nothing. This confirms that there are regional differences when it comes to resource distribution. Sometimes, these differentiations are created when certain areas are targeted for development and government spending while others are neglected or told to wait until funds become available. Areas which do not enjoy such services like North Maragoli suffer the negative consequences, including adverse effects on agricultural production, especially food production.

On the other hand, commercial banks are as wary of lending to farmers as farmers are of borrowing from banks that charge interest rates of between 24-30%. The potential suppliers of credit to farmers are short of liquidity. Thus, the credit promised in policy guidelines has not become available to many farmers.

The Ministry of Agriculture, Livestock Development and Marketing has directed that the National Cereals and Produce Board (NCPB) should first purchase 10 million bags of maize for reserve/stock before farmers can be allowed to sell their produce elsewhere. This directive is unworkable as long as the details of the proposed modalities of maize marketing are not worked out. Uncertainty in agricultural markets, particularly in a resource-poor developing country like Kenya, tends to reduce production. The NCPB buys produce at a much cheaper price than the selling price on the open market and expects to sell at a slightly higher price. Farmers are harassed by the NCPB on the delivery of their maize for example that the moisture content in the maize is too high, that the bags are old, that the maize is not clean or that the maize is not well selected. The result is that farmers are sent back with their produce. This is expensive because most farmers rely on public transport. While the majority of farmers are harassed, the rich and well-connected farmers easily deliver their maize and earn their money. The poor record and the delayed payments from the NCPB have discouraged many farmers to sell their produce to this body and dissuaded some farmers from growing maize for sale altogether. Many farmers require their money immediately after the delivery of the produce to enable them to reinvest and take care of family needs. Some farmers, therefore, prefer to sell their produce to private traders who are willing to pay on the spot although sometimes at a much lower price. The Ministry has tried to control private traders through movement permits, police roadblocks and the provincial administration, but has not been effective. On the other hand, such controls beat the logic when the market is supposed to be liberalized and make questionable the honesty in set policies, objectives and programmes.

The Ministry has been disturbingly silent on fertilizer pricing and supply in spite of the shortages being felt nationwide and sky-rocketing prices. Once in a while, fertilizer donations are received from foreign countries. However, donor fertilizer is sold on the market at the same price as imported fertilizer.

Though the government has announced economic liberalization and privatization, there is still much government intervention. The process has been slow and largely closed to public scrutiny. The producer prices and market prices of many agricultural products and inputs are still set and controlled, negatively affecting free market competition. The government still holds an upper hand in deciding the prices of maize, sugar cane, and milk. Such interventions sometimes hurt the producers, leading to disincentives of the selling price turns out to be lower than the cost of production. In fact, the opening up of the market and competition associated with liberalization is still to be realized in Kenya. 'Liberalization' rather appears to have sparked off the spiralling of agricultural input prices and to have lowered the standard of living of many Kenyans.

7.9 Policy and Gender

As pointed out by Leonard (1991), government policies are the most important way in which a State affects both social and economic development. Development specialists are in substantial agreement that more efficient methods of promoting agricultural production today in Africa, and Kenya in particular, include righting distorted prices, devaluing inflated currencies, and generally decreasing the extent to which the state is extracting resources from the farming sector of the economy. Other experts suggest that macro-economic policies should be formulated and implemented in a way that supports agriculture, because massive nominal devaluations that are not supported by substantial external assistance to relieve the consequent import constraint, can starve agriculture of vital inputs while provoking high rates of inflation (FAO, 1992). The emphasis is often not so much on improving the operations of the State as on keeping its overall policies congenial to agriculture and on finding ways to decrease the negative role it has often played in the past. These are the most important issues in public management because they are among the most critical variables affecting the success and sustainability of agricultural development (Leonard 1991).

Other concerned researchers state that some expansion of production in the high-potential areas should be possible with the conversion of idle lands, currently held for speculation, into productive areas through the distribution of such lands to small cultivators or by the imposition of a land tax (FAO 1992). This policy guideline was already recommended by the ILO (JASPA Report of 1972) and in other studies by Livingstone (1980, 1986) and Hunt (1984, 1985) but has not been considered a policy measure as yet. The same documents recommend a reduction of the disparity between the poor and the rich, equal distribution of land and other resources and the consideration of women in such policy matters as a way of improving the rural agricultural areas. It is also thought that extending policy reforms to the traditional export crop sector, for example the elimination of delays in payments of agricultural products, is necessary in order to improve incentives, expand input use and speed up the adoption of improved varieties (FAO, 1992).

However, most agricultural sector policies in developing countries such as Kenya are oblivious to gender or are gender-blind as has been evidenced in this study. Women are invisible to agricultural policy-makers. It appears that policies that ignore gender and specifically the role of women in agriculture often have important but negative impacts on farm management and overall agricultural production and, worse, on the lives and livelihoods of women and many rural farm-households. Moreover, such policies sometimes fail to achieve their objectives due to the fact that they neglect of the role of women in the processes that they aim to influence. While in the previous chapters we have made an attempt to expose the important multiple roles of women in the small-scale farming sector of Kenya, conceptual awareness by the Government of the issues of gender and development has not resulted in its translation into adequate policy formulation, planning and, most crucial, implementation. Some of the reasons for this are (a) most authorities responsible for development planning have only reluctantly recognized gender as an important planning issue; decision-making powers continue to remain not only male-dominated but also gender blind in orientation, (b) the primary concern of much recent feminist writing has been to highlight the complexities of gender divisions in specific socio-economic contexts, rather than to show how such complexities can be simplified so that methodological tools may be developed enabling practitioners to translate gender awareness into practice and (c) for those involved in planning practice, it has proven remarkably difficult to insert gender into existing planning disciplines (Moser 1989).

In the official statements regarding development and agricultural production, society and households are often treated as an undifferentiated whole. Thus planning and drawing policies on this basis distorts the reality. Another tendency of planners, as already discussed above, is to look at the household as a nuclear family of husband, wife and children and as an aggregated unit of consumption and production. It is not recognized that within the household there is a clear gender division of labour in which the man of the family as 'breadwinner' is primarily involved in production work outside the home, while the woman as the housewife and 'homemaker' takes overall responsibility for the production, reproductive and domestic work involved in the organization and feeding of the household. Implicit in this view is also the assumption that within the household there is an equal control over resources and the power of decision-making between the man and the woman in matters affecting the household's livelihood. Such common oversights not only disregard how policies have a differential impact on various sectors of society and the household, they also overlook the needs and contributions of members in the household, and frequently ignore some of the most important forms of market and non-market activity and expenditure in households which determine the farming system in the small-scale farming areas of Kenya. Policy recommendations often do not take into account the different roles played by household members in the struggle to survive. Women and children are always lumped together and portrayed as the 'weak' and 'vulnerable' groups with respect to health, education, and various social services. However, it is precisely these women who so often keep their families afloat during unprecedented hardships (Aili Mari Tripp, in Beneria et al. 1992, Moser, 1989). Because the burden of supporting the family rests no small degree on women, this makes women especially important, not only as recipients of social services, but also as potential beneficiaries of agricultural extension services, credit, training, and technical assistance in policies geared to improving the small-scale farming sector. In this study, it has been demonstrated in this study that in this respect women's marginalization negatively affects farm management, resulting in low agricultural production as evidenced in many of the farms studied.

The importance of considering the roles of women as paramount when formulating and implementing agricultural policies is not simply because there is concern for the subservient status of women and their marginalization - economically and socially. It is especially explained by the fact that neglect of gender (specifically women) sometimes results in policy outcomes that fall short of policy intentions in the sense that the policies pursued may reduce rather than increase the living standards of those designed to assist (Kandiyoti 1990). In North Maragoli it was found that agricultural production is low in all the farming sub-systems, partly because the policies lack a clear focus in their implementation. Neither do they incorporate of women despite the fact that the Government realizes the important role women play. Many farm-households were of the contention that their agricultural production was declining annually, their food-self sufficiency was a myth, their purchasing power had been eroded, and their standard of life was much lower than ten years ago or in the 1960s. The vexing question that calls for discussion is how can policy guidelines for the rural small-scale farmers be implemented to benefit all members of the society and are such policies workable in the first place?

7.10 Conclusions

The overall problems and the declining performance of the small-scale farming sector to service and provide for Kenya's needs is at least partly a result of unclear or poor policy making, policy inconsistency and instability, an absence of implementation, and the neglect of the role of women in policy. The policies are well written for office shelves but little is done to put them in practice. It would appear that policy success depends on good organization, transparency, accountability and unity among those involved in its translation and implementation. It also seems that arrogance and a 'don't care' attitude by those concerned with its formulation and implementation may contribute to policy failure. The Government's heavy handedness and suspicion results in failure of implementing policies. We can deduce from our discussion on policy that there is in fact an urgent need for new policies to replace some of the outdated and dictatorial ones (for example, the ban on uprooting coffee even when the returns to the farmer are zero) being used to oppress farmers in various parts of the country. We have seen that important issues choking the agricultural sector are still largely ignored or side-lined contributing to the low morale and poor production of many crops. An open question for future discussion is- "What are the prime tasks of the government in agriculture and which ones should be left to private sector?"

It is possible that the lack of success of government efforts to stimulate agricultural development in the past is not so much due to too much intervention but to doing the wrong things, such as the

T&V agricultural extension approach for example where the right group is not targeted or recognized by policy (viz. the female small-scale farmers). However, it is important to avoid interventions which, to be effective, would require considerably more technical and managerial capacity and financial resources than Kenya can afford. Also to be avoided are those policies, which, even if effective involve the risk of distorting markets.

Food problems and cash crop production problems may be a result of poor policy formulation, implementation and the neglect of women in policies designed for the small-scale agricultural sector. Therefore, as we move closer to the twenty-first century, Kenyans should reconsider their agricultural policy again if they hope their farms to be well-managed, food self-sufficiency is to be maintained at the household level and if the country is to be able to feed itself and to expand the export crop production.

The country's agricultural production both for food and for some cash crops (coffee) has deteriorated since 1990 (see FAO 1992 and Kenya Development Plan 1994). The country is now over ten years into its adopted national food policy and livestock development policy programmes and it is obvious that the production of maize and other crops as well as milk production has not increased as planned for in 1980 and 1981. What is clear is that the production of maize in the country is low, and that maize has had to be imported between 1991 and 1994. It is also clear that the acreage under coffee is declining in many areas of Kenya.
All these problems at the national level are keenly felt at district and location levels, such as Vihiga District and North Maragoli Location. This has become abundantly clear in our preceding analysis of the impact of selected government policies and programmes on farm management and agricultural production in relation to gender in our study area.

Notes

1 "A man slashed and killed his mother following a disagreement over money accruing from the sale of tea. The man (name withheld) went home in Mago Village of Sabatia division, and demanded that his mother give him some of the money she had received from the sale of the crop." (Daily Nation, December 8, 1994).

2 Another problem which must be faced when analyzing the effectiveness of extension services is the complex organizational, economic and ecological setting which impinges upon smallholders (Farrington 1994, Moris 1991 and Leonard 1977 and 1982). No one profession or academic discipline encompasses all the major components. Those who see extension as technology transfer will focus on communicating information about new varieties and husbandry innovations to farmers; those who deal with community problems will look at local leadership and how individual farmers participate in their communities; while those who analyze environmental trends will adopt a geographical and zonal approach. The most important organizational group are the farmers and their workers (household members and others) who are in direct contact with crops, livestock and trees, and who react to dynamic ecological factors (Moris 1995). Extension to a large extent does not focus on these people.

3 This limitation of the T&V approach has also been observed in other regions and countries where the approach has been tried, especially in many parts of Asia (see Antholt 1994, Antholt 1991, Nayman 1988 and Moris 1988).

4 Ignorance and lack of understanding of policy guidelines was portrayed by some of the implementing officers (e.g lack of accessibility to the documents or misinterpretation of policy). A few officers proclaimed that the documents are locked up in the higher offices and sometimes labelled 'Government Confidential Documents'. At other times, the single copy of the policy document is to be shared by a number of officers in different offices, at the risk of getting lost in the process of circulation. The few officers who may be interested are frustrated because they lack the resources, facilities and support to implement the policies. Sometimes the policy documents are kept at the headquarters and the officers in the field are not aware of the new changes or may not know how to translate and implement written policies.

Female Farmer Plucking Tea

8

Women's Groups as a Strategy for Agricultural and Rural Development

8.1 Introduction

Development projects, government policies and support services appear to exclude many women and this may partly explain the impoverishment of female-headed and female-managed households in the rural areas. The Government of Kenya has tried many strategies for rural development, including the decentralization of a number of sectors (AFC, Agriculture, KGGCU, KNTC), the district focus for a rural development approach and special rural development projects aimed at improving agricultural production and the general living standard of the people in the rural areas. But neither of these approaches and projects have explicitly incorporated women in their development objectives. Most often women are neither a target group nor recipients of rural development projects and agricultural programmes. This neglect of women has been pointed out in the previous chapters.

Recently, both the government and the NGOs have been hard pressed to find a formula for directing development to women in the rural areas. One of the methods suggested in one of the forums concerned with orientating development to women and overcoming some of the unfavourable characteristics of the situation in which rural women are living is through women groups and other collective organizations with strong networking among the women concerned. It is thought that women group organizations geared to income-generating activities could serve as successful coping strategies for women as they might improve the standard of living of women in rural small-scale areas by generating money and enabling women to invest in agricultural production. It is also hoped that the women's group movements can be used as a strategy for rural development generally and that the government may be able to implement agricultural extension services and credit facilities with more effect through women's groups.

Some writers argue that the existence of such coping strategies may be used as an excuse for policy inaction. Women should be enabled to move from coping strategies to transformation strategies - i.e strategies that help to transform gender relations by meeting both women's basic and strategic gender needs (Beneria et al. 1992).

The community self-help groups and women organizations that women construct as coping strategies might provide a springboard for the transformation of rural life if they are redirected toward women's strategic and pragmatic gender needs. This may be achieved if, besides the organization of collective consumption, questions are raised about the way the contribution of women's unpaid labour is undervalued and how resources within the family are highly unequal in their distribution. Gerhart (1974) emphasizes the importance of moving towards equality in the control of resources between men and women. This will enable women to execute their duties more efficiently as household managers and make them less dependent on men. Any improvement for women should begin from the immediate needs that women themselves express, and the interconnections between those needs and strategic gender needs should be taken into account. Poor women's needs for more resources cannot be adequately campaigned for in terms of abstract demands for equality. But there is a connection between greater resource availability for the poor women and greater equality between women and men and between greater resource availability for poor women and the empowerment of women (Beneria et al. 1992). For many poor women in the rural areas empowerment and gender equality are not a priority because their main concerns are access to resources and a better standard of living. Indeed women would lead better lives and they may be able to improve the farming system if they had the same ownership rights and access to resources and institutional support as men.

This chapter attempts to give an analysis of the women groups movement as a coping strategy for women, as a strategy for the improvement of agricultural production and as a strategy for overall rural development. The chapter briefly describes and examines the debate and history of women groups, their objectives, conditionalities of membership, activities, and aspirations and problems of women groups. In this chapter we shall try to answer our final research question : What is the role of women groups in the development and the improvement of the living standards of rural households and in increasing of agricultural production?

It is important to find out whether the role of women is more recognized in women groups than in the individual households. To come to grips with these aspects of women in the rural areas, it is important to analyze different groups, activities undertaken and limitations that impede the performance of these groups. An important reason for devoting this chapter to the theme of women groups is to reveal the achievements, roles and constraints facing women groups in rural development.

8.2 A Review of the Debate on Women's Organizations

The 1970s witnessed two major events that have affected women studies all over the world and particularly in the Third World Countries. These were the changes that took place in theory and practice of economic development and the United Nations' designation of the Women's Decade, which started in 1975 with the International Women's Year. Realizing that capital and technology transferred from the industrialized countries had not filtered down to the poor in the developing countries and had not spread from modern enclaves to traditional sectors in these countries, development agencies designed new strategies to directly improve the standards of living of the poor. In his address to the World Bank Board of Governors in 1973, Robert McNamara, then president of the Bank, made explicit the need to redirect the investments made by development agencies from those that focused purely on economic growth to those that would also attempt to reduce poverty. International and national relief agencies soon followed suit and initiated their transformation into private development foundations.

The worlds of relief and economic growth grew less apart in the late 1970s and 1980s. In the

1990s, the World Bank has invested heavily in social sectors and took the lead in research on basic human needs, while the private voluntary organizations have professionalised their staffs and implemented small-scale programmes in economic development. These agencies have been slow, however, to shift their orientation from relief to development in their work with women. While women have been a focus of economic development research because of their important role in meeting the family's basic needs, in development action projects concerned with the empowerment of women and research control have lagged behind. In part, it is argued that this can be attributed to the institutional legacy from the approach adopted by the world of relief and social welfare for women (Buvinic, 1986).

The International Women's Year brought to the world attention the concerns of women in industrialized and developing countries, assigned legitimacy to work on women's issues in economic development, and enticed a small but critical budget allocations from the international development agencies to undertake work on the subject. Since then it became appropriate and a selling slogan for development agencies to include in their anti-poverty portfolios projects intended to improve the situation of poor women in developing countries. Some women's organizations that had been in existence since the 1950s, started to revise their aims and were in a position to implement these new projects. However, these organizations were better equipped to implement relief and welfare rather than productive projects for women. The disproportionate growth of 'women and development' agencies, and the conviction that women-only institutions have to expand and do what they know how to do best helps, to explain the welfare orientation of projects for women in the Third World. A major recommendation to governments from the World Bank and the International Monetary Fund (IMF) was development for women and this explains the call for women's organizations or self-help groups in many countries and why these groups have been promised financial support from governments and international organizations.

In Kenya, self-help has led not only to community projects but also to the emergence of a variety of groups oriented towards development, among them women groups. Two types of approaches in the literature on contemporary Kenyan women's organizations may be identified (Sorensen 1992): (1) a functional approach which basically focuses on how women's organizations are used by women, and (2) a systematic approach which concentrates on the underlying conditions for women's organizations and incorporates a historical dimension into the study of women groups. A considerable number of authors have followed the functional approach, the majority in a rather descriptive manner (cf. Mwaniki, 1986; Riria-Ouko, 1985; Wipper, 1982). A few analyses within this category are, however, more analytical (Nelson, 1978; Thomas 1988). Thomas's study of Kikuyu women groups shows that there are differences among women based on their economic resource base, educational standard, marital status, and sometimes religion. Such differences are reflected in their organizations. The roles women groups play and the activities they undertake vary to some extent according to the resource base of the community and the level of affluence of the household. Although membership includes all socio-economic strata in the community, women's organizations are particularly attractive and useful to women in low-income households (Thomas 1988). Therefore, through providing some income-generating and investment opportunities, and offering ways to save and providing other kinds of support, they help to address the needs of poor and single women managing their households under difficult circumstances. According to Thomas, participation in women's organizations is related primarily to women's practical needs in parts of rural Kenya.

The systematic approach is mainly represented by Stamp's (1986) study of Kikuyu women's self-help groups in Mitero, Central Province. She is among the very few who consider both the linkages between former and present types of organization and the effects of changing gender relations on participation in women's organizations. In contrast to Thomas's and other functional analyses of women's organizations, Stamp's study looks at the broader organization of society as providing the basis for women's organizations. Stamp's major aim is to contribute toward a better understanding of the relation between the sex-gender system and the mode of production in Africa. In her analysis she integrates the two concepts 'articulated modes of production' and 'sex-gender systems'. The communal mode of production and the bride-wealth sex-gender system which characterized pre-capitalist Kikuyu society and gender relations are seen as a shaping force in the non-exploitative relations of production. Stamp goes on to analyze the contemporary form of women's self-help groups in Mitero and sees them as successors to Kikuyu women's customary age-grade organizations. She further argues that women's organizations are the source of the most radical consciousness to be found in the Kenyan countryside and views women's organizations as radical and vital organizations in resisting exploitation (1986).

According to Sorensen (1992) however, Stamp has now distanced herself from earlier generalizations, and expresses concern about the lack of attention to differences among women by scholars dealing with the issue of women's organizations. On the basis of her own recent research in Mitero she demonstrates how members of successful women's organizations exploit other women and receive a disproportionate share of the group savings and possibly divert resources intended for community development projects to their own use (1989).

Gladwin (1991) states that organizations of women have a long history on the African continent and have been important in a variety of contexts. However, existing groups have taken on new functions and new types of organizations have come into existence in recent years. Their increasing prominence is, in part, a result of the current economic crisis and structural adjustment policies. Historically, women's organizations in Africa were organized primarily at the local level around a variety of concerns and for a variety of purposes. Some were basically social groups or clubs, others were organized for the purpose of generating savings and providing access to credit (economic groups); others were religious groups and professional groups, especially in the urban areas. A number of scholars state that many of the women organizations were and are typically small and focus on parochial concerns of interest to their members alone. Few are incorporated in any formal sense. Even in those areas where the co-operative movement has played a role, women's organizations have rarely been registered as co-operatives (Sorrenson 1992, Gladwin 1991, Thomas 1988, Stamp 1986 and Were 1985).

The traditional mutual aid system provides firm underpinnings for the activities of women's groups. Traditionally, women have worked together in small groups of two, three, or four sharing tasks according to the agricultural season. Thomas (1985) states that there is a qualitative and a quantitative jump to the kinds of endeavors in which women's groups are involved today, but the basis for mutual cooperation and mutual enterprise can be found in a social tradition. She explains that the existence of women's groups is also supported by the firm dichotomy between male and female solidarity. Women's roles have traditionally involved agricultural labour and related tasks, as well as home and child care. Women's groups are building upon the customary responsibilities and duties of women. The groups do not lure women into completely new and hitherto unrelated interests and concerns. In other words, women groups have not transformed the gender-relations and the perception of women in many rural areas.

A study by Buvinic (1986) in both Latin America and Africa discloses that projects for women and women groups fail in many Third World Countries. The reason is that while policies that incorporate the results of research and stress the importance of economic programmes for women are often firmly in place, a welfare orientation has prevailed in the execution of projects for low-income women in many countries. She claims that many male planners are frequently reluctant or even downright hostile to implementing projects with an economic orientation for women, and this has been used to explain the prevalence of welfare action for women. Such welfare projects are designed to deliver information, education and sometimes free handouts (money, food, technology) to poor women in their role of homemakers, reproducers and child rearers. Examples are projects in material and child health, hygiene, nutrition, home economics and home-based appropriate technologies. Income-generation designs, in contrast, involve teaching a new skill or upgrading the income-generating skills women already have and providing some of the resources needed to use these skills in the production of marketable goods and services. Therefore, many women's projects fail because: (1) emphasis is given to the welfare aspect in the execution of women's projects as seen in specific project characteristics that are shared by a majority of interventions; (2) administrative concerns negatively interfere with the execution of production objectives; and (3) implementation is geared towards social and not economic goals. The underlying reason for this is that agencies which implement women's projects have more expertise in the welfare sector than in the technical and economic aspects of development programmes for women. Another reason is an institutional preference for welfare action because of the lower financial and social costs involved in the execution of welfare versus economic-oriented policies for women in developing countries (Buvinic, 1986). A typical and popular women's project is small-scale, situation-specific, and uses limited financial and technical resources. It is implemented by women, many of whom are volunteers with little technical expertise, and it benefits only women. The intended target group is poor women, but project tasks usually require beneficiaries to volunteer their time and labour, which tends to exclude those with heavy demands on their time and the poorest women who cannot afford to make these investments.

8.3 A History of Women's Organizations in Kenya

The origin of women's organizations in Kenya is very varied. A detailed analysis is given by Were (1985). The mobilization of women in Kenya can be traced as far back as the women's councils of the Kikuyu in the nineteenth century. During the first half of the twentieth century women in areas like Vihiga, Kakamega, Kiambu, Murang'a and Machakos formed mutual assistance groups. Neighbours and relatives helped each other in birth, disease and death. In the late forties women's clubs appeared as a formalized structure and were among the first to be organized in 1951 under the National Women's Organization 'Maendeleo ya Wanawake' (Progress for Women) (Maas, 1991). Maendeleo ya Wanawake which is currently the largest women's organization or association in Kenya was set up by a small group of European women in the early 1950s to promote the advancement of African women and to raise African living standards. Maendeleo was organized following a philanthropic model where white, middle-class women volunteers provided assistance to rural women's clubs in welfare-oriented matters such as teaching home management, child-care, sewing and knitting, embroidery, music for example. Maendeleo ya Wanawake was therefore not concerned with promoting and advocating for national issues such as the education of women, employment of women, women representation in the legislation or political system etc. It was parochial in approach, concerned with making women good housewives without providing the economic base for a good standard of living. Maendeleo ya Wanawake began early on to Africanise its staff but the particular volunteer, welfare and relief legacies remained. Maendeleo ya Wanawake coordinates women's clubs around the country and in 1984 had a membership about of 40,000 women (Buvinic, 1986, Were 1985). Maendeleo ya Wanawake is now affiliated to the ruling party and it has been given a bigger role in co-ordinating most of the women's groups projects in Kenya. In recent years it has expanded its scope of work to include the implementation of income-generating projects using government funds and contributions from international donors.

Apart from Maendeleo ya Wanawake, there are many other women organizations. Some began during the colonial regime as a means of educating women, but have now 'taken root', and become independent and strong African movements. These include the Ismalia Women Association (1926), the National Nurses Association (1958), the Nairobi Business Women (1955), East African Women's League (1917) and the Home Economics Association of East Africa (1958) to name but a few. Some others have emerged as branches of world movements such as Y.W.C.A (1912), and the Kenya Girls' Guides (1920). The churches also initiated some women's organizations, for example, the Mothers' Union (1955) started by the (Anglican) Church of the Province of Kenya.

After Independence the few that there were professional women quickly realized that political independence in Kenya did not translate into economic and social improvement, especially for their immediate families and themselves. This realization led to a change of focus in the organizations formed during the colonial era to tally with the aims of organizations formed after Independence, such as the Kenya Association of University Women (1965), Mfangano Women Groups (1973), Nyeri Women's Association (1974), Breast-feeding Information Group (1973), and the Kenya Women Finance Trust (1981) for example. Many women's organizations now realize that they have political, social and economic roles in the country. However, Maendeleo ya Wanawake continues to be at the apex of women's organization with about 6500 women's groups affiliated to it in 1985 throughout the whole country (Were 1985). The current figure is larger as more groups have been formed during and since the late 1980s and early 1990s.

In the rural setting, the groups extended their activities to saving groups and working parties, mostly in Nyeri and Murang'a Districts of Central Province. They became known as 'women's groups'. The members worked as agricultural labourers and saved their earnings to buy corrugated iron sheets for roofing materials for the houses of group members. The basis for the rapid expansion of these groups is the self-help (Harambee) ideology formulated shortly before and after Independence (Maas, 1986 and 1991).

A major expansion of these groups occurred around 1970-1975, due to the stimulating efforts of the Central Government. The Women's Bureau was established by the Kenya government in 1976, in response to the decision made at the UN World Conference for International Women's Year, held in Mexico in 1975 which suggested that governments of the world and especially in the Third World, should create a government machinery to deal with issues related to women's problems. Women's organizations and women groups have become part of the government planning apparatus. However, a disturbing phenomenon of women's organizations in Kenya is that there seems to be no aggregation in their approach to women's development issues.

The Kenya Development Plan 1994-96 states that one of most significant developments in the efforts of rural women to take affairs into their own hands has been the formation of self-help and income-generating groups. Recent surveys reveal that there are 23,614 women groups in the country, both registered and unregistered, with a membership of 968,941 (5% of the total Kenyan population). Out of these groups 1% were formed in the 1950s, about 4% in the 1960s while over 80% of the groups were formed during the last two decades. The major activities undertaken by the groups in the rural areas are in agricultural and livestock production, sales and services, and handcrafts and social welfare. Out of the 23,614 women groups enumerated, 15,839 are engaged in these four groups of activities. These activities, in fact, account for 67% of the total range of activities the groups are engaged in. Table 8.1 shows the number of women groups that received government support between 1988-1992.

Given the large number of women groups in Kenya, only a handful have been assisted financially by the Government through the Ministry of Culture and Social Services. The amounts involved

Table 8.1
Number of Women Groups Assisted by the Kenya Government (source: Kenya (1994) Kenya National Development Plan 1994-1996).

Province	No of Groups Assisted					Amount Granted Kshs 000				
	1988	1989	1990	1991	1992	1988	1989	1990	1991	1992
Nairobi	10	10	4	7	1	61	61	150	62	10
Coast	60	60	4	10	8	226	238	200	90	40
Eastern	60	60	52	9	6	201	213	420	75	25
N.Eastern	30	30	30	6	4	93	99	460	30	40
Central	50	50	45	16	9	200	210	500	160	36
R.Valley	130	130	36	8	5	468	494	500	50	30
Nyanza	40	40	40	5	3	154	162	600	25	15
Western	30	30	32	6	2	118	124	450	30	10
Total	410	410	243	67	381	521	1601	3280	522	206

have been minimal and have decreased over the years. It is also clear that the crisis of government finance in 1991 and 1992 has had a definite impact on the support for women groups.

In another parallel development, the Kenya Government, within the framework of the 'Harambee' ideology, recognized the collective effort by women's groups and organizations as an integral part of the entire Harambee movement. Consequently, the Department of Social Services in the Ministry of Culture and Social Services was charged with the duty of co-ordinating and controlling all self-help women activities in Kenya.

Women's group development can be differentiated. Brown (1990) claims that there are certain women's groups that act as 'magnets' and that project bias operates whereby assistance is attracted by a small number of favoured groups. These are treated as 'showpiece' projects (Chambers 1983). Groups from certain sub-locations or locations (administrative units) are excluded by development agencies because such agencies may not have access to them for example and because the groups may not have influential political patronage.

8.4 Formation and Objectives of Women Groups in North Maragoli

Formation
Due to their multiplicity of duties, women in the traditional setting of North Maragoli formed groups based on the clan system, age groups or women married within the same family to assist one another with farm work. This assistance wad given on a rotational basis. These groups extended their functions to planting, weeding, harvesting, storing grain and other activities such as collecting water, plastering the walls of houses and serving during celebrations such as weddings, funerals, diseases and 'thanks giving' ceremonies.

As has been pointed out earlier, the social system in this area was completely disrupted by colonization of Kenya. Most men were recruited to work on the large farms and by implication women had even more work to do than before. Therefore, the function of women groups was strengthened as women organized themselves to help one another to work on their holdings and to support each other with mutual and social assistance. At the same time, the colonial government introduced 'women leagues and clubs'. Most of these women leagues were affiliated to the different Christian denominations but the main dominant church in North Maragoli was 'The Friends Church' or the Quakers, supported by sponsors from the U.S.A and other protestant churches. Later on, with the formation of Maendeleo ya Wanawake, the women's groups affiliated to it in addition to their affiliation to the religious groups.

Unfortunately, at their meetings with the sponsors, the women's role as home keepers and wives was enhanced through the type of training given. The women were taught the art of spinning, knitting, cookery and embroidery. This reinforced the ideology that a woman's place was in the home and dissuaded them from venturing into wage employment.

With the attainment of Independence, there was an increase in male migration to the urban areas to search for remunerative jobs. In most of the families in this area education was reserved for sons and this further stimulated the migration of males to urban areas. After Independence, the government did not address itself immediately to the issue of girls education. This has meant that women remained in the rural areas to looking after the farm and the children. Tradition also dictates that a good woman stays in the rural areas to perform her duties as housewife and farmer. But in trying to execute these duties properly, women encounter many problems. In the previous chapters we have already described the main problems facing the majority of rural women. Therefore, it suffices to summarize these problems as follows: (a) weak economic base; (b) labour constraints because the amount of work for women in small-scale farming areas is enormous both in reproduction, production and domestic work; (c) inaccessibility to resources by the majority of women; (d) low educational levels leading to poor agronomic practices, especially where exotic crops are concerned, when the right management methods are not conveyed to women; (e) little support from spouses in the up-bringing of children which includes, feeding children, dressing them, making sure that they go to school, providing school requirements and caring for their health; (f) poor financial support from men or remittances that are insignificant, infrequent or targeted for specific use, especially school fees payment; (g) absence of any clear government policy as to how working men should support families financially and otherwise, and the lack of social welfare facilities or other economic support which would enable women to function independently in the rural areas; (h) culture of having large families or children of a particular sex (especially boys) which wears women out; (i) limited property rights for women - land being inherited by men in accordance with the Maragoli (Luhya) traditions - which further erodes the bargaining power of women as far as the control and use of resources are concerned and degrades them to mere unpaid labourers in cash crop farming on their husbands' land; (j) absence of a clear policy to incorporate rural women into the socio-economic and political structures of the country; and (k) lack of security and sense of belonging experienced by many rural women perpetuated by the cultural and traditional norms. The rural woman is dependent and sometimes desperate due to lack of economic resources of her own and the overwhelming nature of her roles and this makes her vulnerable and fearful of the authority of men and of those perceived to be of higher status in terms of education, employment and government responsibilities. This situation can be easily exploited by men to their own advantage. One of the reasons for the formation of women groups is to try and overcome some of these problems.

The majority of the groups investigated during this study started in the 1980s and early 1990s. Most of the groups were established in response to government initiatives. The women groups in Maragoli can be categorized along the following lines: (1) members of the same church; (2) women residing in the same area, either a sub-location or location; (3) women with common interests and goals (rural teachers, nurses, market women); (4) women married within the same clan; and (5) women who are related. It also appears that women choose each other according to age. Many members in a group appeared to be within the same age cohort. Education may be a criterion for group formation in the future. It was noticeable that young wives with at least the basic primary school education were already joining together to form groups. Marital status seemed to be a condition for joining women groups, more than 95% of group members were married women.

Objectives of the Women Groups in North Maragoli

It was observed above that the formation of women groups, especially groups with welfare objectives is not a new phenomenon in North Maragoli. The aspect that is new for many women in this area is the formation of income-generating groups and the formalization of groups through registration. The encouragement given by the government through the Ministry of Culture and Social Services and local administrators such as chiefs has had a great influence on the number of groups that have been formed in this area. A stimulating factor was the allegation that the government was going to distribute money and materials only to those women who were members of registered groups, who possessed a bank account, and had an on-going project. Many women joined together haphazardly to form groups. The main objectives of these groups are:- (1) to better the lives and living standard of women and their households in the rural areas; (2) to solicit government and NGOs for financial assistance, and to enable members to apply for materials, grants and loans from the government, NGOs and other funding organizations; (3) to contribute money and start income-generating activities to generate money to be a source of income for the members; (4) to enable members to form savings and credit societies (revolving funds) to generate money which members can borrow from when the household is in financial need; (5) to reactivate the welfare aspect of the women's groups to provide labour to members on a rotating basis or at a subsidized price in the different farming sub-systems; (6) to educate one another about different methods of farming; (7) to educate one another in health care and to invite officers from the Ministry of Health to give instructions on family planning methods, child care and good feeding methods; (8) to help illiterate members to learn how to read and write; (9) to enable members to have forums where they can air their views and frustrations and get advice from fellow members, especially the older members; and (10) to assist one another in all aspects of life in the rural areas.

Conditions of Membership to Women Groups in North Maragoli
According to the Ministry of Culture and Social Services, for a group to qualify for registration it must have a minimum number of 10 persons. When an inventory of the self-help groups was taken in North Maragoli in April 1992, 41 groups were found to have been officially registered by the Department of Social Services. By the time the actual interviews took place between September and December 1992 only 37 groups were still operational. These had a total membership of 1,156 persons, representing 20% of all the households in North Maragoli (see Appendix 9). Of these 37 groups, 33 are women's groups, 2 are youth self-help groups and 2 are male groups. We shall concentrate our analysis on the 33 women's groups.

Membership ranged from 15 to 60 persons per group. Total membership for the 33 groups was 1,086 women, an indication that 1,086 households were participating in this movement. North Maragoli had a total of approximately 5,900 households. Therefore, officially registered formal women groups represented 18% of all the rural households in this area. Apart from the formal women groups, there are also some informal women groups. The informal women groups have fewer members and members are normally in the same age cohort. Unfortunately, we did not take an inventory of these groups although a large number of the younger women belong to these informal groups. If these informal groups were to register, it is possible that close to 40% of the rural households would be represented. This would give a strong basis for using women groups as one of the approaches to rural development. There was another category of women who not belong to any group movement. This group comprised of the very old women and those who cited financial constraints as a barrier to joining women groups.

All the women groups have a formal leadership structure. Members choose a committee consisting of a chairperson, a secretary, a treasurer, an assistant to each official, and three ordinary members. These officials are responsible for the organization and functioning of the groups. To become a member of a women group one pays an entry fee, and the group continues to ask for contributions either on a monthly basis or occasionally for specific purposes. The membership fee ranges from Kshs 10 to Kshs 200 depending on the economic base of the members. Members are also required to contribute money towards accepted group projects, apart from the continuous monthly contribution, and this ranges from Kshs 20 to Kshs 100 depending on the group's aspirations and the ability of the members to contribute. Some groups encourage members to purchase shares. These different ways of raising money assist the groups to accumulate money. This initial capital is utilized by groups to start small projects and they hope to get funding to expand such small projects later. This phenomena will be discussed below. The same money facilitates the registration of groups with the Ministry of Culture and Social Services and the opening of bank accounts which are among the conditions for formal recognition. Aside from the contributions mentioned above, there are also rules which the members abide by. As a general rule the women groups aim at realizing some kind of project, for example a poultry farm, a zero-grazing farm, or a grain mill.

Something that was found to be characteristic, unique and interesting in most of the women groups was that they enlisted few male members who were seen as patrons or advisors, and who ran some of the projects which were physically strenuous and traditionally male jobs in this area such as brick-making, bee-keeping, fish farming and timber-sawing.

8.5 Activities and/or Projects of Women Groups in North Maragoli

Many of the members (at least 60%) of women groups are either illiterate or only have basic primary school education. A large number of the women have grown up working on the land and most of their knowledge concerns food production. It is not surprising then to notice that many of the income-generating activities/projects and welfare/service activities are mainly concentrated in the area of agricultural production. A number of women groups start with agricultural-related activities since they have very little knowledge of the other sectors and also a large sum of money may be required for more enterprising sectors (see Table 8.2).

The level of education, age of members and interaction with urban life may determine some of the women activities. Elderly women, who have always lived in the village, have very limited scope when it comes to types of women's activities and they are more comfortable with traditional agriculture-oriented activities or simpler activities such as pottery, group farming, petty business for example. On the other hand, women who are literate and have lived in the urban areas will conceptualize of projects that are supposed to be more income-generating, for instance tailoring, renting houses and grinding mills. For discussion purposes women's activities in North Maragoli are classified into the following types:- (1) Agricultural-related activities- (a) Crop farming- (b) Livestock farming and (c) Tree production; (2) Secondary activities; and (3) Tertiary activities (see Table 8.2). Each type has a number of different activities. Many of the women groups are involved in more than one activity (see Annex 9), with a few activities basically being undertaken by all the groups.

Table 8.2

The Most Prevalent Activities among Women Groups in North Maragoli (source: North Maragoli Women Group Survey 1992).

Classification	Activities and/or Projects	No of Groups	Percent (%)
Primary Sector Activities	Group Farming	33	100
Crop Production	Horticulture	20	61
	Organic Farming	1	3
Livestock Production	Zero-Grazing (Dairy Farming)	6	18
	Poultry Keeping	6	18
	Bee Keeping	4	12
	Rabbit Rearing	3	9
	Fish Farming	1	3
	Pig Keeping	1	3
Tree Production	Tree Nursery	10	30
	Agro-forestry	1	3
Secondary Sector Activities	Handicraft	20	61
	Brick-making	7	21
	Timber Making	4	12
	Tailoring	4	12
	Pottery	2	6
Tertiary Sector Activities	Petty Business	28	85
	Rotating Saving & Credit Association	24	73
	Choir	6	18
	Plot and Rental Houses	4	12
	Making/Selling Energy Saving Stoves	2	6
	Adult Education & Nursery School	2	6
	Hotel and Butchery Business	1	3
	Traditional Midwifery	1	3
	Clinic	1	3
	Hotel and Butchery Business	1	3

8.5.1 Crop Production and Income-Generation

The main activities in crop production are group farming, horticulture and organic farming. Land as a resource is essential for all these activities. There are five ways in which women groups acquire the land they use:- (a) members using part of their kitchen gardens land for the group, (b) land on church compounds, (c) a few husbands of women group members allocate some land,

(d) purchase of small land plots by some women groups, and (e) renting of land in the neighbourhood of the women groups.

Group Farming

All the 33 groups claimed to be involved in group farming and, in fact, many women's groups start off with group farming before diversifying into other areas. Group farming is the practice of members of a group executing some agricultural production tasks together such as planting, weeding and plucking tea, and rotating from one member's farm to another. Some groups also work for money in their neighbourhood. Despite the fact that farm holdings in North Maragoli and Vihiga District are small, labour requirements on these farm holdings are quite large as the small holdings are sub-divided into many small parcels planted with tea, French beans, bananas, maize, coffee, beans, sweet potatoes, cassava, and trees. Labour for many of these crops is supplied mainly by women who are the main permanent residents in many households. During peak labour seasons, especially between mid-February and early May when planting and weeding of the main food crops takes place, coupled with the weeding and picking of tea and the picking of French beans, women find themselves with much farm work and with shortage of labour at household level. The problem is aggravated by women's reproductive and domestic chores within the household.

Some groups are formed to raise labour collectively because this is a fast and cheap method. Labour is contributed on a rotational basis. Members within the group who do not participate in group farming because of old age or illness are free to hire and pay for the group these services at a subsidized price (Kshs 100 per day).

Many of the women's groups are not keen to provide labour for money because of time constraints. Group farming is slowing down and the number of members is shrinking. Collective action appeared to be deteriorating and many members were defecting from providing labour which is the main essence of such groups. The failure of group farming is pegged to the timeliness needed in agricultural activities. Food crops are planted at the same time and grow at the same pace, and so weeding must also be carried out at practically the same time. Rotating from one farm to another takes time and those whose farms are reached last suffer the consequences of late weeding. Members prefer to weed at the right time and those who have money employ hired labour instead of delaying weeding until the group reaches them which may result in delays and in a poor harvest. In fact, the opportunity cost of farming is higher within the functioning of group farming.

Horticulture

Horticulture is the cultivation of fruits and vegetables. In other words horticulture can be referred to as gardening. Vegetables are the main crops grown by the women groups. Vegetable growing is not a new event in Western Kenya and many parts of Kenya because vegetables are used as the main sauce served with food in most homes. Vegetables and fruits have been grown traditionally for many years. However, new methods of growing and managing vegetables, and controlling diseases and pests in horticultural crops have been introduced, and this is especially relevant if groups are cultivating vegetables on an intensive scale for both the market and home consumption. The most common vegetables in North Maragoli are traditional vegetables (these include omurere, likhuvi, lizeveve, tsisaga, lisutsia, elivogoi, and emiro) and others such as kales (sukuma wiki), cabbages, tomatoes, onions; the fruits are mainly bananas, mangoes, guavas, and avocados.

The growing of horticultural crops is popular among women groups as it is an extension of what they do in their kitchen gardens. In North Maragoli, 20 women's groups (61%) of the 33 women's groups interviewed practice horticultural farming to generate income and to improve the family food supply. Groups buy seeds and distribute these to their members. Vegetables are sold on the local market by different group members and the money is deposited in group accounts. Some groups supply vegetables to local institutions. The demand and market for vegetables in this area is very high. One of the groups was buying vegetables from Nandi District and selling these on the local market.

Only two of the twenty groups growing horticultural crops had been visited by agricultural extension officers. The officers instructed the groups on vegetable management and these included the selection of certified seeds, planting in the seed-beds, transplanting, weeding, the use of fertilizers, and disease and pest control. They also discussed the sale of the crops. The two groups that had received agricultural extension support were the most successful in horticultural farming and earned between Kshs 300 to Kshs 500 during the long rains from vegetables while the others earned about Kshs 100 during the same period.

Horticultural farming appears to be a viable project, taking into consideration the small size of farms in North Maragoli. Also soils and climate are conducive to this type of farming. The majority of groups concentrated on cultivating mostly traditional vegetables rather than experimenting with more paying horticultural enterprises such as tomatoes, onions, French beans and fruits. In fact, the initial capital required for horticultural farming is not as high as needed for other types of group projects. It is a pity that 80% of Kenya's vegetable oil needs are imported at a staggering price of Kshs 1.2 billion each year (Kenya 1994 and The Daily Nation, Thursday, July 9, 1992). The irony is that the list of crops which can be used to produce vegetable oil is endless and includes sunflowers, groundnuts, simsim, macadamia nut, rapeseed, safflower, soyabeans, castor bean and others. Many of these crops could be easily grown by women groups if financing, processing and marketing were organized.

To illustrate that horticultural pays well when the right crops are planted and well managed, two examples of independent women who grow tomatoes and cabbages for commercial purposes are discussed here. Neither of these women had to buy land, rather they have used part of the family land for the enterprise. The initial capital invested in 1992 was approximately Kshs 5,000 for the purchase of inputs and labour. Crops were cultivated on a plot of 0.4 hectares. Both women followed the correct management procedure and they consulted agricultural extension officers privately. Two harvests were achieved in 1992. Each of these women earned about Kshs 50,000 from the two harvests. This income is higher than what most households make from tea in North Maragoli.

Many women groups in North Maragoli choose the simplest and most traditional activities because they are afraid of venturing into new activities. Thus the insistence on growing traditional vegetables, cabbages and sukuma wiki. Actually, the groups do not discuss cost-benefit analysis of projects they undertake. Women seem to lack the relevant skills for adequate horticultural management, and they do not have sufficient capital for investment and group land. Many groups appeared ignorant of the importance of the timeliness of activities such as the preparation of seed-beds and planting, the spacing of crops, and the proper use of fertilizer (some groups applied NPK which is used on tea and because it is very acidic resulting in very poor yields).

Many groups grow vegetables mainly during the wet season. This is the time when other farmers also grow precisely the same crops, and therefore an oversupply is created on the local market, thus depressing prices. Vegetables are more marketable during the dry season and it would be beneficial to these women groups if they had water to practice garden irrigation in order to grow vegetables through-out the year. Women groups in Siaya (Bondo women group), South Nyanza and Kisii which are sponsored by NGOs and receive close supervision in the management of the different horticultural crops, perform much better than the groups in North Maragoli where technical knowledge and financial support is limited.

Vegetables are planted on family plots by many women groups members. Problems have been experienced by some women's groups in this regard if husbands object to a small part of their land being converted to group use by wives. A number of cases have arisen where husbands have sold out the crops (especially cabbages, tomatoes, onions and sukuma wiki) and confiscated the money. Groups cannot contest this because the land belongs to the man and no contracts have been signed between groups and land owners.

The incidence of pests and diseases are quite common on many vegetable plots due to poor management skills. Pesticides are rarely used because knowledge of pesticides and pesticide control appeared to be very poor. The limited use of pesticides is also explained by the fact that pesticides are very expensive.

The vegetable plots are neither well prepared or managed. Weeding is done poorly. The same plots are used over and over again for the same horticultural crops, and this leads to the exhaustion of fertility and to poor harvests. Many horticultural crops require spraying and/or dusting to kill pests but this was only being practiced by one group.

The role of women in horticultural production in North Maragoli in particular and in Vihiga District in general cannot be over-emphasized. About 90% of the activities related to horticultural production are carried out by women and youths. The contribution of men to this activity is on a limited scale and this is one of the factors that has contributed to low commercialization of this activity in this area. Men as the main decision makers and owners of the holdings do not give much weight and assistance to these crops and hence vegetables entirely remain 'women's crops' as it is termed in this area. Some men do not allow their wives to use the family holdings to grow horticultural crops on a commercial basis. A case in point is the French beans project. A number of women alluded that they would prefer to grow the crop on a slightly larger plot but their husbands do not allow them any more land.

Women's groups lack funds to invest in horticulture farming and the marketing of products outside the local market where they could make more money. Horticulture could be an alternative in such areas as North Maragoli, Vihiga District and other densely populated districts in Kenya. Farmers may earn more from these crops, be able to buy food for their families and other necessities as compared to maize if optimal management and economic efficiency was taken into consideration at every stage of cultivation.

Women groups members claimed that if supported both technically and financially, they would play a bigger role as promoters and participants. Horticulture requires much commitment, dedication, hard work and an open out-look as far as production, demand/supply and marketing are concerned. Women are known to have these qualities and, backed with proper information, there is no doubt that they could play a bigger role in the management and development of this industry. However, the horticultural industry can only improve if Government gives women the right support: (a) information on horticultural management as this pertains to seed selection, fertilizer use, timeliness in planting, weeding and harvesting and marketing; (b) control of pests and diseases; (c) accessibility to credit and other capital resources; (d) mutual sharing in land use between men and women; (e) better strategies for pricing and marketing products; (f) irrigation possibilities during the dry season; and (f) high quality seeds.

Organic Farming
This is the type of farming where crops are grown without the use of artificial chemical fertilizers and pesticides. It does not degrade the soils, in fact, it helps the soil to build more fertility naturally. Also it is claimed that products from this type of farming are more healthy.
Only one group - the Avugwi women group - practiced organic farming. The group received a grant of Kshs 25,000 from the Ministry of Culture and Social Services. Part of the money was used to purchase a plot. In the initial stages, the agricultural extension staff visited the group and provided it with seeds, fertilizers, pesticides and sprays. At the same time, the group was given instructions on the management of different crops especially horticultural crops. About the same time a course was organized by the Ministry of Agriculture and one of the NGOs on organic farming at Manoa House, Kitale. This group was given a chance to send two members for three months training in organic farming. The group recruited two form four school leavers, children of two members of the group. The two youngsters came back from this training as instructors in organic farming. The two youths embarked on a programme of training all the members of this group using a local church plot for farming demonstrations and manure from individual members' cow-sheds.

The small piece of land at the church compound was divided into different plots. The plots were prepared through double digging whereby the top soil is removed and kept aside. Secondly, the sub-soil is removed and kept aside. The two layers of soils are mixed and placed on the different plots. This brings the soil nutrients close to the surface for the benefit of crops. Farmyard or compost manure is added to the prepared plots and they are left for two weeks so that the soil can break down the manure. After two weeks the soil is mixed with manure and is assumed to be ready for planting crops or vegetables. Vegetables are planted in different styles, some are broadcast and others are planted in rows. This helps the women assess the highest yielding method.

Members have been trained in the making of green, farmyard and compost manure. They prepare one to three holes in which they throw maize stalks, green waste and left over garbage from the kitchen, and add a lot of cow-dung and other wastes from cattle. This mixture is covered with soil, and during the dry season water is added. It is turned and mixed after one month and again left to break down for a further three weeks. It takes about two months for members to produce manure for planting different crops. Some of the members' holdings were visited and indeed their crops looked healthy and definitely gave a better yield compared to other members in the locality who were not practicing organic farming. Organic farming is cheap to produce, although the materials for the production of manure are limited in North Maragoli.

This group also makes pesticides from local materials to spray their vegetables. One method encouraged was dusting vegetables with the use of ash which is collected and kept after every family cooking. The second method is buying or collecting hot pepper from members' farms. This is chopped into tiny pieces. These pieces are put in water and bowled to produce a pepper solution. The solution is sieved to remove the unwanted pepper covers. The solution is cooled and put in a spraying pump ready for spraying vegetables. The third method is to soak tobacco leaves in water for one week, to sieve the solution and then spray it on cabbages and other vegetables. Women also spread some local flowers and onions on their vegetable gardens. The smell from the flowers is thought to dispel pests and therefore, pests fly off or are strangulated by the smell.

The Avugwi women group also makes local ammonia for top dressing maize. They collect cattle waste (cow-dung and urine). This waste is put in a sack which is tied at the top. The sack is immersed in a barrel full of water. This is left standing for three weeks after which the water is squeezed from the sack into the barrel and this solution is used for spraying on the maize.

The group was doing very well. It earned between Kshs 3,000 to Kshs 5,000 on every harvest of vegetables, especially during the wet season. The group also makes vegetable oil from groundnuts, sunflowers and simsim for its members and for sale. The group had diversified into other projects such as the establishment of tree nurseries and zero-grazing which were also quite successful as discussed in the relevant sections. A number of members have benefitted because they now practice organic farming on individual holdings which is a healthier and cheaper method of farming. Families of the women belonging to this group have an adequate supply of vegetables. One of the youngsters who trained as an organic farmer instructor has been employed by an NGO to train women groups in organic farming in Migori and the other has been employed by the Ministry of Agriculture. The two are grateful to this group for having given them a chance to have this training which has led to their present employment and so they come whenever they have time to encourage and instruct the women on new techniques in organic farming. This system of farming is suitable for all women groups since they do not have a source of income to purchase artificial fertilizers which are also environmentally degrading to both soils and humans. The group is succeeding because of the commitment and unity of its members. It also has a strong, dedicated and uncorrupted leadership.

8.5.2 Livestock Production and Income-Generation

North Maragoli is no longer viable for the large herds of livestock that were kept in pre-colonial and colonial times when land was still abundant. Farm families keep an average of two animals as already mentioned. The most important factors that farm families and women groups should take into consideration in regard to the livestock farming sub-system are high milk production and less space utilization. The zero-grazing method has been advocated by the Ministry of Agriculture, Livestock Development and Marketing particularly for those areas with increasing population and shrinking farm sizes, such as North Maragoli, Vihiga and Kisii Districts. Zero-grazing has also been recommended for women groups in order to increase milk for the family, especially for children and to generate an income for women.

Women Groups and the Zero-Grazing Project
At present any agricultural practice in North Maragoli should be intensive and economically efficient if it is to satisfy the needs of the rural households. On many farms it was noted that one or two cows were kept on average for milk production. Zero-grazing is more practical in this area because it does not require large pieces of land, and is economically more productive if the right management is provided. Milk is in great demand in many households.

Many women groups wish to venture into zero-grazing because it is a well paying enterprise. The initial capital for starting or investing in a zero-grazing project is quite high and therefore deters many groups. An initial minimum capital of Kshs 28,000 is required by any group that desired to embark on this project as of 1992: Kshs 18,000 for the purchase of a grade cow and Kshs 10,000 for the establishment of a zero-grazing unit and napier grass. Six groups (18%) are practicing zero-grazing. The groups are Chanderema, Evojo, Mago, Mwenyelitsi, Vokoli and Wekhonye women groups. Members of Evojo, Mago, Vokoli and Wekhonye women groups contributed between Kshs 200 to 300 per member towards the purchase of the cross-breed cows. These four groups have not constructed the approved zero-grazing units due to financial constraints and members were not willing to contribute more. The cross-breed animal of each group is managed by one of the members of the group. It is kept in a small enclosure without the appropriate amount of space for movement. The four groups do not buy the concentrates for the animals but feed them on maize stalks and napier grass. Because of these poor feeding procedures and inadequate food supplies, milk production was low. Instead of the animals producing the standard 12 to 18 litres a day for a cross-breed cow (improved cattle), only 6 to 8 litres was produced. Instead of these groups earning between Kshs 3,500 and Kshs 4,500 per month, they were only making about Kshs 1,920 and Kshs 2,600 a month in 1992. This is a low income taking into consideration the fact that the group has to purchase napier grass, and buy disinfectants and drugs for the animal as well as providing labour.

The other two groups, Chanderema and Mwenyelitsi women's groups had grants from different organizations for their zero grazing projects. Mwenyelitsi received a grant of Kshs 10,000 from the Ministry of Culture and Social Services. With the money two cross-breed cows were purchased for Kshs 5,000 each in 1987. Both animals were kept by one of the officials. It was expected that the sale of milk would bring members a steady source of income, and when the animal calved, other members would receive the heifers. The management of these animals was very poor. The animals often suffered from inadequate food and water and they were seldom disinfected. Both animals contracted 'East Coast Fever' and died. Initially, the official keeping the cows treated them as personal property and was enjoying the total income from the sale of milk. When there were complaints from group members, the animals were purposely neglected and in the end nobody benefitted.

Chanderema women's group received a loan of Kshs 20,000 from an NGO known as Partnership for Productivity based in Kakamega. The group also received a grant from the United Nations Development Fund (U.N.D.P) through an NGO affiliated to GTZ, called 'Africa 2000 Network'. The Kshs 20,000 loan was distributed to members to buy materials to build zero-grazing units. Each member received Kshs 800. The money from U.N.D.P was also distributed among the members and each received Kshs 2,000. The cost of a good zero-grazing unit was Kshs 6,000 in 1991/92. Each member received Kshs 2,800 and so an additional amount (Kshs 3,200) was required per member to complete the construction of the zero-grazing units.

From its inception in 1977, the group had encouraged members to make monthly contributions. The money contributed by the group members had been banked on a group account and each member had accumulated up to Kshs 3,700. Each member obtained Kshs 3,500 from the group's savings to complete the zero-grazing units and to establish napier grass in readiness for the animals.

The group recruited more members and from an initial membership of 25 the group grew to 35. Staff from 'Africa 2000 Network' in-conjunction with staff from the Department of Livestock Development visited all the women to certify that the zero-grazing units had been completed as recommended and that 0.4 hectares of napier grass had been established. 'Africa 2000 Network' purchased and donated 35 Frisians and Ayshires dairy animals from a farm in Nakuru to Chanderema women's group between August 1992 and February 1993. Each member of the group received one cow. The officers from 'Africa 2000 Network' visited the women each month. The NGO has continued its role of supervising and managing the project with help from livestock officers. The group is encouraged to enlist more members so that more families in the vicinity can benefit from this project. This group had an enrolment of 60 members at the end of 1993. The first beneficiaries of the project were obliged to donate their first heifer to a member of the group who had not benefitted. By July 1993 the group had 15 heifers for distribution to other group members.

Because of close supervision, the women have been kept on their toes to ensure that the animals received correct and adequate management and supplies. All the 35 dairy animals had calved and each was producing about 600 litres (800 bottles) of milk a month. Part of the milk is used for home consumption, 45 litres (60 bottles) per month is reserved for the calf, and the rest is for sale. Each woman was earning approximately Kshs 3,500 per month from the sale of milk. Each member contributes Kshs 210 per month to the group savings. The money is saved for emergency purposes such as members having to borrow money to buy drugs and feeds for the animal.

Members of this group now have a steady monthly income of Kshs 2,500 from the sale of milk after deducting contributions to the group and other miscellaneous requirements. Households of women that belong to this group now have enough milk for children to drink and therefore poor nutrition will soon be eradicated amongst these families. This project indicates that women can contribute a great deal to development. However, they need sustained institutional support, sponsorship, technical knowledge, good supervision, encouragement and understanding from the qualified officers.

Though this project seems quite successful, a cost-benefit analysis shows otherwise: the costs in time spent and the miscellaneous expenses by the personnel from 'Africa 2000 Network' and the

Department of Livestock outweigh the benefits at the moment. This is in addition to the free labour supplied by the women. Maybe in the long term the benefits will outweigh the costs after the project has been in operation for many years.

Poultry Keeping
Apart from zero-grazing, poultry keeping is another farming practice that does not require much space and so is suitable for many densely populated areas. Six of the groups (18%) visited had ventured into poultry keeping. These groups are Gavudia, Givudimbuli, Lumavo, Masigolo, Mukuyu and Vokoli. Three groups were keeping broilers and the other three layers.

Poultry is an expensive undertaking and all these groups started with an initial capital of between Kshs 10,000 and Kshs.15,000. The women bought one-day old chicks from suppliers who were mainly based in Kisumu and Nakuru. The cost of one-day old chicks were as follows:- Broilers Kshs 22 per chick and egg pullet day-old chicks (layers) were Kshs 28 each. Gavudia and Masigolo started with 500 chicks and the other four groups with 250 chicks. Two to three members were appointed from each group to manage the chicks. The other members were expected to provide labour and to transport feeds from the buying centres. Once the layers started laying eggs, members were expected to assist in selling the eggs. For those with broilers, all members were involved when it came to the sale of hens when mature. Members are encouraged to learn the management of poultry keeping. Cleanliness is a must and the chicks must be provided with the right feeds. They should also be vaccinated against a number of diseases at different stages. Some members appeared keen to learn because the plan was that poultry should be kept on a rotational basis.

Poultry projects have not been very successful for the following reasons.
(a) The initial cost of starting poultry farming is high for many groups. Each member of each group contributed a minimum of Kshs 250 to enable the project to take-off. In-between members were also required to contribute money towards the buying of feeds and vaccines. This proved quite costly for many members and has prevented more groups from going into poultry keeping. Some members of these groups avoided paying and this has led to the chickens being inadequately fed. This affected production capacity and eventually culminated in the failure of this project among some groups. No profit was realized in any of the groups and in some cases the money invested in the project was not recovered.
(b) Day-old chicks are not always available. The demand for these chicks is higher than the supply. Farmers who want chicks often have to wait from between six months and a year. Such long delays result in some groups giving up altogether. It was claimed that this problem is caused by the institutional denial of licenses to those interested in private breeding.
(c) Feeds are not readily available. On many occasions the chicken feeds are not in stock at the local market centres such as Mudete and Sabatia, which are a long distance and away from women's groups situated in the interior. The women have to make long journeys to either Kisumu, Nakuru or Eldoret with mixed chances of getting feeds or missing them. Chicken feeds are a big problem for people who are involved in poultry keeping all over Kenya. Sometimes the companies manufacturing the feeds produce low quality feeds which affect the health of the hens and the production of eggs. When the feeds are available at the local centres, the price is inhibiting. The situation has deteriorated due to liberalization, and the de-controlling of prices

and the wrong attitude coupled with poor management of some manufacturing companies. Chicken feeds were sold at between Kshs 650 and Kshs 1250 per 50 kilogram bag as of June 1993. Yet, the same bag sold at Kshs 350 as late as November 1992 before prices were decontrolled. The price is therefore very high for rural women and small-scale farmers keeping chicken on a small scale. In fact our analysis revealed that at the end of the project, it was an uneconomic activity because an egg only sells at Kshs 3.50 and a grown grade chicken sells at Kshs 80 to Kshs 100 while the investment may be twice as much.

(d) The chicks and the chicken are vulnerable to many diseases such as Newcastle Disease, Gumboro, Marek, and Hitcher B. An acute problem is shortage, unavailability and the expensive vaccines and drugs. Lumavo and Givudimbuli women's groups lost all their birds because of diseases. The other groups have also suffered large losses.

(e) The women complained of lack of relevant knowledge about exotic poultry keeping. Ninety percent of the women interviewed claimed that livestock extension agents rarely take the initiative to instruct women groups on poultry keeping unless the group is funded by an NGO and staff from the Ministry are seconded to the project. On many occasions women have to take the initiative to seek help and information from the extension staff. The women's groups involved in poultry keeping often are not able to diagnose a problem early enough and so they only call livestock staff when the birds are in a critical condition. This leads to many losses.

(f) Poultry management was poor in almost all the groups. The exotic birds were being reared within the family's residential houses on cold floors, instead of constructing the recommended structures outside the main house. Only the Gavudia women group had constructed the recommended poultry structure. Financial constraint was the main reason impeding these groups from building the right structures for poultry. The sharing of the residential house with poultry posed a health hazard to the birds. Children and other household members walk in and out of the room where the birds are kept without disinfecting their feet and hands. Smoke produced from the daily cooking and noise disturbs the birds. Grade birds appeared to be easily contaminated by the local birds. The birds were not fed on schedule, the containers for feeding the birds and the sleeping area were not kept clean.

(g) The groups keeping poultry are situated far away from the market and urban centres. Marketing of the eggs and chicken appeared difficult and expensive. The groups seemed ignorant of the strategies for getting tenders for their products. Many rural people prefer local birds to the grade birds. Local populations also rarely encourage their families to eat eggs and eggs are considered mainly for hatching or selling. Therefore, due to the high cost of feeds, poor management, and poor marketing strategies, the women groups that ventured into this project have not made any profits and this has been quite discouraging for other groups. However, Gavudia women's group has not given up and they have invited an extension officer to instruct them on all the relevant techniques before they decide to buy more broilers.

Many groups expressed a desire for poultry keeping but financial and labour constraints, and poor knowledge on management prevented them. It is quite clear from our discussion that the economic situation in the rural areas in connection with poultry keeping is very depressing for this type of venture. On the other hand, poultry keeping is assumed to be simple and income-generating. A short survey was carried out in Kakamega and Eldoret using three professional working women (an agriculturalist, a co-operative officer and a teacher) who keep poultry as a part-time business. With an initial capital of Kshs 25,000 in 1993 ((a) Purchase of off-cuts and construction of shed Kshs 4,000; (b) Purchase of pullets (one day old chicks Kshs 6,000; Buying of feeds the first three months Kshs 12,000 and labour, medication and professional fees Kshs 3,000 totalling Kshs 25,000), these women built small raised chicken sheds measuring 10 by 15 m. behind their main living houses and each reared approximately 200 layers. These women used services from livestock staff to the extent of the chicken being checked once a month and provided with the required medication. By the end of the year these women had each made an approximate profit of Kshs 40,000 from the sale of eggs and the final sale of the layers. This demonstrates that poultry can be income-generating under controlled and good management. Thus correct knowledge, availability of support services and inputs and finance are very crucial. This implies that women in the rural areas living under the present difficult financial conditions may not be successful in poultry keeping.

Fish Farming
Fish farming is being encouraged in the rural areas by the Department of Livestock Development to eradicate malnutrition among children. Wadigula women's group and a youth group are involved in fish farming. Officers from the Department of Livestock Development visited this group and held a seminar to demonstrate how to make a fish pond and the subsequent management of the fish and pond.

The initial capital for starting fish farming is low compared to many projects. Members of the group provide labour to dig the ponds after which the ponds are filled with water. The young fish (fishlet) are bought at a government subsidized price of Kshs 500 per lot. Local materials such as left over 'ugali', banana leaves, and other suitable vegetation are used as feeds. At the first harvest, the groups earned Kshs 2,500 and Kshs 3,000 respectively.

Wadigula women's group's fish pond was destroyed during road construction in the area. A lot of soil and rubbish was pushed from the nearby road construction site into the pond filling up the pond and killing the fish. Since then the group has not been able to get the pond cleaned up and the group has not been compensated for by the company that constructed the road for the destruction of their fish pond. The ignorance of the women and the arrogance of the men led to the collapse of this project. Women also pointed out that theft of fish from the pond hindered this activity. They feel that fish farming is more suitable for men since traditionally men have been known to be fishermen in many areas and not women.

Pig Farming
Pig farming is a new farming activity in this area as the Maragoli are not keen pork eaters. The women were encouraged to rear pigs for sale in other parts of Western Province. Chegobero women's group received a grant of Kshs 111,000 from the government of Kenya in 1983 through the Ministry of Culture and Social Services in order to start this project. The group claims that they bought a plot at Kshs 3,000 on which they constructed units for the pigs. Twelve pigs were bought to start the project. The money that remained after the purchase of the pigs and construction of the units was for feeding the pigs. This project lasted about two years and then the money ran out. By 1985, the pigs had multiplied greatly and the group had well over one hundred pigs.

After exhausting the government money the group had no strategy for feeding pigs. Members were requested to contribute foods, banana leaves, cabbages, grass, and maize leaves to feed the pigs. However, these are items were in short supply and in demand on the farm households themselves for family livestock. The pigs started missing foods or were given inadequate meals.

The members had been encouraged to provide free their services to clean out the pig sties. Many members were not willing to do this and, as a result the pigs ended up living in very dirty sties. The older pigs started eating young ones. One morning the women woke up to find that 25 pigs had been stolen. The pigs started dying from diseases, hunger, discomfort and suffocation as they were congested. No marketing arrangements had been organized by the group and it therefore appeared that the women did not have anywhere to sell the pigs.

This group was riddled with disunity, discontent and many other village squabbles among the groups' leadership. Many members were not happy with the way the officials had used the grant from the government. Members started questioning officials who refused to give a full report and record about what had been happening within the movement over the last three years. Due to these conflicts, the leaders sold off the pigs, and shared the money among themselves. This led to the split up of the group and its eventual collapse.

This project would have been well paying, but management of pigs and money was very poor. It appears that the leaders were dishonest, irresponsible, and therefore may have swindled the other members. The officials of the group did not co-operate well and did not have a smooth working relationship. The group also lacked supervision from the Ministry of Agriculture, Livestock Development and Marketing and other parties concerned with rural development at the district level. Women projects need considerable supervision from the officials who give loans and from qualified social development agents (SDA).

Bee keeping and rabbit rearing projects were also investigated. Both did not pay well because the projects had not been properly rated. Women think rabbit rearing is not a marketable business because rabbits are only eaten by men in this area. Bee keeping is quite cumbersome for women. Many women fear bees and so rely on sons or husbands to do everything for them. Yet, bee keeping has been very successful with women groups in Machakos and these groups are supplying honey to Nairobi and Mombasa and many other urban areas in Kenya. Women's groups in this area lack however, the technology and incentive to keep bees and rear rabbits.

8.5.3 Tree Production

About ten groups (27%) had established tree nurseries. They were selling seedlings to members of the community and members of the groups were encouraged to plant trees on their holdings. Groups involved in this project earn about Kshs 100 per month. We will deal with environmental projects more in detail later.

8.5.4 Secondary Sector Activities

Handicrafts

Twenty groups (61%) are in the handicraft industry. These groups make products such as table clothes, knitted sweaters and cloths, embroidery work, table mats and rings, bags, pottery, tie and dye materials, and print. This is made possible by money contributed by members in order to buy woollen threads and materials to make the different articles. Some of these handicrafts have been made traditionally for many years and so the skills are widespread.

One of the problems with the handicraft project is the lack of market, especially at the local level. Many of the groups make the same articles and, as a result the local market is saturated. Rural households lack money to make such purchases which are perceived as non-essential goods. The women groups in North Maragoli have not penetrated into the urban market due to lack of information and poor organizational skills. There is need for bigger organizations and umbrella organizations to assist rural women's groups in marketing of handicraft products. A few groups that have marketed through some middle women end up selling at lower than the production cost which, of course, cripples the groups financially and disenchant many members. Not much money has been earned from handicraft by the different women's groups.

Brick Making

Brick making is an old art among the men of this area. The soil in this area is suitable for making very good bricks for building houses. It is not possible to definitely say when the people in this area started making and using bricks. Our humble guess is that brick-making was introduced by the missionaries in the early 1900s with the establishment of the Friends African Mission of the Quakers at Kaimosi in 1902 and The Church Mission Society of the Anglican Church at Kima Bunyore.

Since brick-making requires special skills and is strenuous work, men are trained in this art. In fact, even among the men, the young and energetic are better than the old. This is the reason why brick making groups whether male or female have to use male labour. Bricks are widely used in the construction industry and therefore it is a very lucrative business. Seven groups (21%) had invested in brick-making. Four of the groups used labour provided free by the group members and the other three groups had employed two permanent workers each. The main work involves collecting water from the streams, molding the soil into bricks and firing the bricks. The women in all the groups supplied all the water required.

A few members in the groups involved in brick-making donate a small portion of their holdings for this business while other members donate firewood for firing the bricks. It takes about three months before the bricks are ready for sale. The groups earned a minimum of Kshs 8,000 and a maximum of Kshs 30,000 every year from bricks. They mould and fire between 3,000 to 10,000 bricks in a year. A single brick sells at between Kshs 2 and Kshs 4/50 depending on the size and quality. Sometimes the bricks are sold at a very low price because many people in Vihiga District are involved in brick-making and there is no clear marketing strategy.

The major problem with this project is that it contributes to environmental degradation. Large depressions or open man holes are left behind after excavating the soil to make the bricks. Also many trees are cut down for firing the bricks. Therefore, although the project may be money making, it is not environmentally sound. In all women's groups this project was mainly dominated and run by the few male members who rarely provided the women an account of its profit-making abilities but only complained of the labour involved and demanded the full participation of women.

Tailoring

Tailoring is one of the money-generating activities that women have been encouraged to do since the colonial period. The different groups encouraging tailoring among women assume that tailoring inherently is easier for women to learn. The Ministry of Culture and Social Services has been supplying sewing and knitting machines to a number of women's groups in Kenya in order to encourage the groups to start tailoring businesses. Three groups (9%) are involved in tailoring. In North Maragoli, two groups Mudete and Mwenyelitsi have benefitted from the generous assistance of the Ministry. Each group received two sewing and two knitting machines. However, in Vokoli women groups contributed money themselves and bought one sewing machine per group.

One of the youth groups is also involved in tailoring. The Ambenge youth group enrolls two students every six months to train them in tailoring. Each student pays Kshs 2,400 to complete the training. This group makes about Kshs 4,800 every six months. The rate of school drop-out among girls is quite high as is common in many parts of Kenya and North Maragoli is no exception. Tailoring is seen as a women's career and, therefore, many girls in this area are on the waiting list to enrol for this course.

These groups are all involved in making school uniforms and school sweaters to sell to the local schools. This has led to the saturation of the local market, resulting in depressed prices. Officials of the groups that received machines from the Ministry all claim that the machines have been stolen and, at the same time, some of the officials have started personal tailoring businesses. Tailoring is not an easy project because its success requires trained people. Many of the members in women groups lack tailoring skills and so many of the machines have never been used by the group members. Indeed, women require to be trained to enable them make presentable articles for sell. Tailoring demands sufficient time and patience from members. The implication is that such groups have to employ professional tailors to compete favourably on the market with other designers.

8.5.5 Tertiary Sector Activities

The tertiary sector comprises of petty trade, commercial services, and welfare activities.

Petty Businesses

Most groups (85%) are involved in petty businesses. With as little as Kshs 200, groups are able to invest in trades like buying bar soaps, tea leaves, paraffin, dried maize, salt, peas, bananas, dried millet, dried sorghum, groundnuts and sardines for example and sell them at a profit. This type of trade is normally done in the afternoons or on the market days. However, very little money is raised from these petty businesses. The accounting system is not easy for the women as they do not keep records.

A few members of women groups are engaged in long distance business buying bananas, avocados, beans and sweet potatoes from the local markets in Maragoli and transporting these commodities to Nairobi for sale. From Nairobi, they buy fruits, and Irish potatoes to sell on the local markets at home (Mago, Sabatia, Mudete and Chavakali). Also a few members of women's groups go to other towns such as Kisii, Migori, Eldoret and Nakuru where they buy assorted wares to sell on the local markets in Maragoli. Small profits are made but the benefits to the group are limited as many of these members eventually consider such businesses as an individual business because of the time and labour involved.

Hotel (eating house) and Butchery
Two groups were involved in the hotel business, Ambenge and Gurugwa. Gurugwa received a grant from the Ministry of Culture and Social Services worth Kshs 50,000. The money was invested in opening up a local hotel (eating house) and starting a butchery business. Members were very enthusiastic about these two projects and voluntarily provided labour in shifts to cut down on costs and to maximize the profits. Some of the food stuffs cooked in the hotel (at the eating house) were provided by the members.

Unfortunately, this group has experienced conflicts and misunderstanding between the members and the leaders because of the poor accounting of funds and the way the hotel and butchery were run. By mid-1993, a few officials had turned both businesses into family business and this led to the collapse of the group.

The Ambenge youth group, however, runs a canteen at the Goibei market centre. This canteen was started with members' contributions. This group consists of youths many of whom are school drop-outs and in the same age category i.e between 16 to 22 years. Their main objective was to make some profit so that their members could benefit. All members are literate and officials are trustworthy on all the affairs of the group. Members of the group participate in shifts to work in the canteen where a small profit of Kshs 300 is made per month.

Plots and Rental Houses
Four groups had already acquired plots at the local market centres, Avugwi, Gurugwa, Mudete and Wadigula. One group has put up a building (Mudete women's group). Mudete women's group received assistance in the form of building materials from the Ministry of Culture and Social Services worth Kshs 157,000 for the construction of the building. Members of the group contributed a further Kshs 18,000 towards the completion of this building. The building contains a central hall, and three side rooms and sanitary facilities outside the building. The main objective of the group was to use the main hall for group meetings and tailoring purposes, one of the three side rooms was to be used as the office for the group officials and the other two rooms were to be rented out to generate monthly income for the group.

Mudete women's group is not earning any money from this building because the area chief is using the building as his office and the central hall is utilized for the 'Chief's Baraza' and other meetings. The chief, being a government officer and the executive head of the location claims to have full rights over certain assets under his jurisdiction and, he further asserts that the building was partly sponsored by the government, so he has the right to its use without paying rent. The two extra rooms in the building that are not used by the chief are occupied by personnel from the Ministry of Culture and Social Services who were the funders of this project and some staff of the Ministry of Health, both groups not paying rent. The complacency of a few government officials frustrates women's efforts and defects the purpose of making women groups independent and income-generating groups. Members of this group worked and participated in all aspects during the construction of this building with the believe that they would be able to get rewards from the project. Members have been disillusioned and bitter that they have lost both money and time invested in the construction and completion of the building. These women suffer because of ignorance about their legal rights and because they fear those in authority. As mentioned above, only a few groups have been fortunate to receive financial and/or material support (see Table 8.3).

Only eleven (33%) out of the 33 women groups had been assisted financially or otherwise by the Government through the Ministry of Culture and Social Services and a few NGOs. Out of these 11 funded groups, two groups received Kshs 800 and Kshs 500 as a token of appreciation from the area Member of Parliament, which is too low to be categorized as financial aid for project work. A total of Kshs 865,300 had been received by 11 groups. This is an average of Kshs 78,663/60 for each of the eleven groups and Kshs 26,221/20 for each of the 33 groups if the money had been distributed to all groups equally. Only a handful of women groups have been financially assisted and so the idea that women groups can be used as a strategy for rural development may not bear much fruit unless more impartial investment, supervision and commitment is provided by the government institutions, international organizations and NGOs concerned.

Rotating Savings and Credit Associations:-
Small and micro enterprise development has a vital role to play in the economy of Kenya. But, micro-enterprise in general has not had access to credit from the formal financial institutions. These require collateral before loans can be secured and only limited amounts of credit had been received from NGO programmes. It is said that it is only in the last few years that NGO credit programmes have evolved to provide credit and savings for individuals with a low income on a far greater scale, at market rates and applying commercial principles (Kresterton, 1992). However, many of these schemes are urban based or found on the peripheries of the larger towns. The traditional savings and credit associations (Merry Go-Round) or ROSCAs in North Maragoli are close to the Grameen model in Bangladesh or the Minimalist approach elsewhere in Kenya but they lack formal financial credit support. In the rotating savings and credit associations, savings contributions are mandatory for all members.

Most groups (73%) are involved in the revolving fund and rotating savings and credit associations. In the revolving fund, members contribute between Kshs 100 and Kshs 200 each month. The money is allowed to accumulate and members borrow from this fund and repay at an interest of between 10-20% on the amount borrowed. This fund has helped many members to pay school fees for their children and to offset many other expenses. Two women members had each borrowed Kshs 10,000 from group savings to buy iron sheets to roof their houses. At the end of each year, all the money is returned to the treasurer for accounting purposes. Members share a certain amount for Christmas and New Year expenses and something is left in the treasury for New Year lending.

Table 8.3
Financial Assistance to Women Groups in North Maragoli (source: Department of Social Service Annual Reports, Vihiga District, 1993).

Women's Group	Amount of Money Received	Materials Received	Source of Money	Year Given
Mwenyelitsi	1000/00		Area Member of Parliament	1989
		4 Sewing Machines	Ministry of Culture and Social Services	1990
	10,000/00		Ministry of Culture and Social Services	1991
	300,000/00		UNESCO*	1993
	50,000/00		Global Fund for Women	1992
Total	361,000/00			
Mudete		2 Sewing Machines and 2 Knitting Machines	Ministry of Culture and Social Services	1982
	75,000/00		Ministry of Culture and Social Services	1984
	82,000/00		Ministry of Culture and Social Services	1985
Total	157,000/00			
Avugwi	25,000/00		Ministry of Culture and Social Services	1990
		Fertilizers and Seeds	Ministry of Culture and Social Services	1990
		Seeds		1991
		Seeds		1992
Total	25,000/00			
Chegobero	111,000/00		Ministry of Culture and Social Services	1983
Total	111,000/00			
Chanderema	20,000/00		Partnership for Productivity	1990
	62,000/00		Africa 2000 Network	1991
		Dairy Grade Cattle (35)	Africa 2000 Network	1992 & 1993
Total	82,000/00			

Gaigedi	800/00	Area Member of Parliament	1992
Evojo	500/00	Area Member of Parliament	1992
Kisangula	10,000/00	Visiting Member of Parliament	1989
Gurugwa	50,000/00	Ministry of Culture and Social Services	1990
Lwenya	10,000/00	Rural Enterprises	1990
Mukuyu	58,000/00	Ministry of Culture and Social Services	1990
Grand Total	865,300/00		

A rotating credit association is one where members of the group contribute money or buy articles for one another in turn. The rule is that members contribute the same amount of money on a monthly basis and members receive money or buy items in rotation. When the cycle is complete, the rotation starts again. This has helped many households to acquire cooking and serving utensils. This is the most common project among many groups as the amounts of money involved are small.

Welfare Activities

Avugwi women's group runs a small dispensary (clinic) because one of the members of this group is a retired nurse. At the same time, the elderly women in this group offer traditional midwifery services to the local community. Not much money is collected from these services but it is a big contribution to the welfare of the local people who may not have money to consult medical doctors.

Many women groups have formed choirs to sing during meetings to entertain themselves and to preserve and pass over some of the old songs to the younger generation. Six groups had formed choirs to entertain on important occasions especially during public holidays or at political rallies. Since most of the singing is for entertainment, these groups receive little money as a token of appreciation and they make between Kshs 500 and Kshs 1000 in a year.

An array of activities and projects among women groups in North Maragoli have been discussed. A few of the projects can be termed successful but many are failures. The activities considered most successful are rotating savings and credit associations, revolving funds, organic farming and zero-grazing. Individual group members and households of group members benefitted directly from these projects which could explain why they were more successful. Better organization was also noticed in these groups.

8.6 The Role of Women's Groups in Agro-forestry and Environmental Conservation in North Maragoli

Conservation is "the management of human use of the biosphere so that it may yield the greatest sustainable benefit to present generations, while maintaining its potential to meet the needs and aspirations of future generations. Thus conservation is positive, embracing preservation, maintenance, sustainable utilization, restoration, and enhancement of the natural environment. Conservation, like sustainable development, is not only relevant to nature itself, but also to people. While development aims to achieve human goals largely through use of the biosphere, conservation aims to achieve them by ensuring that such use can continue" (WCS, IUCN, 1980 International Union for Conservation of Nature and Natural Resources).

The reliance of rural communities on living resources is direct and immediate, and unless these resources are conserved, there is no prospect of improving living standards in the long run. Women make up a considerable majority of the rural communities involved in active agricultural production and therefore, they are the people who should be primarily concerned with resource conservation (especially soils, water and forests) since resource depletion will impoverish them and create work (labour hours) for them. Women should also have information about the right inputs and how such inputs should be used. Fertilizer is a case in point which needs proper handling as it can be harmful both to the human body and the soils.

Changing people's awareness about the environment may require a virtual revolution. But what is more difficult for this area is changing attitudes towards women and the environment. Women play an important role as conservationists and sustainers of the environment. However, as their role remains informal, it is easy for governments, International agencies, and even NGOs to gloss over their importance and to ignore their potential usefulness as a force for sustainable development (Dankelman and Davidson 1988).

The perception of forest ecosystems as having multiple functions for satisfying diverse and vital human needs for air, water, medicine, shelter and food has been replaced by a sizeable shift to 'scientific forestry' (Shiva et al. 1985). This 'scientific forestry' is no more than a calculation of timber yields to serve commercial and industrial demands. This is clear in Maragoli where women have problems securing firewood for cooking while the men keep tree resources for sale.

Natural forests have been converted into mono-cultures of such species as eucalyptus and blue gum, which, as we have seen, appear to be destroying the water balance, soils, ecological diversity, and the capacity to produce fodder and organic materials.

Within 'environmental issues' in North Maragoli, afforestation is the most important activity found in the women's group organizations. Women's groups are actively involved in Kenya Energy Non - Governmental Organization (KENGO) activities, a platform of more than 200 NGOs dealing with energy issues. Although their main objective is to safeguard energy supplies through, for example, information about fuel-saving stoves, another programme deals with reforestation using indigenous trees. After collecting information from local women on the medicinal, cultural, ecological and economic value of trees, KENGO passes it on to women groups and others by means of district workshops, exhibitions, radio programmes, newspaper articles and publications. It also sponsors tree seed projects, in which people are trained in collecting, handling and pre-treating indigenous seeds for planting. In all these ways, KENGO hopes to promote individual and communal reforestation with indigenous trees (Musumba, 1985).

The Kenya Woodfuel Development Programme (KWDP) has systematically analyzed fuelwood supplies in Kenya, and, on the basis of its conclusions, developed strategies to alleviate Kenya's fuelwood shortage. For Kakamega and Vihiga Districts, a self-sustaining system of tree planting was developed to contribute to fuelwood supplies. Using the findings of surveys of agro-forestry activities and the cultural backgrounds of the districts' inhabitants, the KWDP has designed an approach that accommodates indigenous expertise with its traditional beliefs and taboos rather than attempting to impose potentially unacceptable solutions on the people. In this process, women in particular are encouraged to formulate solutions, so that these will not conflict with traditional values. Technical help from trained local extension staff is available whenever needed. The project is linked to an awareness-raising programme, using popular mass-media techniques such as drama, pamphlets and posters (Chavangi et al. 1985). However, this is dependent on the acquisition of the above materials and the number of literate women in the village who had time to sit down to read the materials provided.

Both the groups that have worked in Kakamega and Vihiga state that women's participation in agro-forestry and conservation is constrained in many ways: competition for land, problems over land tenure, lack of time left after domestic duties, cultural taboos and lack of familiarity with forestry. Moreover, women do not have legal institutional and organizational support.

North Maragoli women are concerned with both agro-forestry (the practice of combining woody perennials, trees, and shrubs for example with agricultural crops and sometimes animals within a unified production system) and the conservation of their land and soils. In one way or another all the groups in the survey were contributing positively towards the conservation and management of their environment.

Ten groups (30%) were involved in the establishment of tree nurseries. Some of the trees planted in the tree nurseries are Leucaena (Lusina) which is a leguminous shrub that grows to some 20 metres high. It grows at high altitudes and thrives in a rainfall range of between 600-1700 mm. The women also grow Sesbania Sesbania ('Shikhule or omosabisabi'). Traditionally, this tree is inter-cropped with both cash and food crops. When transplanted, it grows to a height of about 5 metres. The Sesbania Sesbania variety fixes atmospheric nitrogen, therefore improving the soil through green manure. Both Leucaena and Sesbania Sesbania are also used in the study area as fodder for cattle.

Other types of trees planted in the nurseries by the women groups are Calliandra (Calothyrsus) and Eucalyptus. The establishment of tree nurseries has encouraged members to plant their own trees for fuelwood. Ninety percent of the households where women belonged to a women group have sufficient trees to provide firewood for household consumption. The trees also conserve the soil and providing fodder for livestock.

Some women are now discarding some of the traditions and cultures that have perpetuated the social and economic underdevelopment of women in the rural areas for a long time. Women in Maragoli were not allowed to plant certain trees as trees are viewed as permanent assets or property. There are many traditional superstitions associated with women planting trees. To give a few illustrations: it was believed that if a woman planted a tree then any of the following things could happen: (a) the woman would become barren; (b) if the woman was married, the husband would die; (c) the woman would be seen as a social outcast and disgraced and therefore a person of ill omen; (d) a tree that is planted by a woman cannot be used for house construction or for cooking because if it were used the house would burn down mysteriously; (e) if a woman planted a tree her children would die; and (f) if the woman is married and she planted a tree, she would eventually be divorced by the husband and clan, but if the woman was not married, then such a woman would be excommunicated by her family as she was looked upon as an outcast.

These taboos were encouraged by men and because of their fear of the repercussions of disobeying the traditional laws, women internalized the above beliefs and these became part of their lives. The younger generation of women are now waking up to the consciousness that some of these taboos are retrogressive and lead to hardships for women. Many women have realized that being the main suppliers of food at the household level, the dissipation of forests has resulted in more work for women because they have to walk long distances looking for firewood if that is still possible or otherwise buy firewood from the market.

A number of women are now planting some traditional trees and eucalyptus trees. This was unthinkable in the past. The women who first started planting these trees did not suffer from the ill luck as they had been led to believe. Most women, especially those who belong to women groups now plant trees to provide fodder and firewood for their own families. Women group members now encourage other local people in the community to plant trees. The elderly women above 50 years have not changed their attitudes and beliefs and so this group of women use their sons or employed labour to plant trees for them.

The eucalyptus tree is the commonest tree in the district and in North Maragoli. It is an exotic species that was introduced into Kenya in the 1920s. The tree grows very fast and can provide timber for many purposes such as building/construction work, fencing logs, making furniture and firewood. The tree is therefore very good for commercial purposes. This species is planted mostly along river valleys and in the homestead. Unfortunately, it is said that the eucalyptus absorbs large quantities of surface water, so that women as the water collectors are now faced with lowering groundwater levels and the drying up of their water sources (Shiva et al.1985). From our field observation this seems to be completely true. The areas where eucalyptus plantations are established are the former wet lands where special traditional crops like arrow root, yams, bananas and indigenous trees (forests) were found. These were the water collecting points and homes of rare species of birds and animals as some very old informants explained to the writer. During drought years such areas were utilized as refuge areas for agricultural production and families did not suffer much.

However, the wet lands have retreated and the water table is much lower nowadays. What were formerly large streams are now very small they dry up during severely dry years. In some areas, where these streams were very small, the water has dried up. Many women complained that the soils in such areas are no longer suitable for farming as the soils look like a dry pan. The solution to this problem is to encourage women's groups and other people in the rural areas to plant more indigenous trees and trees that do not deprive the soil of water and other food nutrients. Eucalyptus trees should be planted in select areas that are already dry and unsuitable for agriculture if the trees can grow there. For instance areas that are left behind after the molding and firing of bricks.

The women groups that have established tree nurseries are encouraging and influencing the community (especially women) to buy seedlings from them for planting. Environmental protection awareness is being created by the women groups and if these groups are given encouragement they can contribute to the protection of the environment in the rural areas.

A few of extension staff especially from soil and water conservation, and forestry departments are now finding ways to reach large communities through women groups. Environmental, soil and water conservation officers have visited such groups as the Mudete, Gaigedi, Wadigula, and Avugwi women's groups. However, these are only four groups out of 33 and so the impact is very limited. These groups have been instructed by these officers about how to make tree nurseries, the types of trees to plant and the management of the trees, making terraces and cut-off drains, 'Fanya Juu' terraces, building of ridges, the importance of crop rotation and mulching, making seedbeds and vegetable gardens. Gaigedi and Avugwi women's groups benefitted from a donation of tree seedlings and other assorted seeds from the forest officers. Members of women's groups use the information received at group level also on family holdings. Many women's groups are spreading the information on environmental conservation to other members of the community. This is one example of how women's group organizations may be used to forge development in the rural areas.

Since the soil is a very valuable resource in this area due to the small size of farms, nobody would like to part with or see his/her soil being washed away into another person's farm or into a river. This coerces the farmers in this area to be concerned with issues of environmental conservation and protection without incentives or a push from the government officers. Many women (90% of those interviewed) had established terraces on their farms. Along the terraces and ridges they plant napier grass to hold the soil together and so prevent soil erosion. The napier grass is also used as fodder for livestock. Cowdung or waste from cattle plus all the waste and other household rubbish is placed in a pit for making compost and green manure in a number of homesteads.

To further enhance a sustainable environment, three groups are involved in making energy-saving stoves referred to as the 'Maendeleo Jiko'. The Maendeleo ya Wanawake women, KENGO and the German Technical Agency (GTZ) have been responsible for this project. These improved stoves are being promoted because they cut down on energy consumption and prevent the destruction of forests. Two members from each group were trained in the construction of the stoves, which are built around clay liners. The clay liners cost between Kshs 40 to Kshs 60 in 1992 depending on the size, but by the end of 1993 their cost had risen to Kshs 90. The members trained in building energy-saving stoves (jikos) have instructed the other members of their groups. About 60% of the women belonging to these three groups had already acquired the energy-saving stoves in their kitchens. These groups were also building energy-saving stoves for other women in the community at a fee. Many women interviewed were interested in owning such a cooking stove but its price was prohibitive.

A number of women are aware that the 'Maendeleo Jiko' is environmentally friendly. They confirmed that the stove uses very little firewood and see that this reduces the number of trees felled by households. The stove also emits very little poisonous smoke into the atmosphere.

Fencing is encouraged, various farmers have fenced their holdings using local fencing plants. Many farmers reinforce the fences annually to prevent their neighbours' animals from straying into their fields.

Avugwi women's group is involved in double digging and organic farming. The women of this group are aware that chemical fertilizers and pesticides may degrade the environment and may poison soils as discussed above. The women are using green manure for planting crops, and they make pesticides from natural plants.

As discussed above, brick-making is the only activity that women are involved in that is environmentally degrading. About seven groups are making bricks for sale. These groups are excavating and opening up big depressions in search of soil to mould bricks. The excavated areas are left in a state of disuse and are mosquito breeding grounds. The groups also use a lot of wood to fire the bricks and this has led to the cutting down of trees in areas where these groups operate.

There are also a few groups who are involved in logging, sawing-timber for sale and also the sale of firewood and charcoal. All these groups need technical advice so that they can stop degrading the environment. This advice should be given by the government officers in charge of community development and by the environmental, soil and water conservation officers. These officers should help the women in the choice of activities for the groups to engage in.

8.7 The Role of Women Groups in Farm Management

Farm management in small-scale farming, especially in food production in Africa, and in Kenya in particular, is carried out by women. All the women groups in the survey were in one way or another involved in agricultural production. They started off as farming associations. Through these associations, a few women's groups started to assist one another on a rotational basis in the weeding of crops, or in the plucking and transporting of tea to the leaf-buying/collecting centres. Women groups and youth groups are also contracted by non-members, especially for weeding maize and tea, and sometimes for tea plucking. Over 60% of the women groups are involved in horticultural farming.

We have presented a seeming success story of the Chanderema women's group's involvement in an intensive zero-grazing project. The construction of zero-grazing units and the establishment of napier grass by women for dairy cattle is a manifestation of their role in rural development. Mambai and Chanderema Sub-Locations where this women's group is situated are now almost self-sufficient in milk production and supply during certain periods of the year.

Groups such as Avugwi, Wadigula, Gavudia, Mago and Mwenyelitsi also own one or two dairy cows. Many women's groups have the ambition to adopt zero-grazing because of the related advantages of this sub-system. Women consider zero-grazing as an important activity because it provides a source of income. The women who already own dairy cows are learning the techniques of feeding, disinfecting, milking and the care for calves quite quickly. Many have constructed drains from the zero-grazing units to cemented holes in order to collect the waste from the dairy cattle which is then turned into manure. Some women have the idea of producing biogas for heating, cooking and lighting if both technical and financial support can be made available to them.

In a nutshell, women's groups are very much involved in farm management but still receive inadequate institutional technical and financial support. This frustrates the efforts of improving their situation, especially in agricultural production and income-generating enterprises. Most women concurred with the idea that if their husbands and traditions would allow it, they would sell off the local zebu breeds whose milk production is very low and buy one cross-breed or one dairy cow.

The Avugwi women's group is involved in organic farming and double digging as discussed above. Many members of the group are now aware of the dangers that chemicals pose to the human body and the soil. Members of the group are also know how to make green manure and control diseases and pests using local natural plants/materials. This type of technology is relevant for many rural farmers who lack the capital to purchase artificial chemicals and fertilizers. The technology is more appropriate because farmers use indigenous knowledge and local materials, giving themselves an opportunity of participating in the whole process of its implementation. The fields of maize, vegetables, napier grass, and bananas belonging to members of this group appeared to have a healthier crop than the other farmers in the neighbourhood. An overwhelming percentage (98%) of women in the survey remarked that they lagged behind in both farm management and in being self-supporting because they have been suppressed and exploited by the males, including their fathers, husbands and elders for a long time. The culture and traditions in this area do not encourage women to move out of the home and learn to be self-reliant. Women rarely take the initiative to question the rules and the set laws.

In this area as in Western Province as a whole, men are considered the bread-winners and heads of households. This norm is associated in many households with the idea that the decision-making process should be handled mainly by males, whether such decisions are good or bad. The women and children are obliged to follow these decisions without question. Women's lives are therefore dictated by men. This may partly explain the fact that women, without protests, combine heavier roles in farming with household duties. This as we have seen results in low labour productivity and labour inefficiency.

Some women now realize that, in fact, many men in the rural areas do not deserve to be regarded as bread-winners because a good number of them do not provide for their families. If anything, there are many women who could be considered bread-winners and who should be given the necessary institutional and policy support. It is noticeable that some change is also taking place among women, both socially, culturally and economically. This change is speeding up because of the difficult economic situation in the country. Women are forced by prevailing conditions to devise strategies for survival. Some women can now make their own decisions in farm management, sometimes ignoring decisions made by their husbands who live in the urban areas.

A number of officials of women's groups have attended seminars and gone on tours to other areas like Kisii, Nyanza, Uganda and Tanzania. The fact that a few rural women can now go to seminars for a number of days outside the home area is an implication of willingness and urge to be incorporated in development agendas, especially of resource management. On these tours, women discover that their counterparts in other areas are more progressive because they have discarded some of the traditions that are detrimental to women's life, self improvement and development. However, this creation of awareness, empowerment and the building of consciousness is limited to very few women. The leaders or officials of women's groups appear more knowledgeable and exposed than other group members.

8.8 Women Groups and Agricultural Extension Services and Credit

Agricultural extension as discussed in Chapter 7 is a pre-requisite for widespread and sustained agricultural development. Extension and training through communication channels like 'barazas', funerals, churches, radio, Farmers Training Centres, posters, publications, and visits by extension officers not only arouse the farmers' awareness about new technologies in farming, but also enhances the adoption and implementation of new technologies and practices resulting in better farm management and agricultural production (Saito et al 1990).

Women's groups, and to a lesser extent mixed groups, provide an effective way of channelling and providing feedback on extension services, and offer women the means of overcoming social and educational disadvantages. For a long time the Kenya Government has found it expensive and difficult to reach all the farmers involved in small-scale farming. One problem has been the characterization of the farmer as a male, a concept introduced during the colonial time when policies in agriculture were persistently oriented towards men. The official recognition of women groups by the government should provide a strategy for extension officers to reach women who are the main resource managers in the rural areas. In the past the agriculture and livestock extension staff have advanced reasons why female farm managers have not been able to benefit from extension. It has been alleged that women are shy and are not free to have discussions with male staff in the absence of their husbands. Our field observation indicates that women are not shy and can effectively participate in discussions and learning situations. The Training and Visit (T&V) approach also greatly limited the number of farmers that extension agents can work with. Women asserted that they are ready and willing to learn regardless of the gender of the one who is providing the information, but the system has been slow to acknowledge the role of women as farmers and managers of resources in the rural areas.

It is our experience from our field survey, and it is an experience confirmed by Feldman (1983 in Bahemuka 1987), that home economists who are the officers designated to visit women and women groups are, in most cases, not trained in agricultural or livestock production. These officers are sometimes handicapped in providing information on crop and animal husbandry. Many of these officers appeared ignorant in pest and disease detection and control while they are expected to be specialists in horticultural farming. Many of the home economists who visit women and women groups preferred to discuss nutrition, cleanliness in the home and family planning.

The Kenya Government has realized that the marginalization of women in the rural areas is partly responsible for slow or poor rural development. This may explain the inclusion of women in the current development plan and the verbal insistence that Government Ministries should approach women groups to provide information relevant to rural development. This challenge is being accepted by the different Ministries in theory but empirically progress is slow.

Our study showed that only six women's groups (18%) had been visited by extension officers, and officials of only 12 groups (36%) had been to a Farmers' Training Centre, seminars, workshops or on tour. The total membership of women groups visited was 1,086 members of which 378 members (35%) were interviewed. Only the officials of women groups go to seminars, on tours and to Farmers' Training Centres. Seventy members (6%) had either been to a Farmers' Training centre, seminars and/or workshops. This is a small percentage compared to the population of women in our area.

Some of the officials have been to Bukura Farmers' Training Centre, Manoa House - Kitale, and the Multi-Purpose Training Institute at Kakamega. Instruction concentrated on the following subjects: (a) zero-grazing where the emphasis was on the establishment of zero-grazing units, the selection of cattle, feeding, treating, milking and the care of calves; (b) poultry keeping; (c) bee-keeping; (d) horticulture and double digging; (e) home economics; and (f) family planning and family life education.

Women's group officials who have been to Farmers' Training Centres pointed out that more time is devoted to home economics than to any other type of training. It was a general opinion among women that this was a waste of time for they believe that they understand their role as wives better than the young people who are giving them instructions, although our observation is that this may not be true for all women.

Only four officials from the 33 groups have been to the Lake Basin Development Authority to learn fish farming. A number of members have been to Aldai in Nandi District to visit women groups involved in dairy farming and others have been to South Nyanza and Kisii districts visiting groups involved in both crop and animal production. These visits which have been organized by Young Women Christian Association (YWCA) and the Ministry of Culture and Social Services have assisted women to learn what other groups are doing. The officials have been very good in instructing their members.

The women's groups involved in zero-grazing receive many visits from assistant livestock production officers and officers from the Africa 2000' Network, UNDP and the International Fund for Agricultural Development (IFAD). The women's groups that had access to extension support were ahead of the other groups in the management of livestock and milk production. Milk production was relatively high and the members enjoyed a small steady source of income which leads us to conclude that their management is better than other groups and this has been influenced by the technical information received, institutional support and the good organization of the group.

Two groups had been visited by officers from the Department of Forestry and Environmental

Protection, resulting in awareness raising and an increase in agro-forestry and soil conservation campaigns in the area by women groups.

What has become clear is that women can be good educators. Any information one of the members received was conveyed and demonstrated to other members of the group. During this survey, the women vividly recounted the information that they had acquired.

However, much still remains to be done. The women who are in dairy farming complained that they have to pay for livestock extension services. The extension section is yet to incorporate the use of women groups as one of the strategies for communicating agricultural messages to farmers in small-scale farming. However, this may be the only way out as it is cheaper and information reaches more individuals responsible for managing farms. Indeed, if women received more extension and credit support, there may be an overall improvement in farm management and agricultural production.

It has been suggested that access to financial services (credit) may be needed simply to assist an income-generating to activity continue, but, if applied well, the can also serve as a catalyst, facilitating working conditions and encouraging increases in the income generated (Hilhorst et al. 1992). It has been established that the access of poor women and rural women's groups to formal finance is either unavailable or very limited. The blame for this rests on the barriers and the stereotypes that have prevented banks from seeing women, and particularly poor rural women as prospective clients (Hilhorst et al. 1992). No formal organized government or NGO credit system is in operation for women's groups in this area. The women's groups do not know who they can borrow money on credit from for group programmes. Many women groups thought that money would be supplied by the government. Indeed many groups appeared illiterate as far as drafting project proposals or application for sponsorship are concerned.

8.9 Views of Women Groups on the General Socio-Economic Life of Women in the Rural Areas

The survey of women groups revealed that women in the rural areas face demands which exceed their time and economic situation. They confirm Kariuki's views stating that women are aware that their lack of education has affected the direction of their lives and limited their life opportunities. (Kariuki 1985). Most of the women felt quite strongly about the negative effects of this omission and the way it has affected the story-line in their histories.

Women in the rural areas claimed they experienced many problems. They raised a number of issues that affected their rural life. Both reproductive and productive responsibilities were too demanding given their lack of money. Women do not mind doing the household work but would prefer that their husbands be more responsible and concerned with farm work and overall production for the household. Child care is the work of the women with very limited or no support from men. Child care requires much time. Planning and management of time seemed to be a dilemma for many women. A number of men living in the rural areas hang about shopping centres throughout the day, or go out drinking and come home in the evening only to demand food. The women suggested that if men in Kenya worked half as hard on the farms as the women there would be far fewer food shortages in their households.

Women would prefer to have smaller families. This would lessen the burden of child care and also free more hours for farm work. But the men insisted that the women have more children, especially boys. Many men discouraged their wives from using family planning devices. Women feel that having large families is another form of oppression and a way of subordinating women to men. Men in the western part of Kenya feel that they are born polygamous and therefore it is quite natural for them to have extra-marital relations and neglect their legally married wife. However, because of this neglect of women, men feel threatened and are suspicious believing that if women used family planning then they might also decide to have relationships outside marriage. A few women who had gone for family planning had problems with their husbands later. As a result the husband takes on another wife or brings home children from his affairs outside marriage to punish his wife and increase her responsibilities.

The women confirmed that most women in the rural areas who are 45 years old or older have cases of hypertension (high blood pressure), depression, arthritis, and ulcers brought on they believe, by frustrations and stress caused by their husbands, family worries and responsibilities and the large amount of work they have to do.

Because rural women have to play many roles, farm work is often poorly done (poor farm management practices) and this affects production. 'At any one moment women are usually juggling three or four tasks at once' thus increasing their inefficiency and need for labour-reducing technology (Bryceson:1994:3). Women argued that if they were accorded institutional support to enjoy the same benefits as men, this would lead to improved farm management and enhanced agricultural production which could be ploughed back into the household and to the larger community.

Women appeared to have a feeling of being enslaved by their husbands (in their own households). This feeling emanated from the labour constraints women experience and the total poverty suffered by many women in the rural areas. It has been rightly pointed out by Bryceson that historically women in this part of the world have tended to be controlled rather than being the controllers of labour as far as men are concerned (Bryceson, 1994). This explains the expectation that men and society have about women: they are there to provide all the labour and food for the household. Women criticized the drinking prevalent among some men in this area. Men who drank much alcohol spent most of their money on this habit at the expense of their families, resulting in household impoverishment. Drinking is also associated with conflicts between wives and husbands and poor child rearing practice. Many rural women are obliged by prevailing circumstances to be tolerant and to cope with married life. Drinking affects many households economically and the interpersonal relationships between members of the household suffers. A break-down in communication between the man of the house and his wife and children is experienced in such households. Such men often hardly talk to their wives and children except when giving orders or asking for something. When a break-down like this occurs in interpersonal household relationships it negatively affects farm and household management, resulting in poor agricultural production as illustrated by the case of Iminza, a female in a male-headed households (see 6.3).

As rural women spoke about the changes taking place in the communities around them, they showed their frustrations and feeling of inadequacy. They repeatedly stated that the change from a subsistence economy had resulted in stress within the family and increased disparities between men and women in all areas of life - physical and social mobility, economic and political status, and interpersonal relationships.

The current economic conditions and structural adjustment programmes affect women directly and are salient in their minds and daily lives. The women see a deterioration in the economic environment to be the cause of increased complexity and difficulties in their lives. They stated that lack of money is their main economic problem followed by the difficulties they have in distributing household resources. These inadequacies directly contribute to the negative farm management, the poor agricultural situation and the poverty we have analyzed in this study. These same conditions have also been observed in the role of women in other parts of Africa and Asia.(Agarwal 1985, 1994, Bryceson, 1993, 1994 & 1995, Gianotten et al. 1994, Gladwin 1991, Baud 1989, and Lycklama a Nijeholt 1987).

In many ways, women perceived themselves as having the primary responsibility for the economic well-being of their families. They feel themselves responsible for solving their economic needs, provided that society gives them the opportunity and institutional support (financial, technical and legal) to participate in income-generating sustainable economic activities. Women attribute their underprivileged situation to their unequal share of new options and the ways in which new economic opportunities have been controlled by men due to the patriarchal system and the inheritance of the colonial set-up. Apart from the disparity in social, economic and resource ownership that exists between men and women, women in North Maragoli also pointed out that they are neglected by the elite women in the urban areas and so the marginalization of the rural women is double gendered. They are neglected both by men and by elite women.

They suggested that the government and other international organizations involved in women issues should offer alternatives and improve rural women's living situations and management skills.
These alternatives were formulated as follows: (a) the government should pass a policy whereby working husbands should be made to send at least 50% of their earnings to their wives every month. Of course, this sounds unrealistic since many of the men can barely survive on their meagre salaries in the urban areas. Many of the men rely on food supplies from the rural areas. However, it may be difficult for the government to arbitrate in what are the domestic affairs of the household. However, the women insist the government devise ways of financially supporting women in rural areas to enable them to be effective farmers and child bearers. Some kind of social welfare scheme for rural non-wage earners, especially for women.
(b) The government should incorporate women into the control of resources through property ownership. Resource ownership determines the form of exploitation and affects working relationships. In the case of Vihiga District, property, which is seen in rural areas in terms of land and cattle is owned by men. The women recommend that the government should change the law on the ownership of property, especially where land is concerned, so that once a man marries a woman then the share of land that he inherits from his parents should be registered in both the

man's and woman's names. This effectively means that both names appear on the land title deed and that they are both shareholders. The women think that with both names appearing on the land title deed, men will not just sell off land without consulting their partners, men will not dictate and make unilateral decisions on land use and farm incomes, and will not marry at will imagining they own the property and therefore would share it out as they deem fit. The women envisage that being shareholders of land will confer on them a feeling of security, permanence and belonging where they are married and this would enhance better farm-household management. This situation would facilitate women's access to loans as they would have collateral. Women will be an equal partner to husbands and, therefore, can make farming decisions without fear of intimidation. This is not the case at the moment. The women suppose that property ownership within the home should be equally shared between boys and girls. They stressed that the government should step in to provide equal opportunities for boys and girls so that girls are not discriminated against when it comes to education and job opportunities in the country in the future.

(c) The government should look into ways of improving the extension system so that more women can benefit from practical training and farming skills. Women have remained behind in all spheres of life due to lack of access to technological knowledge. This is seen as the main hinderance to good farm management and agricultural production. This has been pointed out by Bifani, (1985), Kariuki (1985), Saito et.al (1990, 1994) and particularly the ILO report of 1972 that stated that women's knowledge is confined to aspects of a process or parts of a technological device instead of the whole process or the whole production structure. Also in North Maragoli, women lack basic scientific knowledge and most of them depend on their traditional knowledge which cannot help them in the production of cash crops and the good management of the total farming system. Indeed, this was lucidly demonstrated with poor farm management and low agricultural production in many female-headed and female-managed households.

(d) Rural women are disadvantaged because of legal ignorance. They challenged urban women organizations and NGOs to interact with the rural women and help create an awareness of many legal issues on matters such as marriage, property ownership, business, access to credit, education, and access to technology.

(e) The women complained of excessive political control over resources allocation and use, in terms of policy formulation. They, therefore, recommended that the government should formulate policies that would incorporate women into all spheres of social, economic and political development and this should also be implemented. Women would prefer to be considered as full members of the community and not as a special group lumped together with children requiring welfare at best, open slogans and future promises. This delays rural development and further impoverishes the rural households.

(f) Women have also realized that the cultural patterns of control over and use of resources is a factor that limits their management. They recommend a change in the cultural patterns of resource control so that both men and women can benefit equally.

(g) In general women feel that it is the responsibility of women to change the pattern of bringing up children so that there can be a change in the present rigid system of labour division and gender relations. They recommend that parents should make children learn from the early age that both boys and girls are equal, should respect each other and should perform all the tasks that have to be done without discrimination. Of course, the women suggested this but they are well aware that it will take many years for this to be realized. However, they believe that women who live in the urban areas can start this process.

8.10 Problems and Aspirations of Women Groups

From the first UN Decade's Women Conference held in Mexico City in 1975, to the most recent one held in Beijing in September 1995, concern about the plight of women, especially rural women in Third World Countries has increased. It has been repeatedly announced that women should be given equal opportunities with men and discrimination should stop. Women should have the freedom to choose and decide on what is right for them. However, despite all these declarations and the promises of world leaders that they will better the situation of women, not much seems to have been accomplished in the rural areas. Often shouldering the major burden of agricultural production without adequate compensation, rural women in North Maragoli and in Kenya share with rural African women everywhere a list of needs that includes improvement in water quality and supply systems, better housing and sanitation, health care services and more educational opportunities. Many rural women live in conditions of servitude.

Many women groups in the rural areas of North Maragoli and Kenya have been formed in response to the UN Decade of Women Conference's recommendations that the position and role of women should be recognized and that projects should be created for women, especially poor women in the developing countries. This forced the Kenya Government and NGOs to assist women and the use of projects carried out through women groups was selected as an approach. These groups are seen as a new approach to rural development, a way of incorporating women into the country's economic development and leading eventually to an improvement in the resource base of rural women. Women's groups were encouraged to go into income-generating projects so that women in rural areas could find means of generating capital. This would improve general conditions in rural households and rural communities at large. In other countries and areas the approach to project development for women has progressed from the 'welfare approach' that dominated the 1950s and 1960s, through the 'anti-poverty approach' of the 1970s, the 'women and development approach' of the 1980s, to the present 'autonomy and empowerment approach' (Gianotten 1994 and Lyckama a Nijeholt 1987). This progression indicates the strengths and weaknesses that have been found at every stage of the development and incorporation of women into the socio-economic and political structures of different countries. In North Maragoli, our study has shown that the development of women is still seen in terms of the 'welfare' (women are still considered as passive beneficiaries in the development process, emphasizing their reproductive role) and 'anti-poverty' approaches (related to the basic needs strategy limited to producers in the context of self-sufficiency (Gianotten 1994). Thus, women groups in North Maragoli are still at the first stage in the phase of women's development.

This survey brought to light a multifarious number of problems women groups are facing. Below we discuss these problems not necessarily in order of importance.
(a) Financial Problems: Without exception all the groups visited lacked capital to initiate meaningful projects. The amount of money that each member contributes towards the group, Kshs 10 to Kshs 20 per month, is too little for the women to venture into projects that will be productive enough to give every member a monthly income. In fact, most groups, as explained above, had started projects that gave minimal returns.

When the government sent out staff to sensitize rural women to cooperate and form groups, the women claimed that they were promised financial support in terms of grants and loans from the Government, NGOs and international organizations. Most groups formed with the hope of receiving funding. The fact that many of these groups have not received any financial grants or loans is affecting their performance. Quite a number of groups are in the process of collapse. A few groups have benefitted, and some have received financial aid more than once which makes the other groups feel that there is favouritism in the giving out of grants and loans to women's groups. Almost all the women groups that have received financial assistance have strong political and administrative backing and association. Some groups were formed through the instigation of men with a hope of influencing funding to such groups and eventually benefitting the men. It was noticed that a certain amount of corruption and dishonesty was involved in deciding on the groups to be funded. Some of the groups that received aid had only a few literate members and these turned out to be the group's officials. The illiterate members in many of these groups have not benefitted from financial or material assistance as such assistance has been 'personalized' by the officials. A number of men seized the opportunity to form groups thinking it was an opportunity to get easy money from the government. Indeed, the men groups studied were planning big projects such as building houses for rent, running a petrol station or a wholesale shop. The men groups were therefore just waiting for the grants instead of getting something underway.

Whenever there are earnings from the projects, they are generally very small and do not provide members with a steady income. More than 90% of the groups visited had not reached the stage where they could provide members with an income on a monthly or an annual basis. The officials always ask for further contributions from the group and no substantial incomes or profits were made. Without the promised rewards many members withdraw their participation. The management and administration of projects by rural women is still a problem. Accounting and giving a clear record of savings and earnings has not been mastered by many of the groups.

The Ministry and other organizations interested in the women movement both in the urban and rural areas have not gone out of their way to meet all the women groups and train them in how they should function. The organization and management of many groups was very poor. Up to 60% of the groups visited during the course of this study indicated that they had not been visited by any officer. Group members therefore develop a feeling of being disconcerted, and there was lassitude and fatigue in the women movements. It is important that the field staff should visit all groups and encourage them by giving them the right advice about which projects to start before these are considered for funding. When funding is available the money or materials should be distributed to as many groups as possible.

(b) Political or Administrative Problems: The groups are either affected positively or negatively, depending on their level of affiliation with and submission to the local politics. The running and functioning of most groups is very much influenced by the political administration of the area. To a considerable extent a number of women groups are controlled by men operating in the background and some decisions are made by the local politicians. Cases of groups being frustrated and denied funding because the administration imagines that they side or sympathize with the political opposition have been cited, and this has resulted in the disbandment of such groups.

Most groups had choirs as one of their activities. The choir is specifically to entertain guests during political rallies. This means that when the chief or area Member of Parliament has a meeting, the women have to abandon all duties to go and entertain them. Groups complained of being used by the administration for selfish ends and that this took too much of their time.

A few groups had leaders who are wives, mothers, sisters or relatives of a Member of Parliament, Locational Chairman of the ruling party, area chief or a prominent person in the society. Such groups benefitted quite a lot from grants, especially grants from Government. A few of the leaders of these groups had enriched themselves by converting whatever grant had been made available into personal money. Members of such groups have nowhere to go to voice their grievances as they are intimidated by the leaders. It was also quite noticeable that such groups were being run by the men in the background. Some of these men are patrons in some groups, and during our interviews it happened quite often that a few men attended and closely monitored the discussions. The groups that are politically strong were not willing to discuss their financial matters or the amount of money they had in their bank accounts. It was also quite evident that the members were not free as the members kept saying that they agreed with leaders or else had nothing to add. Other groups were therefore, frustrated because they felt that only groups with strong political backing received loans and grants. Indeed, we have found a bias for certain groups among some officials working with women groups because only the politically strong groups were mentioned or visited. This is an indication of either fear or financial corruption by these officers.

(c) Management Problems: Rural women have very few years of schooling. Therefore, most of these women are illiterate or semi-literate. Lack of literacy is often one of the major reasons for poor performance. However, also groups with literate officials show poor management performance and this can be explained by the greed of the officials. Poor management leads to the collapse of projects and to lack of a confidence of members in their leaders. Therefore, members and leaders need training and supervision.

(d) Social and Cultural Problems: Most women are socially inhibited and this prevents them from questioning their leaders and contributing effectively to the movement. This also leads to a reliance on men. Examples are brick-making or timber-sawing activities which cannot be carried out without the involvement of men as women believe these to be typical male activities. Many groups consulted men in banking matters and when it came to filling out official documents which indicates that they are not freeing themselves from men.

(e) Problems of inadequate technology: Women only have traditional knowledge, and this is concentrated above all in farming. This has led many women to invest primarily in agricultural enterprises and petty businesses. It is important for the women to be trained in relevant modern technology in order to be successful in their projects.

(f) Problems of marketing and pricing of goods: There is much duplication in what women groups are doing. Most of these are active in petty businesses, selling the same type of goods. Others are active in making table clothes, knitting and embroidery work, and handicraft (pots, robes, rings), making the same type of goods. It, therefore, becomes difficult for them to sell their goods. Sometimes they sell at throw-away prices, making no profit at all. It is important for women's groups to know each other, to know what each group is doing and to be encouraged to specialize in their projects. Those concerned should try to find markets for the goods produced by women groups and to assist them in pricing their goods. It has been observed that women's projects tend to be similar in both type, content and production, resulting in a lack of marketing outlets which eventually culminates in the collapse of groups. It is assumed, quite correctly, that the failure of many women's group projects lies with the programmes and not with the women (Lycklama a Nijeholt, 1987). In North Maragoli and Kenya as a whole rural women were urged to officially start and register women groups as a government programme without a clear plan, policy or data on how the groups were going to be supported and how the groups would be used in rural development. In fact, the incorporation of women groups into rural development is conceived of in terms of women providing free labour to government community projects (for example, soil and water conservation, tree planting, and the improvement of rural access roads). This does not help women but burdens them further given their already constrained time budget.

(g) Transport problems: Transportation of goods by women groups is a big problem. Most areas have no public transport, and where there is public transport, it is very expensive.

(h) Time problems: Many women already experience time constraints in the provision of labour for their tasks on their farms and homesteads. Therefore, finding time for women's group activities is quite a problem. This explains why women are careful to spend their limited time only on activities that are rewarding. Many of the activities as listed earlier in Table 8.2 are time consuming and need much dedication. Absenteeism was found to be quite high among many groups despite the fact that a fee was levied for absenteeism in some of them. This fee was in any case never paid.

(i) Problems of staff and unsuitable government interventions: It was apparent that government staff mainly visit groups that receive government grants, ignoring the other groups. Many of the projects are not break-throughs in economic efficiency (producing the maximum valued output possible, using cost-minimizing techniques of production and considering effective market demand). No planning was done as to how each of the members of the group benefits or how long they have to wait for the rewards. Money given as grants is very limited for really good business. It was discovered that there was very little follow-up by the Ministry staff. The locational and sub-location-level community staff seem to lack authority and skills in running and organizing women's groups. Many of them appeared ill-trained in handling women issues and questions of group organization and psychology. There is no organized supervision of this staff and no audit by Ministry staff of projects sponsored by the Ministry.

(j) Problems of parochial orientation: Many groups deal with petty issues. Most women in the study area are concerned with their immediate household's survival needs and do not concern themselves with the larger problems facing women in Maragoli, Vihiga, Kenya and at the global level. They claimed that such problems were beyond their control. The groups in North Maragoli are not at all radical.

(k) Problems of male intervention: Men were found to contribute significantly to the problems in women's groups movement in this area. Many groups are patronized by men and male local government administrators and so women are not free to make their own independent decisions on projects and money issues. Many of the men who belonged to women group organizations as 'advisors' were found to be crafty and sly, and out to swindle money from these groups. This eventually caused the collapse of a number of groups. The main problem noticed in many female members of the women groups was their reliance and dependence on men for many things.

(l) Problem of legislation: Though women's groups have been officially encouraged, and are formally registered under the Ministry of Culture and Social Services, no institutional legislation exists for women's groups that could safeguard operation and protection. It appears that many rural women's groups operate in a void so that when wronged they cannot use the law to protect themselves.

(m) Finally, despite the existence of a women's bureau, the support of the government has been primarily verbal, focusing on staffing at the district level, rather than specific resources which actually reach local women. As pointed out by Thomas (1985), the Ministry of Culture and Social Services has placed a female officer in most districts to foster programmes and to assist women in their organizational efforts. However, these officers are hampered in their work by extremely limited budgets and resources and their lack of the relevant skills. Chiefs and Assistant Chiefs have been encouraged to support women groups. Such support most frequently consisted of exhortation to form women associations for development purposes. Often Chiefs and local leaders have used these groups for political and personal purposes.

The above discussion indicates that women groups' projects and, at a higher level women groups as an approach to rural development in North Maragoli are far from successful. Much more planning, management and supervision is required at the moment. Some of our findings about the shortcomings of women groups projects have also been pointed out by other researchers. It has been observed that while the aims and philosophical underpinnings of women's projects through women groups have steadily evolved, the actual form and forum of donor and government support and external intervention has not shown much change (Bryceson 1993:1). It may take a number of years before women groups become a successful instrument for rural development. This success will depend on the Government's, NGOs and International Organizations' commitment and support for this movement.

We conclude this section with an observation of Bryceson that "despite the diversity of approaches and forms of projects, two features remain common to all. First, the objective of project intervention is to raise women's status vis-a vis men. Second, as projects, these efforts are geographically localized and of limited duration. Their direct effect is usually restricted to relatively small numbers of women, whereas their indirect effect is difficult to measure. In theory these projects are high-minded and status conscious about women while in practice they are small and in need of status raising themselves." (Bryceson 1993:8).

8.11 Conclusions

In the current development debate it is often claimed that the women's groups movement appears to be an important strategy of getting women involved in agricultural and rural development. Being a member of a self-help group is considered to be an important coping strategy at the household and community level. Many women have joined such groups in the hope of benefitting financially and this hope has been reinforced because of the Government's encouragement for the formation of women's groups. Many groups have the objective of carrying out income-generating activities. However, our findings illustrate that the most women's groups members have yet to realize this objective. At the same time, the groups are involved in the welfare activities of supporting one another. Many of the projects seem to be focused on local affairs with no thought given to the major problems affecting women in Kenya in terms of employment, labour, education, economics and politics. The groups are disaggregated in nature with little or no linkages with urban women's organizations. The groups try to rely on external funding which is only limited to a few groups with political backing. This discourages the other groups and eventually results in non-participation of many members, in 'dormancy' of some groups and in the collapse of others. Most groups engage in projects that do not generate large sums of money on benefit the members because of limited investment capital and management problems.

We still believe that women's groups can be important instruments for agricultural and rural development. These groups can be used as a focus for agricultural extension services, credit and development projects in the rural areas by the Government and International Organizations. Using women groups rather than using the conventional 'contact farmer' approach may prove to be more efficient.

It has been argued in this Chapter that the present approach to women groups through existing government institutions can be characterized as top-down development, where funded projects are often selected by the sponsors and government agents. Although the women's groups programme has made some progress, success has been constrained by the failure to involve rural women as the target group in the decision-making process and the biased way in which some women's groups are selected for funding. A broad-based, participatory, development approach should be applied to all women's groups which implies the participation and empowerment of those involved in the decision-making process, the availability of capital, and continued and committed institutional support.

A consistent and coherent policy of 'women and development' is a necessary condition to ensure that the position of women and gender relations are taken into account. However, it is not a sufficient condition. Such a policy should be supplemented by a thorough analysis of the role of women in any project area to prevent biased assumptions about that role from becoming the starting point for project interventions.

At a more practical level there is an urgent need for experts on women's studies in Kenya, in Africa and in other Third World Countries who can suggest ways and means of removing the impediments facing rural women in order to achieve faster and more effective agricultural and rural development and to improve the poor standards of living of rural households. It is our contention that given a favourable environment with public support and increasing economic opportunity, women's groups will be able to address the needs and interests of individual women, as well as families and communities. Women's groups can enable women to participate both individually and collectively in development projects, and in doing so contribute to the process of modifying key economic and social relationships in the rural areas of Kenya (see also Thomas, 1985).

Plate 1 - Mother and Children Weeding their Maize Field

Plate 2 - Types of Food Crops: Yams and Onions in the Foreground, 'Sukuma Wiki' in the Middle, Maize and Bananas in the Background

Plate 3 - Healthy Maize

Plate 4 - Maize Badly Affected by Hailstorm

Plate 5 - Tea Fields

Plate 6 - Plucking Tea

Plate 7 - Coffee Trees with Food Crops

Plate 8 - Organic Farming of Vegetable Plots

Plate 9 - Zero-Grazing Unit with Grade Cow

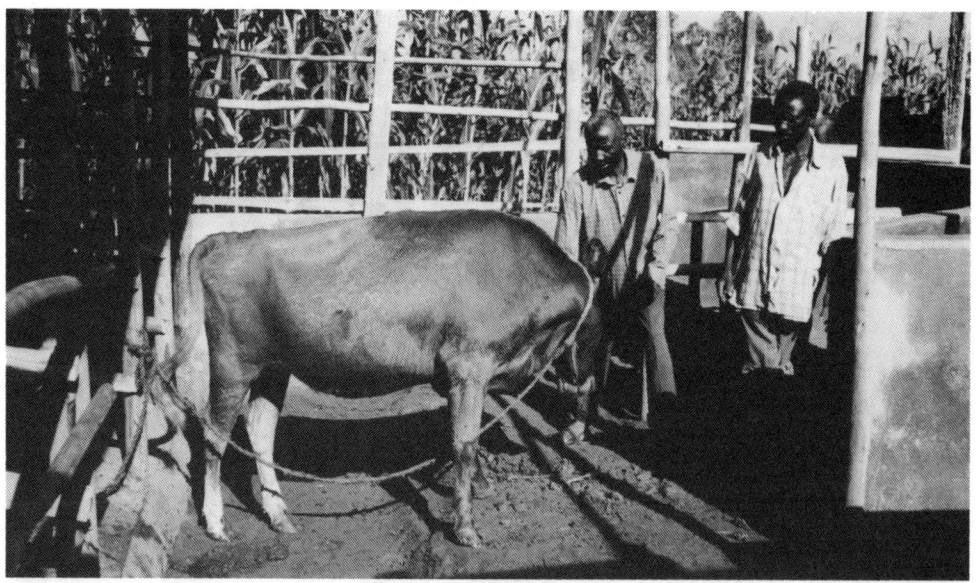

Plate 10 - Feeding a Cross-Bred Calf on the Homestead

Woman Groups as a Strategy

Plate 11 - Eucalyptus Groves in Marshy Valley and Tea Fields on Slopes

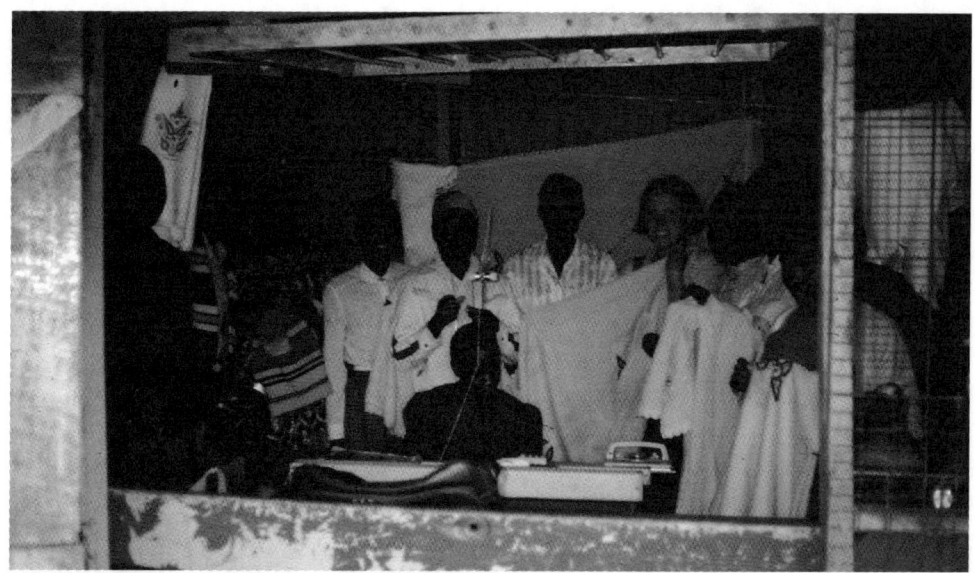

Plate 12 - Women Group Engaged in Tailoring

Plate 13 - Women Group Displaying Home-made Pottery

Plate 14 - Women Choir Practising

Plate 15 - Future Farmers

Notes

1 The population of women group members and the number of groups could be higher because no systematic, organized and supervised countrywide census and inventory has been done. This means that most likely more than the present 5% of the population are enrolled in women groups.

9

Conclusion

9.1 Introduction

This study has tried to improve our understanding of the complex position of small-scale farmers, especially female farmers who are confronted with multifarious constraints in both farm and household management. The study has found that farm management is poor in many farm-households and this contributes greatly to low agricultural production in the study area. In this concluding chapter we give a synopsis of the major problems in small-scale farming and the role of women in farm management. Finally, the principal findings of this study are again reviewed and compared with the research hypotheses.

9.2 The Major Problems in Small-Scale Farming

Agriculture has been the main source of livelihood in Kenya for many generations. However, small-scale farming is now considered by many to be a last option for employment and income generation. People's attitude to agriculture has changed from seeing it as an honourable way of life that can assure families of their livelihood. Today, agriculture is considered by many to be a dirty and non-remunerative activity. The reputation of agriculture will have to be rebuild if Kenya hopes to continue to depend on agriculture in the years to come and if it is to maintain its place as one of country's most important economic activities. The dilemma facing Kenya is the low agricultural production in many parts of the country particularly in small-scale farming.

Kenya's economy precariously relies on the small-scale farm sector with all the risks involved. More than half of the agricultural exports and slightly more than 60% of the food is produced by small-scale farmers. Moreover, the small-scale sector absorbs a large labour force, viz. 26% of those in wage employment and 80% of those who are self employed. It is therefore all the more serious that agricultural production is on the decline. What are the major problems that afflict agriculture in Vihiga District and in North Maragoli that may be generalized for Kenya as a whole?

The major problems identified contributing to poor agricultural production are first and foremost a gender bias and the consequent marginalization of many female farmers - a problem which is taken up again in the next section. Secondly a land tenure problem which negatively affects increasing numbers of people in rural areas. Indeed, with declining farm sizes many families are threatened in their subsistence and are becoming paupers. Many of the small-scale farmers in our study area do not qualify for credit because of their small farm sizes. The survival of small-scale farming is not only a problem for the farmers but also for the Kenyan Government. Shrinking farm sizes have forced many males to migrate to the urban areas. The Government is therefore faced with the double problem of providing enough foodstuffs on the market not only for the large mass of the unemployed in the urban areas but also for the rural households. The vexing question is what way forward is there for the growing numbers of small-scale farmers with holdings that are often less than one hectare in size and which are poorly managed and therefore unable to produce sufficient food for their households? These mini-farms are even threatened with further subdivision and fragmentation because of inheritance customs, thus worsening farm management and agricultural production still further.

Apart from a gender bias and a land problem, there is the problem of inadequate institutional support. Farm inputs were found to be too expensive for more than three-quarters of the small-scale farmers in our study area. Some of the farm inputs, especially the improved seed varieties are of poor quality and more often than not inputs are delivered late or there are perpetual shortages. Under such unfortunate circumstances farmers cannot plan farming activities properly, which again results in operations not being carried out at the right time and consequent low production.

Many farm managers are illiterate or semi-illiterate which explains the important role of agricultural extension. But our findings indicate that agricultural extension services suffer from many shortcomings such as ill-equipped staff and a narrow, top-down approach which means that few farmers are reached. Institutional support for cash crops such as coffee leaves much to be desired and the same goes for hybrid maize. The result is poor management and low agricultural production.

The question of agricultural labour has received very little attention from agricultural policy-makers and planners. It is assumed that family labour is abundant and able to provide all the labour a farm needs. However, many households experienced shortages of labour, another reason for poor farm management and declining agricultural production. Many smallholders can no longer meet their food needs and financial obligations through farming.

The agricultural sector is deteriorating because of these problems. It offers no security and has lost prestige partly because of the low and delayed payments for agricultural products. Lack of adequate facilities such as roads, and electricity in the rural areas contribute to delays in the transportation and processing of agricultural products. Many small-scale farmers are desperate and not sure how they will survive the next day. Agriculture is becoming increasingly associated with poverty, illiteracy and women. How long will rural households in areas with decreasing farm sizes continue to consider agriculture as a means of survival when it does not compare well with other sectors of the economy?

Many small-scale farmers have diversified their sources of income and the relative importance of agriculture has declined, even if agriculture is still the major source of income for many rural households. Income is often invested in off-farm activities which tend to be more profitable. Education of the children, the purchase of food for the household and off-farm businesses are considered priority investments. The families that can be considered well-off by local standards in our study area are those who did not consider agriculture the main basis of their households' economy. The low investment in farm improvements threatens the stability and sustainability of the farming system in Vihiga District and in Kenya in general.

9.3 The Role of Women in Farm Management

To understand and influence patterns of farm management, agricultural production, consumption levels and investments at the farm-household level, we must move beyond the conventional conception of the household as a production and consumption unit, with the male as head and decision maker. First, we have to identify three different types of households in the rural small-scale farming areas of Kenya: male-headed, female-headed and female-managed households. Second, we must look at the 'inner workings' of these households and take account of intra-household interests and conflicting relationships. This is because gender-related work patterns are critical to the system of farm and household management and to agricultural production with women being the principal managers on the farms regardless of the type of household. The key to improving agricultural production are policies which are attuned to the needs of the farm manager with due consideration of his/her type of household.

This study has attempted to show that knowingly or unknowingly the State does not always operate in the interest of women, especially those in the rural areas. In fact, the government plays a major role in perpetuating social, economic and ideological processes that subordinate women. Most agricultural sector policies in Kenya are oblivious to gender and women are invisible to policy makers. Consequently, these policies fail to improve farm management and agricultural production. Women are frequently treated as dependents of men in resource ownership and use, in access to agro-support services (extension and credit) and in administrative procedures rather than as persons in their own right. The government and the cultural setting upholds patriarchal family forms in which women do not have the same control and access to resources as men. All this points to a marginalization of women and explains the poor performance of agriculture, especially in the smallholder sector where women dominate.

Any programme or policy aimed at accelerating the development of small farms should be based on an awareness of their relevance and of their likely impact on both women and men in any given household. For instance, price policies which provide production incentives should take full account of the costs of production and the farmer's responsiveness. Programmes for institutional change, aimed at making resources and new knowledge more readily available to farmers are doomed to fail if they are not guided by an understanding of existing institutions, current practices and farmer's attitudes, especially those of women. Above all, research and the development of new technology should be directed towards meeting farmer's objectives, especially those of women farmers, overcoming their constraints and making use of indigenous technology where relevant. This implies that researchers and agricultural planners should have a thorough knowledge of what farmers do and why they do it. Such knowledge can only be obtained from the farmers themselves and so field studies and the collection of data are essential. A closer interaction between farmers, regardless of gender, and agricultural research stations, personnel in these research stations, their findings and the agricultural staff is vital.

To reduce household poverty farmers adopt the strategy of diversification of sources of income and individualization of economic activities. Both men and women engage in a diversity of on and off farm income generating activities on an individual basis, thereby protecting the household from the risks of crop failure, changes in market supply and demand, and government intervention. However, a big difference was found between the various types of household with regard to these activities. The male-headed households had a more stable means of incomes from outside farming. Men also appeared to spend their incomes on resource investment. Women appeared to be more involved in less remunerative activities like petty trade and were found to spend all their money on consumptive needs. Therefore, a need exists for more government and non-governmental support to women both in the field of technology and capital.

Women groups appear to be a promising strategy for mobilizing rural women. Such groups may be used as forums for creating awareness and bringing about improvements in the social and economic lives of women. Many women are joining and forming these groups all over the country regardless of their resource disposition. Therefore, the strengthening of the women's groups movement as a strategy for household, agricultural and rural development is essential. Although much has been theorized nationally and internationally on women and women movements since the 1970s, not much input has been directed towards them. Indeed, no legal basis has been drawn on which women groups may petition for support or defence. Their neglect is a further confirmation of the marginalization of women and the lip service paid to gender problems in policy documents. The government and the many NGOs involved in assisting women's groups have not abandoned their very centralized structure which prevent them from reaching most rural women's groups. The few groups in our study area that were found to benefit from government support appeared to have a strong connection with politicians. Women's groups as a strategy for agricultural and rural development will only be successful if there is de-linking from the politics of favouritism and corruption and if they become actively involved in decision-making with respect to agricultural and rural development.

9.4 The Research Hypotheses Revisited

The subject of this study were the problems of small-scale farming in densely populated Vihiga District, a case study of North Maragoli Location and in particular the problems women face as farmers and managers in various types of households. Four hypotheses were proposed to guide the research. In this section an attempt is made to revisit and answer the four hypotheses: (1) female-headed and female-managed households have less control over land and less access to agro-support services than male-headed households, which negatively influence their farm-management practices; (2) female-headed and female-managed households experience more labour constraints than male-headed households which leads to poor farm management practices; (3) female-headed and female-managed households benefit less from non-farm sources of income than male-headed households with negative effects on farm management; and (4) because of poor farm management, female-headed and female-managed farms compare unfavourably with male-headed ones in investments, crop yields and farm income. Indeed, if one has to give a single answer to these hypotheses, the answer is in the affirmative in that female-headed and female-managed households face all the mentioned constraints and more so than male-headed ones.

Control Over Land and Access to Agro-Support

Apart from economic reasons, other factors such as differences in education levels, the prevailing socio-cultural norms, the legacy of the colonization process which incorporated men in particular into the evolving money economy, the continuation of gender biased policies, and the declaration of males as the heads of households and as bread-winners have contributed to the migration of many men from the rural small-scale farming areas to urban centres. In doing so men leave behind wives as secondary managers on the family farms and in the households. A situation has arisen where at least 55% of the households and farms in North Maragoli and Vihiga are managed and headed by women. But land and other properties (cows, trees, buildings) among the Luhya can only be inherited by males because this is a patriarchal society. This also applies to the land tenure system in Kenya in general. Therefore, women have no control over land in small-scale farming but have access to it by virtue of being married or related to men. Major decisions on land use and the distribution of land for the different farming sub-systems are made by men. This leaves women in an economically marginal position only providing farm labour and producing food crops for the household. In the commercial farming sub-system men are recognized cash crop farmers and the members of co-operatives which market these crops, despite the fact that women provide most of the labour. A high percentage of men thus continues to earn from these crops but do not provide the labour and cash required. Collateral in small-scale farming is associated with a land title deed. Women are handicapped in farm improvement as they have no access to credit since they lack collateral. The extension services recognize men as the farmers and therefore men receive more agro-support than women. Without extension many women are handicapped in farm management, especially when they grow such crops as coffee, tea, hybrid maize or have improved or grade cattle that require specialized management. Many women experience insecurity as their access to land is not assured because of polygamy and the fact that their spouses can sell land without consulting the women. Therefore, farm management practices are not only negatively influenced in female-headed and female-managed households. In male-headed households as well, farm management is negatively affected because the wife's labour is inadequately remunerated and, conflicts arising about farm incomes between husband and wife, further there is insecurity for the woman, and she lacks of ownership over resources and has little access to agro-support services by the woman.

Labour Constraints

Women's participation and involvement in agricultural production is increasing continually, especially with growing land shortages and male migration in search of wage employment, and women's low opportunity cost in the labour market among other factors. Many rural households have no male head because the husband is working elsewhere or has died, and the children above the age of six are at school. This often only leaves women as the sole providers of agricultural labour. Labour in the agricultural sector is thus added to the women's household chores and their reproductive roles in the domestic sector. It is impossible for them to provide the labour required in both these sectors. Due to the scarcity and lack of money in female-headed and in many female-managed households it is also impossible to supplement this labour shortage with hired agricultural labour. This results in conflicting labour demands on the female-head or female household manager between her duties on the farm and in the household. It also leads to low labour productivity as the women have to rush to try to accomplish their farm work done. In fact, on female-headed household farms, the women often have to hire out their labour in order to raise money to buy food, thus reducing even further the scarce labour supply on their own farms. On the other hand, in male-headed households, the men are able to contract labour, especially for the commercial farming sub-system, and this explains the higher level of farm management and agricultural production on these farms. Gender therefore plays an important role in the supply and use of labour at the household level.

Non-farm sources of income
More than 90% of rural women are not in any formal employment. Farming is considered to be employment, but many do not derive any earnings from their agricultural work. Therefore, many female-headed households benefit the least from non-farm income. Female-managed households often rely on the remittances from spouses which are often unsubstantial and very irregular. On the other hand, many men in male-headed households often have a non-farm source of income - either a regular salary, pension, or earnings from a business - in addition to their farm income. Women therefore lack the money necessary to invest in agriculture. Because of money shortage the use of hybrid seeds, fertilizers and hired labour is lowest on female-headed and female-managed household farms, thus negatively affecting farm management and agricultural production.

General Farm Situation
Due to the many constraints that women face, their farm management is poor. Female-headed and female-managed farms especially were found to suffer from low investments, poor crop yields and small farm incomes and therefore compared unfavourably with male-headed households.

Women are the main force in small-scale farming. Therefore any improvement in this sector can only take place when institutions incorporate women into their programmes and involve them in decision-making. Our findings have shown that women have a positive attitude towards agriculture and are capable of good management, provided they have access to the same resources as men. Despite the multifarious constraints facing women farmers, women, as this study has shown, still manage to keep on farming and maintaining their households. In this respect they are working wonders.

References

Acland, J.D. 1971. East African Crops. Rome: FAO, Longman.

Adams, W.M. 1992, Wasting the Rain: Rivers, People and Planning in Africa. London: Earthscan Publications.

Afshar, H. & C. Dennis (eds). 1993. Women and Adjustment Policies in the Third World. London: Macmillan.

Agarwal, B. 1992. Gender Relations and Food Security: Coping With Seasonality, Drought, and Famine in South Asia In Unequal Burden: Economic Crises, Persistent Poverty, and Women's Work edited by Beneria, L. and Feldman, S. (1992). Boulder. San Francisco. Oxford: Westview Press.

Agarwal, B. 1994. A Field of One's Own: Gender and Land Rights in South Asia. Cambridge: Cambridge University Press.

Akong'a, J.J. 1994. Indigenous and Native: a False Dichotomy in State Formation in Africa. A Paper Presented at the CERES Summer School/CEDLA Workshop on "Popular Culture: Beyond Historical Legacy and Political Innocence." Amsterdam: University of Amsterdam.

Alexander, M.C. and Scott, J.P. 1975. The Implications of Group Credit for Rural Development in Malawi - in Wilson, F.A. and Amann, V.F. (eds) - Financing Rural Development : Kampala: Eastern Africa Agricultural Economics Society and the Makerere Institute of Social Research.

Allan, W.E. 1965. The African Husbandman. New York: Barnes and Noble.

Amara, H. and T. B.F, (ed) (1990), African Agriculture: The Critical Choices. New York: Zed Books Ltd, United Nations University Press.

Andreasen, J. 1990. Urban-Rural Linkages and Their Impact on Urban Housing in Kenya in : Baker, J. (ed) 1990. Small Town Africa: Studies in Rural-Urban Interaction. Seminar Proceedings No. 23. Uppsala: The Scandinavian Institute of African Studies.

Anker, R. & Knowles, J.C. 1983. Population Growth, Employment and Economic-Demographic Interactions in Kenya: Bachue-Kenya. Gower. New York: St. Martins Press.

Antholt, C.H. 1994. Strategic Issues for Agricultural Extension in Pakistan: Looking Back to Look Ahead. In the Proceedings from Seminar on "Productivity Through Agricultural Extension in Pakistan: Isamabad. March 1990.

Antholt, C.H. 1994. Getting Ready for the Twenty-First Century: Technical Change and Institutional Modernization in Agriculture. World Bank Technical Paper No. 217. The World Bank: Washington, D.C.

April, A. & Gordon, L. (eds). 1992. Understanding Contemporary Africa. Boulder. London: Lynne Rienne Publishers.

Bahemuka, M.J. 1987. Socio-Cultural Practices Related to Grain Storage in Western Kenya. A Field Manual for Agriculture Extension Personnel. Nairobi: Development Farming and Research Association.

Barnes, C. 1983. Differentiation by Sex Among Small-Scale Farming Households in Kenya. Rural Africana. 15-16 pp 41-63.

Barret, M. 1987. Women's Income-Generating Initiatives in Kenya: Self-Report Perceptions of the Need for and Value of Women's Groups. African Urban Quarterly Vol. 2 No. 4 Nov. 1987.
Bates, R.H. 1989. Beyond the Miracle of the Market: The Political Economy of Agrarian Development in Kenya. Cambridge: Cambridge University Press.
Baud, I.S.A. & P. Pennartz. (eds). 1986. Women as Workers. Report on the IRENE-EADI Seminar on the Information of Women's Labour. Tilburg: University of Brabant. May 1986.
Baud, I.S.A. 1989. Forms of Production and Women's Labour: Gender Aspects of Industrialization in India and Mexico. Ph.D Thesis. Eindhoven: Technical University of Eindhoven.
Bay, E.G. (ed) 1982. Women and Work in Africa. Westview Special Studies on Africa. Boulder. Colorado: Westview Press.
Becker, G.S. 1981. A Treatise on the Family. Cambridge. Massachussets: Harvard University Press.
Beneria, L. & S. Feldman. (eds). 1992. Unequal Burden: Economic Crises, Persistent Poverty and Women's Work. Boulder. San Francisco. Oxford: Westview Press.
Beneria L. 1981. Conceptualizing the Labour Force: The Underestimation of Women's Economic Activities. In N. Nelson (ed). 1981. African Women in the Development Process. London: Cass. pp 10-28.
Bernstein, H. 1976. African Pleasantries: a Theoretical Framework. The Journal of Peasant Studies. 6: 4, pp. 421-443.
Bernstein, H., B. Crow, M. Mackintosh, & C. Martin. 1990. The Food Question - Profits Versus People? London: Earthscan Publication Ltd.
Bernstein, H., B. Crow, & Johnson. 1992. Rural Livelihoods: Crises and Responses. Oxford: Oxford University Press.
Bevan, D., P. Collier, & J.W. Gunning. 1989. With Arne Bigsten and Paul Hornsnell. Peasants and Governments. An Economic Analysis. Oxford: Clarendon Press.
Bifani, P. 1985. Women and Development in Africa: A Tentative Approach Through Scenario Building in : Were, G.S. (ed). 1985. Women and Development in Africa. Nairobi, Kenya: Gideon S. Were Press.
Bindlish, V. & R. Evenson. 1993. Evaluation of the Performance of T&V Extension in Kenya. Agriculture and Rural Development Series Number 7, Technical Department, Africa Region, March. The World Bank: Washington, D.C.
Blaikie, P. 1978. The Theory of the Spatial Diffusion of Agricultural Innovations: A Specious cul-de-sac. Progress in Human Geography. 2, pp. 268-95.
Blaikie, P. 1985. The Political Economy of Soil Erosion in Developing Countries. New York: John Wiley & Sons Ltd.
Blaikie, P. and H. Brookfield. 1987. Land Degradation and Society. New York: Methuen & Sons.
Blaikie, P. 1989. Explanation and Policy in Land Degradation and Rehabilitation for Developing Countries. New York: John Wiley & Sons Ltd.
Blalock Jr. M.H. 1979. Social Statistics. Lisbon. London. Mexico. Paris. Tokyo: McGraw-Hill International Edition.
Boserup, E. 1965. The Conditions of Agricultural Change. The Economics of Agrarian Change Under Population Pressure. London: Allen and Unwin.
Boserup, E. 1970. Woman's Role in Economic Development. London: Earthscan Publications Ltd.

Bradley, P. 1984. The District Resource Analysis as Applied to Kakamega District. KWDP Working Paper No. 2, Nairobi: The Beijer Institute.

Bradley, P. 1987. District Resource Analysis for Woodfuel Planning: An Overview. Nairobi: Beijer Institute.

Brack, B. 1977. Political Administration and Rural Development in Kenya...Lessons from Settlement and other Rural Development Programmes. Ph.D Thesis. Syracuse: Syracuse University.

Bradshaw, Y.W. 1990. Perpetuating Underdevelopment in Kenya: The Link Between Agriculture, Class, and State. African Studies Review, Vol. 33, No. 1, (1990).

Briggs D. & F. Courtney. 1985. Agriculture and Environment, The Physical Geography of Temperate Agricultural Systems. Singapore: Longman Singapore Publishers (Pete) Ltd.

Brokensha, D.W. & P.D. Little. (eds). 1988. Anthropology of Development and Change in East Africa. Boulder. London: Westview Press.

Brown, K. 1990. Women's Farming Groups in a Semi-Arid Area of Kenya, A Case Study from Tharaka, Meru District. Unpublished Ph.D Thesis, Nottingham: University of Nottingham.

Brown, K. 1991. Assessing Poverty among Women Farmers in Tharaka, Meru District, Kenya. Discussion Paper Presented at an Agricultural Economics Conference, June 1991.

Brown, R.P.C. 1992. Migrants' Remittances, Capital Flight, and Macroeconomic Imbalance in Sudan's Hidden Economy. The Journal of African Economies, Vol. 2, No. 1 1992, pp. 87-108.

Brundtland, G.M. et al. 1987. Our Common Future: The World Commission on Environment and Development. Oxford. New York: Oxford University Press.

Bryceson, D.F. 1990. Food Security and the Social Division of Labour in Tanzania, 1919-85. New York: St. Martin's Press.

Bryceson, D.F. 1993. Easing Rural Women's Working Day in Sub-Saharan Africa. Working Paper Vol. 16/1993, Leiden: African Studies Centre.

Bryceson, D.F. & M. McCall. 1994. Lightening the Load: Women' labour and Appropriate Rural Technology in Sub-Saharan Africa. Working Paper Vol. 21/ 1994. Leiden: African Studies Centre.

Bryceson, D.F. 1995. "Wishful Thinking: Theory and Practice of the Western Donors' Efforts to Raise Women's Status in Rural Africa in: Bryceson, D.F. (ed). Women Wielding the Hoe: Lessons from Rural Africa for Feminist Theory and Development Practice. Oxford: Berg Publishers.

Buchanan. I.W., J.A. Errington, & K.A. Giler. 1982. The Farmer's Wife. Her Role in the Management of the Business. Study No.2. Reading: University of Reading. Department of Agriculture.

Bullock, R.A. 1971. Population and Food in West and Central Kenya. Studies in East African Geography and Development. Edited by S.H. Ominde. London. Nairobi. Ibadan: Heinemann Publishers.

Bulow, D. von. & A. Sorensen. 1993. Gender and Contract Farming: Tea Outgrower Schemes in Kenya. Review of African Political Economy No.56: 38-52.

Burfisher, M. & N.R. Horenstein. 1985. Sex Roles and Development Effects on the Nigerian Tiv Farm Household. Rural Africana, 21 (Winter 1985), pp. 31-49.

Burger, C. P. J. 1994. Farm Household, Cash Income and Food Production: The Case of Kenyan Smallholdings. Amsterdam: Free University of Amsterdam.

Buvinic, M. 1986. Projects for Women in the Third World: Explaining their Misbehaviour. World Development, 14, No.5. pp. 653-664, 1986.

Casley, D.J. & D.A. Lury. 1987. Data Collection in Developing Countries. Oxford: Clarendon Press.

Chambers, R. 1975. Two Frontiers in Rural Management: Agricultural Extension and Managing the Exploitation of Communal Natural Resources. Institute of Development Studies at the University of Sussex, Brighton, England. IDS Communication 113.

Chambers R. 1983. Rural Development. Putting the last First. Longman Scientific Technical Section. New York: Co-Publisher John Wiley & Sons Inc.

Chambers, R. & B.P. Ghildyal. 1985. Agricultural Research for Resource-Poor Farmers: The Farmer-First and Last-Model. Agricultural Administration, No.20.

Chambers, R. & Jiggins, J. 1987. Agricultural Research for Resource Poor Farmers : Transfer of Technology and Farming Systems Research. The University of Sussex, Brighton, England. Agricultural Administration No.27.

Chambers, R., A. Pacey. & L.A. Thrupp. 1989. Farmer First, Farmer Innovation and Agricultural Research. London: Intermediate Technology Publications.

Chavangi, N. 1984. Cultural Aspects of Fuelwood Procurement in Kakamega District. KWDP. Working Paper No. 4. Nairobi: The Beijer Institute.

Christenen, C. et al. 1981. Food Problems and Prospects in Sub-Saharan Africa: The Decade of the 1980s. Washington D.C.: Department of Agriculture, FAER No. 166.

Clark, C.M. 1980. Land and Food, Women and Power, in Nineteenth Century Kikuyu. Africa, 50 (4), 1980, pp. 356-369.

Clark, M.H. 1983. Woman Headed Households and Poverty: Insights from Kenya in: Gelpi, B.C., N.C.M. Hartsock, C.C. Novak & M.H. Strober. (eds). 1983. Women and Poverty. Chicago. London: The University of Chicago Press. pp. 103-119.

Clark, M.H. 1985. Household Economic Strategies and Support Networks of the Poor in Kenya. A Literature Review. Discussion Paper. Report No. UDD-69. Water Supply and Urban Development Department Operations Policy Staff. Washington D.C.: The World Bank.

Clayton, E.S. 1964. Agrarian Development in Peasant Economics: Some Lessons from Kenya. Oxford: Pergamon Press.

Clayton, E.S. 1970. A Comparative Study of Settlement Schemes in Kenya. University of London, Wye College, Agrarian Development Unit. Occasional Paper No.3.

Clayton, E.S. 1970. Agrarian Reform, Agricultural Planning and Employment in Kenya. International Labour Review, Vol. 102, No.5 November 1970.

Clayton, E. S. 1983. Agriculture, Poverty & Freedom in Developing Countries. London: The Macmillan Press Ltd.

Cloud, K. 1977. Sex Roles in Food Production and Food Distribution Systems in the Sahel. Washington D.C.: Agency for International Development.

Cloud, K. 1984 Women's Productivity in Agricultural Systems: Considerations for Project Design. In Overholt et al. (eds) pp 17-56. West Hartford: Kumarian Press.

Cohen, J.M. 1980. Land Tenure and Rural Development in Africa: In Agricultural Development in Africa, Issues of Public Policy. edited by Bates R.H. and Lofchie, M.F. New York: Praeger Publishers and Eastbourne U.K. Hort Saunders Ltd.

Collier P. and D. Lal. 1986. Labour and Poverty in Kenya 1900-1980. Oxford: Clarendon Press.

Collionson, M.P. 1978. Demonstration of an Interdisciplinary Approach to Planning Adaptive

Agricultural Research Programmes. Nairobi: Report No.3. CIMMYT Eastern Africa Research Programme.
Conelly, T.W. 1988. Insect and Weed Control in Subsistence Farming Systems: Western Kenya. Anthropology of Development and Change in East Africa (1988). Edited by Brokensha, D.W. and Little, P.D.
Sunday Nation, December 15, 1991. Nairobi: Nation Newspapers Ltd.
Sunday Nation, March 8, 1992. Nairobi: Nation Newspapers Ltd.
Daily Nation, Tuesday, May 5, 1992. Nairobi: Nation Newspapers Ltd.
Daily Nation, Friday, May 22, 1992. Nairobi: Nation Newspapers Ltd.
Daily Nation, Thursday, July 9, 1992. Nairobi: Nation Newspapers Ltd.
Daily Nation, Tuesday, July 14, 1992. Nairobi: Nation Newspapers Ltd.
Sunday Nation, August 23, 1992. Nairobi: Nation Newspapers Ltd.
Daily Nation, Monday, August 24, 1992. Nairobi: Nation Newspapers Ltd.
Daily Nation, Friday, July 24, 1992. Nairobi: Nation Newspapers Ltd.
Daily Nation, Tuesday, January 3, 1995. Nairobi: Nation Newspapers Ltd.
Sunday Nation, January 8, 1995. Nairobi: Nation Newspapers Ltd.
Daily Nation, Wednesday, January 11, 1995. Nairobi: Nation Newspapers Ltd.
Dale, P.F. & J.D. Mclanghlin. 1988. Land Information Management: An Introduction with Special Reference to Cadastral Problems in Third World Countries. Oxford: Clarendon Press.
Dankelman, I. & J. Davidson. 1988. Women and Environment in the Third World. London: Earthscan Publications Ltd.
Davison, J. 1985. Achievements and Constraints Among Rural Kenyan Women: A Case Study. in Were, G.S (1985), Women and Development in Africa. Nairobi, Kenya: Gideon S. Were Press,
Davison, J. 1987. Without Land We Are Nothing. The Effect of Land Tenure Policies and Practices Upon Rural Women in Kenya. Rural Africana, 27 pp 19-33.
Davison, J. (ed). 1988. Agriculture, Women, and Land: The African Experience. Boulder. London: Westview Press.
De Wilde, C.J. 1967. Vol.II. Agricultural Development in Tropical Africa. Vols.I and II. Baltimore. Maryland: John Hopkins Press.
Deere, C.D. 1985. Rural Women and State Policy : The Latin American Agrarian Reform Experience. World Development, 13, No.9, pp. 1037-1053, 1985. Great Britain: Pergmon Press.
Desai, G. 1988. Policy for Rapid Growth in Use of Modern Agricultural Inputs. Agricultural Price Policy for Developing Countries, eds. J.W. Mellor and R.U. Ahmed, Ch. 12. Baltimore: Johns Hopkins.
Dey, J. 1981. Gambian Women: Unequal Partners in Rice Development Projects. Journal of Development Studies, 17.
Dey, J. 1984. Women in Food Production and Food Security in Africa. Rome: FAO.
Dey, J. 1985. Women in African Rice Farming Systems. In Women in Rice Farming, International Rice Research Institute (IRRI), Ch. 23. Aldershot: Gower.
Dietz, A. 1986. Migration To and From Dry Areas in Kenya. Amsterdam: Published by the Royal Dutch Geographical Society KNAG.
Dietz T. 1987. Pastoralists in Dire Straits: Survival Strategies and External Interventions in a

Semi-Arid Region at the Kenya/Uganda Border: Western Pokot 1900-1986. Amsterdam: Institute of Social Geography, University of Amsterdam.
Dietz, A.J., J.M. Haastrecht. and S.J. Dima. 1994. You Don't Learn By Dodging! Report For ETC Kenya. Eldoret: School of Environmental Studies, Moi University.
Dillion, J.L. 1980. The Definition Of Farm Management. Journal of Agricultural Economics. Vol.31 (1980) No. 2.
Dixon-Mueller, R. 1985. Women's Work in Third World Agriculture: Concepts and Indicators. Geneva: ILO.
Dongen, R.P.M. & J. Frenkel. 1990. Land Evaluation in Kakamega District: A Qualitative and Quantitative Land Evaluation for Soy and Lumakanda Locations, West Kenya. Utrecht: University of Utrecht, Faculty of Geographical Sciences, The Netherlands.
Doorenbos, J., B. Haverkort. & J. Jiggins. 1988. Women and the Rationalisation of Smallholder Agriculture. Agriculture, Administration and Extension, Vol. 28, 1988, pp. 101-112.
Dryson, J.C. 1981. Women and Agriculture in Sub-Sahara Africa: Implications for Development. In Nelson (ed). 1981. African Women in the Development Process. London: Frank Press. pp 29-46.
Due, J.M. 1988. Intra-Household Gender Issues in Farming Systems in Tanzania, Zambia and Malawi. In Poats et al. pp 331-343.
Due, J.M., E. Sikapande. & F. Magayans. 1988. Does T&V Extension Assist Female-Headed Families? Some Recent Evidence From Zambia. East African Economic Review. 1988.
Edalia, J.F. 1980. Hybrid Maize Production in South Maragoli: Farm Size and Yields. Unpublished B.A. Dissertation. Nairobi: University of Nairobi.
Egsmose, R.K. 1990. Women and Agricultural Production: Strategies for Extension Services in Kenya. Danidas Dokumentation Central.
Ekholm, E.P. 1976. Losing Ground, Environmental Stress and World Food Prospects. New York: Norton.
Ellis, F. 1993. Peasant Economics: Farm Households and Agrarian Development. 2nd edn. Cambridge: Cambridge University Press.
Ellis, F. 1994. Agricultural Policies in Developing Countries. 2nd edn. Cambridge: Cambridge University Press.
Elson, D. 1992. From Survival Strategies to Transformation Strategies: Women's Needs and Structural Adjustment. In Unequal Burden: Economic Crises Persistent Poverty, and Women's Work edited by Beneria, L. and Feldman, S. (1992). Boulder. San Francisco. Oxford: Westview Press.
Errington, A. 1986. The Farm as a Family Business. (An Annotated Bibliography). Reading: Reading University.
Europa. 1990. Africa South of the Sahara 1991- Twentieth Edition. Europa Publications Limited.
Evans, A. 1989. Gender Issues in Rural Household Economics. IDS Discussion Paper 254.
Evans, A. 1991. Gender Issues in Rural Household Economics. IDS Bulletin 22.1 pp 50-59.
Evans, H.E. & P. Ngua. 1991. Rural-Urban Relations, Household Income Diversification and Agricultural Productivity. Development and Change (SAGE, London, Newbury Park and New Delhi), Vol. 22 (1991) pp. 519-545.
Evans, H.E. 1992. A Virtuous Circle Model of Rural-Urban Development: Evidence from a Kenyan Small Town and Its Hinterland. The Journal of Development Studies, Vol. 28, No. 4, July 1992, pp. 640-667.

FAO, 1976. Soil Conservation and Management in the Developing Countries. FAO Soils Bulletin 33, FAO: Rome.
FAO, 1984. Tillage Systems for Soil and Water Conservation. FAO Soils Bulletin 54, FAO: Rome.
FAO, 1989. Farming Systems Development, Concepts, Methods and Applications. Food and Agriculture Organization of the United Nations. Rome. 1989.
FAO, 1990. Farming Systems Development. Guidelines for the Conduct of a Training Course in Farming Systems Development. Food and Agriculture Organization of the United Nations. Rome. 1990.
FAO, 1992. The State of Food and Agriculture 1991. FAO Agricultural Series. No. 24. Food and Agriculture Organization of the United Nations. Rome. 1992.
FAO, 1992. The State of Food and Agriculture. Agricultural Series No.25. Food and Agriculture Organization of the United Nations. Rome. 1992.
FAO, 1993. The State of Food and Agriculture 1993. FAO Agriculture Series, No. 26. Food and Agriculture Organization of the United Nations. Rome. 1993.
Fapohunda, E.P. 1987. The Nuclear Household Model in Nigerian Public and Private Sector Policy: Colonial Legacy and Socio-Political Implications. Development and Change. Vol. 18, pp. 281-294.
Farrinton, J. 1994. Public Sector Agricultural Extension: Is There Life After Structural Adjustment? London: ODI Publication. Natural Resource Perspectives. No. 2, Nov.1994.
Feldman, R. 1984. Women's Groups and Women's Subordination: An Analysis of Policies Towards Rural Women in Kenya. Review of African Political Economy 27/28: 67-85. 1984.
Feldstein, H.S. & S.V. Poats. 1990. Conceptual Framework for Gender Analysis in Farming Systems Research and Extension. In Feldstein, H.S. and S.V. Poats (eds), Working Together. Gender Analysis in Agriculture. Volume 1: Case Studies. Boulder: Westview Press. 1990, pp. 7-37.
Fogelberg, T.C. 1990. Women as a Motor in Agricultural Development: Experiences from Western Sub-Saharan Africa. In Beyond Adjustment: Sub-Saharan Africa (Ministry of Foreign Affairs, Directorate General for International Cooperation, Project Group Africa). The Hague: Ministry of Foreign Affairs (cop. 1990).
Foibre, N. 1984. Household Production in the Philippines: a Non-Neoclassical Approach. Economic Development and Cultural Change 32 (1984), pp. 303-330.
Francis, E. & J. Hoddinott. 1993. Migration and Differentiation in Western Kenya: A Tale of Two Sub-Locations. The Journal of Development Studies. Vol. 30, No. 1. October 1993, pp. 115-145.
Franck, G.M.A. (ed). 1990. Establishing a District Information and Documentation Centre: Experiences from the Kakamega District Focus Project. Utrecht: University of Utrecht, Netherlands.
Fresco, L. 1985. Food Security and Women: Implications for Agricultural Research. In Bisilliat, J. (ed). Femmes et Politiques Alimentaires: Actes du Seminaire International sur la Place des Femmes dans I'Autosuffisance et les Strategies Alimentaires, ORSTOM-CIE, Paris, 1985, pp. 94-106.
Fresco, L.O. & S.V. Poats 1986. Farming Systems Research and Extension: An Approach to Solving Food Problems in Africa. In Hansen A. and D.E. McMillan (1986), Food in Sub-Saharan Africa. Boulder. Colorado: Lynne Rienner Publishers, Inc.

Fresco, L.O. 1988. Farming Systems Analysis, An Introduction. Tropical Crops Communication No. 13 (1988). Wageningen: Department of Tropical Crop Science. Agricultural University, Wageningen.

Fresco, L.O., H.G.J. Huizing, H. Keulen, H.A. Luning & R.A. Schipper. 1992. Land Evaluation and Farming Systems Analysis for Land Use Planning. FAO Working Document Jan. 1992. Rome.

Fuglesang, A. 1982. About Understanding: Ideas and Observations on Cross-Cultural Communication. Uppsala: Dag Hammmarskjoeld Foundation.

Garfield, E. (ed). 1978. Women in Kenya. Nairobi: Central Bureau of Statistics, Ministry of Finance and Planning. Republic of Kenya. Consultant Report.

Garfield, E.B. 1979. The Impact of Technical Change on the Rural Household: Evidence from the Integrated Agricultural Development Program. A Research Proposal and Literature Review. Working Paper No. 358. I.D.S. Nairobi: University of Nairobi.

Geisler, G. 1993. Silence Speak Louder Than Claims: Gender, Household, and Agricultural Development in Southern Africa. World Development, Vol. 21, No. 12, pp. 1965-1980. 1993. Great Britain: Pergamon Press.

Gelder, B. & P. Bradley. 1984. The Search for Viable Development Options. KWDP Working Paper No. 5. Nairobi: The Beijer Institute.

Gelder, B. & P. Kerkhof. 1984. The Agroforestry Survey in Kakamega District. KWDP Working Paper No. 3. Nairobi: The Beijer Institute.

Gerhart, J.D. 1974. The Diffusion of Hybrid Maize in Western Kenya. Ph.D Dissertation. Princeton: Princeton University.

Gerhart, J.D. 1986. Farming Systems Research, Productivity, and Equity. in Moock, J.L. and B.N. Okigbo (eds). (1986). Understanding Africa's Rural Households and Farming Systems. Boulder. London: Westview Press.

Ghai, D. & S. Rahwan. (eds). 1983. Agrarian Policies and Rural Poverty in Africa. Geneva: ILO

Gianotten, V. et al. 1994. Assessing the Gender Impact of Development Projects. U.K: Intermediate Technology Publications.

Gilbert, E.H. et al. 1980. Farming Systems Research: A critical Appraisal. A.S.U Rural Development Paper No 6. East Lansing (Ohio).

Gladwin, C.H. (ed). 1991. Structural Adjustment and African Women Farmers. Centre for African Studies. Florida: University of Florida Gainesville: University of Florida Press.

Gordon, A.A. 1992. Women and Development. in Gordon A.A et al. (eds). (1992). Understanding Contemporary Africa. Boulder. London: Lynner Rienner Publishers.

Grigg, D.B. 1976. Population Pressure and Agricultural Change. Progress in Geography. Vol. 8, pp. 133-176 by C. Board, R.J. Chorley, P. Hagget and D.R. Stoddart. London: Arnold.

Groenenboom, M. & M. Drift. 1991. Soil Survey in West Kenya. A Semi-Detailed Soil Survey and Qualitative Land Evaluation for North Maragoli Location (Kakamega District). Utrecht: University of Utrecht, Faculty of Geographical Sciences. The Netherlands.

Groot, C. de. 1991. Labour Migration: A Structure to Development. M.A. Thesis. Utrecht: University of Utrecht, The Netherlands.

Grosh, B. 1991. Public Enterprise in Kenya: What Works, What Doesn't, and Why. Boulder. London: Lynne Rienner Publishers,

Grosh, B. 1994. Contract Farming in Africa: An Application of the New Institutional Economics. Journal of African Economics, Vol. 3, No. 2, 1994.

Grosh, B. 1994. Kenya: A Positive Politics of Parastatal Performance. in Grosh, B. & R.S. Mukandala. (eds). (1994). Africa: State-Owned Enterprises in Africa. Boulder. London: Lynner Rienner Publishers.

Grosh, B. & R.S. Mukandala. (eds). 1994. Africa: State-Owned Enterprises in Africa. Boulder. London: Lynner Rienner Publishers.

Gugler, J. 1989. Women Stay on the Farm No More: Changing Patterns of Rural-Urban Migration in Sub-Saharan Africa. The Journal of Modern African Studies. 27, 2 (1989), pp. 347-352.

Gugler, J. 1991. Life in a Dual System Revisited: Urban-Rural Ties in Enugu, Nigeria, 1961-87. World Development, Vol. 19, No. 5, pp. 399-409, 1991. Great Britain: Pergamon Press.

Gustafsson, B. & N. Makonnen. 1990. Poverty and Remittances in Lesotho. 48 Journal of African Economies, Vol.2, NO.1, 1993.

Guyer, J.I. 1986. Intra-Household Processes and Farming Systems Research: Perspectives from Anthropology. Moock J.L. and B.N. Okigbo (1986). Understanding Africa's Rural Households and Farming Systems. Boulder. London: Westview Press.

Hans- Joachim Fuchs. 1989. Tea Environments and Yield in Sri-Lanka. Tropical Agriculture, Margraf Scientific Publishers.

Hansen, A. & D.E. McMillan (eds). 1986. Food in Sub-Saharan Africa. Boulder. Colorado: Lynner Rienner Publishers, Inc.

Harrison, P. 1987. The Greening of Africa: Breaking Through in the Battle for Land and Food. London: Paladin Grafton Books.

Harts-Broekhuis E.J.A. & A.A. De Jong. 1993. Subsistence and Survival in the Sahel. Utrecht: University of Utrecht, The Netherlands.

Hay, M.J. 1976. Luo Women and Economic Change During the Colonial Period. In Hafkin and Bay, (1976). Women in Africa : Studies in Social and Economic Change. Stanford: Stanford University.

Hayami, Y. V.W. Rutten. 1985. Agricultural Development: An International Perspective. Second Edition. Baltimore. London: John Hopkins University Press.

Hettne, Bjorn. 1990. Development Theory and the Three Worlds. England. New York: Longman Scientific and Technical, and John Wiley & Sons, Inc.

Heyer, J. 1975. The Origins of Regional Inequalities in Smallholder Agriculture in Kenya, 1920-73. East African Journal of Rural Development. Vol.8, Nos. 1 & 2. 1975.

Heyer, J., J.K. Maitha & W.M. Senga. (eds). 1976. Agricultural Development in Kenya: An Economic Assessment. Nairobi. Lusaka. Dar es Salaam. Addis Ababa: Oxford University Press.

Heyzer, N. 1987. Women Farmers and Rural Change in Asia: Towards Equal Access and Participation. Kuala Lumpur: Asian and Pacific Development Centre.

Hilhorst, T. & H. Oppenoorth. 1992. Financing Women's Enterprises: Beyond Barriers and Bias. Amsterdam: Royal Tropical Institute - The Netherlands. UK, UNIFEM-USA. Intermediate Technology Publications.

Hinderink, J. & J.J. Sterkenburg. 1979. Agricultural Commercialization and Rural Development in Sub-Saharan Africa: an Overview, Diskussiestukken 5, Vakgroep Sociale Geografie Ontwikkelingslaan, Geografisch Instituut, Rijksuniversiteit Utrecht.

Hinderink, J. & G.J. Tempelman. 1979. Development Policy and Development Practice in Ivory Coast: A Miracle or a Mirage? Diskusiestukken 4, Vakgroep Sociale Geografie Ontwikkelingslanden, Geografisch Instituut, Rijksuniversiteit Utrecht.

Hinderink, J. & J.J. Sterkenburg. 1982. The Aims, The Methods, and The Means: Agricultural Production Performance and Agricultural Policy in Selected African Countries, N0. 16, March 1982, Vakgroep Sociale Geografie Van Ontwikkelingslanden, Utrecht.

Hinderink, J. & J.J. Sterkenburg. 1987. Agricultural Commercialization and Government Policy in Africa. London: Kegan Paul Ltd.

Hinderink, J. & E. Szulc-Dabrowiecka. (eds). 1988. Successful Rural Development in Third World Countries. Netherlands Geographical Studies 67, Amsterdam/Utrecht.

Hinderink, J. & J.J. Sterkenburg. 1988. The Tenant and the Peasant: A Comparative Analysis of Production Increase and Income Improvement in Southern Sudan and Swaziland. In Successful Rural Development in Third World Countries, edited by Hinderink, J. and Szulc-Dabrowiecka, E. (1988). The Netherlands Geographical Studies 67, Amsterdam/Utrecht.

Hobley, C.W. 1903. British East Africa. Anthropological Studies in Karivondo and Nandi. Journal of the Royal Anthropological Institute 33:325-49.

Hobley, C.W. 1929. Kenya from Chartered Company to Crown Colony. London: H.F.G. Witherby.

Hombergh, H. 1993. Gender, Environment and Development. A Guide to the Literature 1993. International Books (Utrecht, The Netherlands)

Horenstein, R.N. 1989. Women and Food Security in Kenya, Population and Women Resources Department. Policy, Planning, and Research Working Papers 232, Women in Development. The World Bank: Washington, D.C.

Howell, J. 1988. Training and Visit Extension in Practise. Agricultural Administration Unit. Occasional Paper No. 8. Overseas Development Institute.

Hulsebosch, J. 1990. Decision Making of Women in Female Headed, Joint and Migration Households. Agricultural University, Wageningen, The Netherlands.

Hunt, D. 1984. The Impending Crisis in Kenya. The Case for Land Reform. Aldershot, United Kingdom: Gower Publishing Company Ltd.

Hunt, D. 1985. The Impending Crisis in Kenya. The Case for Land Reform. United Kingdom: Gower.

ILO. 1972. Employment, Incomes and Equality: A Strategy for Increasing Productive Employment in Kenya. Geneva: International Labour Office.

Ipara, I.H. 1991. Socio-Economic Factors Affecting the Participation of Women in Agroforestation Activities in Sabatia Division, Kakamega District, Kenya. Unpublished M. Phil Thesis. Eldoret: Moi University.

Jackson, C. 1993. Doing What Comes Naturally? Women and Environment in Development. World Development, Vol. 21. No. 12, pp. 1947-1963, 1993. Great Britain: Pergamon Press.

Jacobson, J.L. 1992. Gender Bias: Roadblock to Sustainable Development. WorldWatch Paper 110. Washington, D.C: Worldwatch Institute.

Jaetzold, R. & H. Schmidt. 1982. Farm Management Handbook of Kenya Vol.II. Natural Conditions and Farm Management Information. Part A : Western Kenya (Nyanza and Western). Nairobi: Ministry of Agriculture, Kenya.

Jazairy, I., M. Alamgir & T. Panuccio. (eds). 1992. The State of World Rural Poverty. An inquiry into its Causes and Consequences. London: The International Fund for Agricultural Development.

Johnston, B.F. & J.W. Mellor. 1961. The Role of Agriculture in Economic Development. The American Review Dec. 1961. No. 5, pp. 566-593.

Johnston, R.J., D. Gregory & D.M. Smith. (eds). 1994. The Dictionary of Human Geography. Third Edition. Oxford: Basil Blackwell Inc.

Jones, C.W. 1983. The Mobilisation of Women's Labour for Cash Crop Production: A Game Theoretical Approach, Paper Presented at the Conference on "Women in Rice Farming Systems," Held at the International Rice Vol. 18, pp. 39-59.

Jones, C.W. 1986. Intra-Household Bargaining in Response to the Introduction of New Crops: A Case Study from North Cameroon. In Understanding Africa's Rural Household Farming Systems, ed. J.L. Moock & B.N. Okigbo, Ch. 6. Boulder and London: Westview Press.

Jong, S.M. & H. Th. Riezebos. 1990. The Use of Multispectral Satellite Images for Land Use and Land Cover Mapping in the Kakamega District, Kenya: A Feasibility Study. Utrecht: University of Utrecht, Netherlands.

Jungheim, K. 1989. Access to Land in Kakamega District Kenya: Kakamega District Focus Project. Utrecht: Dept. of Geography of Developing Countries, University of Utrecht, Netherlands.

Kandiyoti, D. 1990. Women and Rural Development Policies: The Changing Agenda. Development and Change, 21. No.1. 1990.

Kennedy, E.T. & B. Cogill. 1987. Income and Nutritional Effects of the Commercialization of Agriculture in Southwestern Kenya. International Food Policy Research Institute, Research Report 63.

Khasiani S.H. (ed). 1992. Groundwork: African Women as Environmental Mangers. Nairobi: African Centre for Technology Studies (ACTS) - Nairobi, Kenya.

Killick, T. 1992. Explaining Africa's Post-Independence Experiences. London: Overseas Development Institute.

Kiros, F.G. (ed). 1985. Challenging Rural Poverty: Experiences in Institutional-Building and Popular Participation for Rural Development in Eastern Africa. Trenton. New Jersey: Africa World Press.

Kitching, G. 1980. Class and Economic Change in Kenya, The Making of an African Petite-Bourgeoisie. New Haven. London: Yale University Press.

Kitching, G. 1989. Development and Underdevelopment in Historical Perspective: Populism, Nationalism and Industrialization. London: Routledge.

Kliest, T.J. 1985. Regional and Seasonal Food Problems in Kenya. Food and Nutrition Planning Unit, Ministry of Finance and Planning, Nairobi, Kenya: and Africa Studies Centre, Leiden, Netherlands.

Kongstad, P. & M. Monsted. 1980. Family, Labour and Trade in Western Kenya. Centre for Development Research. Publication 3. Uppsala: Scandinavian Institute of African Studies.

KTDA. 1971. The Kenya Tea Development Authority-Tea Cultivation Manual. Nairobi: KTDA.

KTDA. 1992. Tea Annual Reports. Nairobi: KTDA.

KTDA. 1994. Tea Annual Reports. Nairobi: KTDA.

Label, G.G. & H. Kane. 1987. Sustainable Development: A Guide to Our Common Future - The Report of the World Commission on Environment and Development. Geneva: Switzerland.

Lamb, G. & L. Muller. 1982. Control, Accountability, and Incentives in a Successful Development Institution: The Kenya Tea Development Authority. World Bank Staff Working Paper No.550. Washington D.C.: World Bank.

Lavrijsen, J.S.G. 1984. Rural Poverty and Impoverishment in Western Kenya. Utrecht: University of Utrecht, Netherlands.

Lele, Uma. J. 1973. The Role of Credit and Marketing Functions in Agricultural Development. Prajnan. Quarterly Journal of the National Institute of Bank Management, Bombay, Vol. 11. No. 2. pp 129-130.

Lele, Uma. J. 1974. Designing Rural Development Programme: Lessons from Past Experience in Africa. Second International Seminar on Change in Agriculture, England. (a discussion of integrated rural development, with a section on the problems and effects of land policy).

Lele, Uma. J. 1975. The Design of Rural Development: Lessons from Africa. Baltimore. London: John Hopkins University Press. Published for the World Bank.

Leonard, D.K. 1977. Reaching the Peasant Farmer: Organization Theory and Practice in Kenya. Chicago & London: The University of Chicago Press.

Leonard, D.K. 1991. African Successes: Four Public Managers of Kenyan Rural Development. Berkeley. Los Angeles. Oxford: University of California Press.

Leonard, D.K. 1993. Structural Reform of the Veterinary Profession in Africa and the New Institutional Economics. Development and Change (SAGE, London, Newbury Park and New Delhi), Vol. 24 (1993), 227-267.

Lipton, M. 1974. Towards a Theory of Land Reform. In Lehmann, D. (ed) (1974), Agrarian Reform and Agrarian Reformism, Studies of Peru, Chile, China, and India.

Little, D.P. & M.M. Harowitz. with E.A. Nyerges. 1987. Land at Risk in the Third World: Local Level Perspectives. Boulder. London: Westview Press.

Livingstone, I. 1974. Creating Employment in Kenya: The ILO Mission Report. Journal of Administration Overseas. April.

Livingstone, I. 1981. Rural Development, Employment and Incomes in Kenya. ILO/JASPA. Addis Ababa. Gower.

Livingstone, I. 1986. Rural Development, Employment and Incomes in Kenya. ILO/JASPA: Gower.

Livingstone, I. 1991. A Reassessment of Kenya's Rural and Urban Informal Sector. In World Development, Vol. 19, No. 6, pp 651-670, 1991.

Lofchie, M.F., S.K. Commins & R. Payne. (eds). 1986. Africa's Agrarian Crisis: The Roots of Famine. Boulder. Colorado: Lynne Rienner.

Lofchie, M.F. 1994. The New Political Economy of Africa. Political Development and the New Realism in Sub-Saharan Africa edited by David E. Apter and Carl G. Rosberg. Charlottesville (etc): University Press of Virginia.

Low, A. 1986. On-Farm Research and Household Economics: in Moock, J.L. and B.N. Okigbo (eds) (1986), Understanding Africa's Rural Households and Farming Systems. Boulder and London: Westview Press.

Luning, H.A. (n.d) Agricultural Pricing Policy, mimeo.

Luning, H.A. (n.d) Agricultural Land Use and Agro-Ecological Potentials in Kakamega District. mimeo.

Luning, H.A. & J.J. Sterkenburg. 1987. Kakamega District Focus Project. Kakamega: Unpublished Mission Report.

Luning, H.A. 1988. Kakamega District Focus Project. Kakamega: Unpublished Mission Report.

Luning, H.A. 1989. Kakamega District Focus Project. Kakamega: Unpublished Mission Reports (May and October 1989).

Luning, H.A. & H. Th. Riezebos. 1991. Land Evaluation and Farming System Analysis for Rural Development. Paper presented at a Regional Development Research Seminar 1991, 10th December 1991. University of Utrecht.

Lyclklama a Nijeholt, G. 1987. The fallacy of Integration: The UN Strategy of Integrating Women into Development Revisited. Netherlands Review of Development Studies, Vol 1: 23-37.

Lycklama a Nijeholt, G. (ed). 1991. Towards Women's Strategies for the 1990s: Challenging Government and State. The Hague: Institute of Social Studies (ISS).

Lycklama a Nijeholt, G. 1992. Women and the Meaning of Development: Approaches and Consequences, Silver Jubilee Paper 7. The Hague: Institute of Social Studies (ISS).

Maas, M. 1986. Women Groups in Kiambu, Kenya: It is always a good thing to have Land. Research Reports No. 26/1986. African Studies Centre, Leiden / Netherlands.

Maas, M. 1991. Women's Social and Economic Projects: Experiences from Coast Province. Report No. 37/1991. African Studies Centre. Leiden, Netherlands.

MacArthur, J.D. 1966. Some Thoughts on Future Trends in Farm Employment - in Education, Employment and Rural Development - Report of the Kericho (Kenya) Conference 25th September to 1st October 1966, James R. Sheffield (ed) - Published for University College, Nairobi.

Mackenzie, F. 1986. Local Initiatives and National Policy: Gender and Agricultural Change in Murang'a District, Kenya. Canadian Journal of African Studies 20.3 pp 377-401.

Mackenzie, F. 1991. Political Economy of the Environment, Gender, and Resistance under Colonialism: Murang'a District, 1910-1950. Canadian Journal of African Studies, Vol. 25, No. 1 1991, pp.226-256.

Mandi, W.J. 1971. Kenya Tea Development Authority. (Tea Cultivation Manual). Nairobi: Regal Press.

Mbote. K.P. 1992. African Women as Environmental Managers. Nairobi: African Centre for Technological Studies (ACTS).

Mellor, J.W., C.H. Delgado & M.J. Blackie. (eds). 1987. Accelerating Food Production in Sub-Saharan Africa. Baltimore. London: The John Hopkins University Press.

Mickelwait, D.A., M.A. Riegelman & C.F. Sweet. 1976. Women in Rural Development: A Survey of the Roles of Women in Ghana, Lesotho, Kenya, Nigeria, Bolivia, Paraguay and Peru. Boulder. Colorado: Westview Press.

Midamba, B.H. 1993. The Legal Basis of Gender Inequality. African Journal of International Comparative Law (1993), Vol. 5, pt 4, pp 850-868.

Moock, J.L. 1976. The Migration Process and Differential Economic Behaviour in South Maragoli, Western Kenya. University Microfilms International, Ann Arbor, Michigan, U.S.A. London, England.

Moock, J.L. 1978. The Content and Maintenance of Social Ties Between Urban Migrants and Their Home Based Support Groups: The Maragoli Case. African Urban Studies, 3 (Winter 1978 - 1979) 15.

Moock, J.L. & B.N. Okigbo. (eds). 1986. Understanding Africa's Rural Households and Farming Systems. Boulder: Westview Press.

Moock P.R. 1973. Managerial Ability in Small-Farm Production, An Analysis of Maize Yields in The Vihiga Division of Kenya. Ph.D Thesis, University Microfilms International, Ann Arbor, Michigan, U.S.A. London, England.

Moock, P.R. 1986. The Efficiency of Women as Farm Managers: Kenya American Journal of Agricultural Economics 5815 pp 831-835.

Moris, J. 1991. Extension Alternatives in Tropical Africa. Agricultural Administration Unit, Occasional Paper 7. London: Overseas Development Institute.

Morss, E.R., J.K. Hatch, D.R. Mickelwait & C.F. Sweet. 1976. Strategies for Small Farmer Development, An Empirical Study of Rural Development Projects in the Gambia, Ghana, Kenya, Lesotho, Nigeria, Bolivia, Colombia, Mexico, Paraguay and Peru. Vols I and II. Boulder. London: Westview Press.

Moser, C.O.N. 1989. Gender Planning in the Third World: Meeting Practical and Strategic Gender Needs. World Development, 17, No.11, pp. 1799-1825, 1989. Great Britain: Pergamon Press.

Muchena, F.N. & R.M. Kiome. 1995. The Role of Soil Science in Agricultural Development in East Africa. Kenya Agricultural Research Institute, National Agricultural Research Laboratories. Wageningen: Elsevier, Geoderma 67 (1995) 141-157.

Muro, A. 1985. Women Commodity Producers and Proletariats: The Case of African Women. In Kiros, F.G. (ed). (1985). Challenging Rural Poverty : Experiences in Institutional - Building and Popular Participation for Rural Development in Eastern Africa. New Jersey: Africa World Press.

Mutoro, B.A. 1985. Selected Environmental and Socio-Economic Factors Affecting Agricultural Production in South Maragoli Location in the Western Province of Kenya. Unpublished M.A. Thesis. Nairobi: University of Nairobi.

Muzorewa, B.C. 1975. Security for Loans in Rural Development - in Wilson, F.A. and Amann, V.F. (eds). 1975. Financing Rural Development. Kampala: Eastern Africa Agricultural Economics Society and The Makerere Institute of Social Research.

Mwaniki, N. 1986. Against Many Odds: The Dilemmas of Women's Self-Help Groups in Mbeere, Kenya. Africa, 56 (2), 210-28.

Nasimiyu, R. 1985. Women in the Colonial Economy of Bungoma: Role of Women in Agriculture, 1902-1960. In Women and Development in Africa, edited by Were, G.S. (1985). Nairobi, Kenya: Gideon S. Press.

Nelson, N. (ed). 1981a. African Women in the Development Process. London: Frank Case.

Norman, L. & R.B. Coote. 1971. The Farm Business. London: Longman Group.

Nzomo, M. 1992. Beyond Structural Adjustment Programs: Democracy, Gender, Equity, and Development in Africa, with Special References to Kenya. in Beyond Structural Adjustment in Africa: The Political Economy and Democratic Development, edited by Nyang'oro, J.E and Shaw, T. M. (1992). New York. Westport. Connecticut. London: Praeger.

Ochieng', W.R. 1975. An Outline History of the Rift Valley of Kenya Up to A.D 1900. Kampala. Nairobi. Dar es Salaam: East African Literature Bureau.

Ochieng', W.R. (ed). 1990. Themes in Kenyan History. Nairobi: Heinemann Kenya and James Currey Publishers.

Odingo, R.S. 1971. The Kenya Highlands: Land Use and Agricultural Development. Nairobi: East African Publishing House.

Odingo, R.S. 1971. Settlement and Rural Development in Kenya. In S.H. Ominde (ed) - Studies in East African Geography and Development. London. Ibadan: Heinemann Educational Books.

Odingo, R.S. 1973. A Study of Agrarian Change in Kenya with Special Reference to Kericho District. A Special Study Prepared for FAO, Rome.

Odingo, R.S. 1984. Agricultural Resettlement Programmes and their Contribution to Rural Development in Kenya. Paper Presented to the 25th International Geographical Congress, Symposium No.8. Commission on Rural Development, Neuchatel, Switzerland, August, 1984.

Odingo, R.S. 1985. The Dynamics of Land Tenure and of Agrarian Systems in Africa, Land Tenure Study in Kenya. Rome: FAO.
Ogada, F. 1969. The Role of Maize Research in Stimulating Agricultural Progress in Kenya. Paper Presented at the East African Agricultural Economics Society Conference 1969.
Okoth-Ogendo, H.W.O. 1976. African Land Tenure Reform. In Heyer, J. et al (eds). (1976), Agricultural Development in Kenya: An Economic Assessment. Nairobi: Oxford University Press.
Okoth-Ogendo, H.W.O. 1981. Land Ownership and Land Distribution in Kenya's Large-Farm Areas. In Papers on the Kenyan Economy: Performance, Problems and Policies, edited by T. Killick, 329-38. Nairobi: Heinemann.
Okoth-Ogendo, H.W.O. 1991. Tenants of the Crown: Evolution of Agrarian Law and Institutions in Kenya. Nairobi, Kenya: ACTS Press.
Okuku, T.M. 1991. Rural Development Policies And Programmes: the Impact of Rural Development Fund at District Level in Kenya. Enschede, The Netherlands.
Ominde, S.H. 1971. Rural Economy in West Kenya. in Studies in East African Geography and Development - edited by S.H. Ominde, London. Ibadan. Nairobi: Heinemann Educational Books.
Oosten, C. 1989. Farming Systems and Food Security in Kwale District, Kenya. Food and Nutrition Planning Unit, Ministry of Planning and National Development. Nairobi, Kenya; and African Studies Centre, Leiden, Netherlands. Report No. 30 / 1989.
Orna', S., H. Anders & M.A. Salih. (eds). 1989. Ecology and Politics: Environmental Stress and Security in Africa. Uppsala: Scandinavian Institute of African Studies.
Osogo, J. 1966. A History of the Baluyia. Nairobi. London. New York: Oxford University Press.
Ostergaard, L. (ed). 1992. Gender and Development: A Practical Guide. London: Routledge pp. 220.
Oyugi, W.O. 1985. Central-Periphery Linkages in the Development Process: An Assessment of the Kenyan Experience. In Challenging Rural Poverty: Experiences in Institutional Building and Popular Participation for Rural Development in Eastern Africa (1985), edited by Kiros, F.G.
Pala, A. O., T. Awori & A. Krystal. 1978. The Participation of Women in Kenya Society. Nairobi: Kenya Literature Bureau.
Pala, A.O. 1978. Women's Access to Land and Their Role in Decision-Making on the Farm: Experiences of the Joluo of Kenya. Discussion paper No. 263, IDS. Nairobi: University of Nairobi.
Pala, A.O. 1980. Daughters of the Lake and Rivers: Colonization and the Land Rights of Luo Women. In Etienne, M. and Leacock, E. (eds) (1980), Women and Colonization: Anthropological Perspectives. pp 186-213, New York. NY: Praeger.
Pala, A.O. 1983. Women's Access to Land and Their Role in Agriculture and Decision Making on the Farm - Experience of the Jaluo of Kenya. Journal of Eastern African Research and Development 13 pp 69-87. Also, University of Nairobi, Institute for Development Studies, Discussion Paper No. 263.
Palmer, I. 1985. Women's Roles & Gender Differences in Development: The Impact of Male Out-Migration on Women in Farming. West Hartford: Kumarian Press.
Parpart, J.L. & K.A. Staudt. (eds). 1989. Women and the State in Africa. Boulder. London: Lynne Rienner Publishers.
Pearce, D.W. & J.J. Warford. 1993. World Without End: Economics, Environment, and Sustainable Development, A Summary. Washington D.C.: The World Bank.

Pepels. E. 1991. Tarmacking in Nairobi. Labour Migrants from Kakamega District. M. A. Thesis. Utrecht: University of Utrecht.
Poats, S.V., M. Schmink & A. Spring. (eds). 1988. Gender Issues in Farming Systems Research and Extension. Special Studies in Agricultural Science and Policy. Boulder: Westview Press.
Potash, B. 1985. Female Farmers, Mothers - In-Law and Extension Agents: Development Planning and Rural Luo Community. Working Paper #90.
Rathgeber, M.E. & B. Ketter. Women's Role in Natural Resource Management in Africa. International Development Research Centre.
Reijntjes, C., B. Haverkort & A. Waters-Bayer. 1992. The Farming for the Future. An Introduction to Low-External-Input and Sustainable Agriculture. London: Macmillan Press Limited.
Rempel, H. 1970. Rural to Urban Migration: A Tabulation of the Response to the Questionnaire Used in the Migration Survey. Discussion Paper no. 92. Institute for Development Studies. Nairobi: University of Nairobi.
Rempel, H. 1971. Labour Migration into Urban Centres and Urban Unemployment in Kenya. Ph.D. Dissertation. Wisconsin: University of Wisconsin.
Rempel, H. & R.A. Lobdell. 1978. The Role of Urban to Rural Remittances in Rural Development. Journal of Development Studies, 14 No.4 pp. 324-341.
Republic of Kenya. 1965. African Socialism and Its Applications to Planning in Kenya. Sessional Paper No. 10, 1965. Nairobi: Government Printer.
Republic of Kenya. 1966. Kenya Population Census, 1962 Volume III African Population. Statistics Division, Ministry of Economic Planning and Development. Nairobi: Government Printer.
Republic of Kenya. 1968. Report of the Commission on the Law of Succession. Nairobi: Government Printer.
Republic of Kenya. 1971. Kenya National Population Census 1969. Nairobi: Government Printer.
Republic of Kenya. 1979. Statistical Abstract. Central Bureau of Statistics, Ministry of Planning and Community Affairs. Nairobi: Government Printer.
Republic of Kenya. 1979. National Development Plan 1979-1983. Nairobi: Government Printer.
Republic of Kenya. 1979. Ministry of Co-operative Development. Plan of Operation for Farm Input Supply Scheme Phase II. Nairobi: Government Printer.
Republic of Kenya. 1980. Kenya National Population Census 1979. Nairobi: Government Printer.
Republic of Kenya. 1980. National Livestock Development Policy. Ministry of Livestock Development. Nairobi: Government Printer.
Republic of Kenya. 1981. National Food Policy - Sessional Paper No. 4 of 1981. Nairobi: Government Printer.
Republic of Kenya. 1981. Statistical Abstract. Central Bureau of Statistics, Ministry of Planning and Development. Nairobi: Government Printer.
Republic of Kenya. 1983. Paths to Rapid Development: Policy Guidelines on the Department of Social Services. Nairobi: Government Printer.
Republic of Kenya. 1984. Kakamega District Development Plan 1984-1988. Ministry of Finance and Planning. Nairobi: Government Printer.
Republic of Kenya. 1986. Economic Management for Renewed Growth: Sessional Paper No. 1 of 1986. Nairobi: Government Printer.
Republic of Kenya. 1986. Revised Edition. Laws of Kenya: The Agricultural Act: Chapter 318. Nairobi: Government Printer.

Republic of Kenya. 1987. Agricultural Information Centre: Major Crops Technical Hand-book. Agricultural Information Centre. Nairobi: Government Printer.
Republic of Kenya. 1987. Ministry of Agriculture. Fertilizer Use Recommendation Project (Phase I). Final Report Annex III. Description of the First Priority Sites in the Various Districts. Vol.7. Kakamega District. Nairobi: Government Printer.
Republic of Kenya. 1988. Reflection on Livestock Production Strategy Under the District Focus Strategy for Rural Development 1988 and Beyond. Ministry of Livestock Development. Nairobi: Government Printer.
Republic of Kenya. 1988. Ministry of Agriculture. Fertilizer Use Recommendation Project (Phase I). Main Report. Methodology and Inventory of existing information. Nairobi: Government Printer.
Republic of Kenya. 1988. Rural Labour Force Survey. Nairobi: Government Printer.
Republic of Kenya. 1989. Kakamega District Development Plan 1989-1993. Nairobi: Government Printer.
Republic of Kenya. 1990. Economic Survey, Central Bureau of Statistics, Ministry of Planning and National Development. Nairobi: Government Printer.
Republic of Kenya. 1990. Statistical Abstract. Central Bureau of Statistics. Ministry of Planning and National Development. Nairobi: Government Printer.
Republic of Kenya. 1992. Economic Survey, Central Bureau of Statistics, Ministry of Planning and National Development. Nairobi: Government Printer.
Republic of Kenya.1994. Vihiga District Development Plan 1994-1996. Office of the Vice President and Ministry of Planning and National Development. Nairobi: Government Printer.
Republic of Kenya. 1994. National Development Plan 1994-1996. Nairobi: Government Printer.
Republic of Kenya. 1994. Kakamega District Development Plan 1994-1996. Nairobi: Government Printer.
Republic of Kenya. 1994. Kenya Population Census 1989, Volume 1. Nairobi: Government Printer.
Republic of Kenya. 1994. Kenya Population Census 1989, Volume 11. Nairobi: Government Printer.
Riria-Ouko, J.V.N. 1985. Women Organizations in Kenya in : Were, G.S. (ed) (1985),Women and Development in Kenya. Nairobi: Gideon S. Were Press.
Robertson, C. & I. Berger. (eds). 1986. Women and Class in Africa. New York. London: African Publishing Company.
Roche, Jr.D. 1992. Quarttro Pro 4 Quick Start. (The step by step approach). Carmel: QUE.
Royal Tropical Institute. 1990. Environmental Management in the Tropics (an annotated bibliography). Amsterdam: Royal Tropical Institute: Department of Information and Documentation.
Rundquist, F.A. 1984. Hybrid Maize Diffusion in Kenya. CWK Gleerup.
Rusell, M. 1993. Are Households Universal? On Misunderstanding Domestic Groups in Swaziland. Development and Change, 1993, 24, pp. 755-785.
Ruthenburg, H. 1971. Farming System in the Tropics. Oxford: Oxford Scientific Publication.
Ruthenberg, H. 1980. Farming System in the Tropics. Oxford: Claredon Press.
Rutten, M.M.E.M. 1992. Selling Wealth to Buy Poverty: The Process of Individualization of Landownership Among the Masaai Pastoralists of Kajiado District, Kenya, 1890-1990. Ph.D Thesis KUN, Nijmegen Studies in Development and Cultural Change 10.

Safilios - Rothschild, C. 1985. Policy, Theoretical and Methodological Issues Regarding Statistics on Women in Agriculture. Workshop on Improvement of Statistics on Women in Agriculture. 21 October - 31 October 1985, FAO, Rome, Italy, pp. 1-21.

Safilios - Rothschild, C. (n.d). A Typology of Men's and Women's Farming Systems in Kenya. The Hague: The Ministry of Development Cooperation of the Netherlands.

Safilios-Rothschild, C. & E.K. Mburu. 1987. Men's and Women's Agricultural Production and Incomes in Rural Kenya. Paper Presented at the Seminar on `Agricultural Development, Population and the Status of Women, Nyeri, Kenya, September 2-3.

Safilios - Rothschild, C. 1989. Farming Systems and Gender Issues: Implications for Agricultural Training and Projects. Ministry of Agriculture and Fisheries of the Netherlands and International Agricultural Centre. : in Walsum, E. ed. (1989), Gender Issues in Agriculture. Dept of Gender Studies in Agriculture, Wageningen Agricultural University.

Safilios - Rothschild, C. 1990. Women as a Motor in Agricultural Development: Lessons Learned from Eastern and Southern Sub-Saharan Africa. Africa Seminar. Maastricht, The Netherlands. Beyond Adjustment. Sub-Saharan Africa. Ministry of Foreign Affairs. Directorate General for International Co-operation, Den Haag, 1990, pp. 153-164.

Saito, K.A. & C.J. Weidemann. 1990. Agricultural Extension for Women Farmers in Africa. World Bank Discussion Papers. Washington, D.C.: The World Bank.

Saito, K. A. et al. 1992. Raising Productivity of Women Farmers in Sub-Saharan Africa. PHRWD, Women in Development Division. Washington, D.C.: The World Bank.

Saito, K.A. 1994. Raising the Productivity of Women Farmers in Sub-Saharan Africa. World Bank Discussion Papers, No. 230. Washington, D.C.: The World Bank.

Sangree, W. 1966. Age, Prayer and Politics in Tiriki, Kenya. London: Oxford University Press.

Savane, M.A. 1986. The effects of Social and Economic Changes on the Role and Status of Women in Sub-Saharan Africa in: Moock, J.L. and B.N Okigbo (eds) (1986), Understanding Africa's Rural Households and Farming Systems. Boulder. London: Westview Press.

Schrijvers, J. 1991. Dialectics of a Dialogical Ideal: Studying Down, Studying Sideways and Studying UP. In Constructing Knowledge: Authority and Critique in Social Science. Edited by Nencel, L. and Pels, P. (1991). London. Newbury Park. New Delhi: SAGE Publications.

Schrijvers, J. 1995. Participation and Power: A Transformative Feminist Research Perspective. In Power and Participatory Development: Theory and Practice. Edited by Nelson, N. and Wright, S. (1995). London: Intermediate Technology Publications 1995.

Shaw, D.P. 1987. Rural-Urban Linkages in the Malawian Context. African Urban Quarterly, September 1987.

Schwartz, L.A. & J. Kampen. 1992. Agricultural Extension in East Africa. World Bank Technical Paper No. 164, Washington, D.C.: The World Bank.

Shipton, P. 1992. Debts and Trespasses: Land, Mortgages, and the Ancestors in Western Kenya. Africa- Journal of the International African Institute, Vol. 62, No. 3, 1992, pp. 357-385.

Shiva, V. 1989. Staying Alive: Women, Ecology and Development. London. New Jersey: Zed Books.

Singh, R.D. 1988. Economics of the Family and Farming Systems in Sub-Saharan Africa. Development Perspectives. Boulder. London: Westview Press.

Smith J.F. 1972. Farm Business Management. The Decision Making Process. London: Macmillan Publishing Co, Inc.

Sorensen, A. 1992. Women's Organizations Among the Kipsigis: Change, Variety and Different Participation - African, Journal of the International African Institute Vol. 62 No. 1 1992.

Spring, A. 1986. Women Farmers and Food in Africa: Some Considerations and Suggested Solutions in: Hansen A. and D.E. McMillan (eds), (1986), Food in Sub-Saharan Africa. Boulder. Colorado: Lynne Rienner Publishers, Inc.

Spring, A. 1988b. Using Male Research and Extension Personnel to Target Women Farmers. Poats et al.

Stahl, M. 1993. Land Degradation in East Africa. Ambio No. 8. pp. 505-508.

Stamp, P. 1990. Technology, Gender, and Power in Africa. Ottawa: International Development Research Centre IDRC- TS3e

Staudt, K. 1974. The Effects of Government Agricultural Policy on Women Farmers in Kenya : A Research Proposal. IDS Working Paper 200.

Staudt, K. 1975. Women Farmers and Inequalities in Agricultural Services. Rural Africana 29 pp 81-94.

Staudt, K. 1978. Agricultural Productivity Gaps: A Case Study of Male Preference in Government Policy Implementation. Development and Change (SAGE, London and Beverly Hills), Vol. 9 (1978) 439-57.

Staudt, K.A. 1978. Administrative Resources, Political Patrons, and Redressing Sex Inequalities: A Case of Western Kenya. The Journal of Developing Areas 12 (July 1978), pp. 399-414.

Staudt, K. 1985. Agricultural Policy Implementation: A Case Study from Western Kenya. West Hartford: Kumarian Press.

Staudt, K. 1987. Uncaptured or Unmotivated? Women and the Crisis in Africa. Rural Sociology, Vol. 52, No.1, pp.37-55.

Sterkenburg, J.J. 1988. Views on Rural Development and Rural Development Policy. In Successful Rural Development in Third World Countries, edited by Hinderink, J. and Szulc-Dabrowiecka, E. (1988). Netherlands Geographical Studies No. 67.

Sterkenburg, J.J. 1990. Livestock and Livestock Policy in Kakamega District - Kenya. Kakamega District Focus Project. District Information and Documentation Centre, Kakamega District.

Stichter, S. 1982. Migration Labour in Kenya: Capitalism and African Response 1895 - 1975. England: Longman.

Streeter, C.P. 1973. Reaching the Developing World's Small Farmer. New York: The Rockerfeller Foundation.

Suda, C.A. 1990. Labour and Land Resources in Small-Holder Agriculture in Siaya and Kakamega Districts: Analysis of the Factors Which Influence Farm Labour Supply. Kenya Journal of Sciences, Series C, Social Sciences 2(2).

Suda, C.A. 1992. Towards an Understanding of Small-Holder Agriculture in Western Kenya: An Analysis of the Factors Influencing Crop Production. Discovery and Innovation Vol. 4 No. 4 December 1992.

Swartz, M. 1968. Local Level Politics. Chicago: Aldine.

Swynnerton, R.J.M. 1955. A Plan to Intensify the Development of the African Agriculture in Kenya. Colony and Protectorate of Kenya. Nairobi: Government Printer.

Tazelaar, B. 1989. Maize Production and Productivity in Kakamega District, Kenya. Utrecht: Department of Geography, Utrecht of University.

Thomas, P.T. 1985. Politics, Participation and Poverty: Development Through Self-Help in Kenya. Boulder. London: Westview Press.

Thomas, M.F. & W.G. Whittington. 1989. Environment and Land Use in Africa. Metheun & Co. Ltd.

Tidrick, G. 1979. Kenya: Issues in Agricultural Development, Ministry of Economic Development and Planning, Nairobi, May.

Tiffen, M., M. Mortimore & F. Gichuki. 1994. More People, Less Erosion, Environmental Recovery in Kenya. Chichester. New York. Brisbane. Toronto. Singapore: John Wiley & Sons.

Tiffen, M. 1995. Population Density, Economic Growth and Societies in Transition: Boserup Reconsidered in a Kenyan Case-Study. Development and Change Vol. 26 (1995), 31-66.

Timberlake, L. 1985. Africa in Crisis: The Causes, the Cures of Environmental Bankruptcy. London: Earthscan Publications LTD.

Tivy, J. 1971. A Study of Plants in the Ecospace. Oliver and Boyd (Division of Longman Group).

Tivy, J. 1990. Agricultural Ecology. Longman Scientific and Technical. New York: Co-publisher John Wiley and Sons. Inc.

Todaro, M.P. 1994. Economic Development. 5th Edition. New York. London: Longman.

Tripp, A.M. 1992. The Impact of Crisis and Economic Reform on Women in Urban Tanzania in Unequal Burden: Economic Crises, Persistent Poverty, and Women's Work edited by Beneria, L. and Feldman, S. (1992). Boulder. London: Westview Press.

U.S.AID. 1982. Kenya Rural Roads, A.I.D Project Evaluation Report No 26. U.S.AID.

Udvardy, M. 1988. Women Groups Near the Kenyan Coast: Patron-Clientship in the Development Arena. Anthropology of Development and Change in East Africa (1988), edited by Brokensha, D.W. and Little, P.D.

United Nations. 1993. Statistical Yearbook 1990/91. New York: United Nations.

United Nations. 1988. Statistical Yearbook 1985/86. New York: United Nations.

United Nations. 1990. Statistical Yearbook 1987. New York: United Nations.

University of Utrecht. 1989. Location Development Profile, South Kabras: Kakamega District Focus Project. Department of Geography of Developing Countries. Utrecht, Netherlands.

University of Utrecht. 1989. Location Development Profile, Central Marama: Kakamega District Focus Project. Department of Geography of Developing Countries. Utrecht, Netherlands.

University of Utrecht. 1989. Location Development Profile, Central Kisa: Kakamega District Focus Project. Department of Geography of Developing Countries. Utrecht, Netherlands.

University of Utrecht. 1989. Location Development Profile, Lumakanda: Kakamega District Focus Project. Department of Geography of Developing Countries. Utrecht, Netherlands.

University of Utrecht .1989. Location Development Profile, North Maragoli: Kakamega District Focus Project. Department of Geography of Developing Countries. Utrecht, Netherlands.

University of Utrecht. 1989. Location Development Profile, Soy: Kakamega District Focus Project. Department of Geography of Developing Countries. Utrecht, Netherlands.

University of Utrecht. 1989. Location Development Profile, North Marama: Kakamega District Focus Project. Department of Geography of Developing Countries. Utrecht, Netherlands.

University of Utrecht. 1989. Location Development Profile, South Marama: Kakamega District Focus Project. Department of Geography of Developing Countries. Utrecht, Netherlands.

Upton. M. 1973. Farm Management in Africa. The Principal of Production and Planning. London: Oxford University Press.

Upton. M. 1976. Agricultural Production Economic and Resource Use. London: Oxford University Press.

Upton, M. 1987. African Farm Management. Cambridge: Cambridge University Press

Vasthoft, J. 1968. Some Experiences in East Africa with Special Reference to Kenya, Africa. Studien 53, JFO - Institut fur Wirtsckaftsforschung, Munich, Weltforum.

References

Velzen, J. van. 1960. Labour Migration as a Positive Factor in the Country of Tonga Tribal Society. Economic Development and Cultural Change 8:265-78.

Von Pischke, J.D. 1975. A Critical Survey of Approaches to the Role of Credit in Smallholder Development - in Wilson, F.A. and Amann, V.F. (eds) 1975, - Financing Rural Development. Produced jointly by Eastern Africa Agricultural Economics Society and the Makerere Institute of Social Research, Kampala, Uganda.

Wagner, G. 1949. The Bantu of North Kavirondo. Vol.1. London. New York. Toronto: Oxford University Press.

Wagner, G. 1956. The Bantu of North Kavirondo. Vol.II Economic Life. London. New York. Toronto: Oxford University Press.

Walsum. V.E. 1989. Gender Issues in Agriculture, Wageningen: Wageningen Agricultural University.

Walsum. V.E. (1991), Gender Methodology in Agricultural Projects. Wageningen: Wageningen Agricultural University.

Wanjiku, M. B., J.A. Oduol & M. Nzomo. (eds). 1993. Democratic Change in Africa. Women's Perspectives. ACTS, Nairobi, Kenya.

Were, G.S. (ed). 1985. Women and Development in Africa. Journal of Eastern African research and Development Volume 15, 1985. Nairobi: Gideon S. Were Press.

Wilson, F.A. 1975. The Role of the Non-Specialist Credit Agency: The Commercial Bank in Zambia - in Wilson, F.A. and Amann, V.F. (eds) - Financing Rural Development. Kampala: Produced Jointly by Eastern Africa Agricultural Economics Society and the Makerere Institute of Social Research.

Wisner, B. 1988. Power and Need in Africa. London: Earthscan.

World Bank. 1976. Appraisal of Rural Access Roads Project, Republic of Kenya. Nairobi: Regional Projects Department Eastern Africa Regional Office.

World Bank. 1979. Kenya Small-holder Coffee Improvement Project. Staff Appraisal Report. Volume II: Annexer. Nairobi: Eastern Africa Region Central Agriculture Division.

World Bank. 1990. Poverty: World Bank Development Report 1990, World Development Indicators. Oxford. New York. Toronto: Oxford University Press.

World Bank. 1992. Development and the Environment: World Development Report 1992, World Development Indicators. Oxford. New York. Toronto: Oxford University Press.

World Bank. 1994. Agricultural Extension: Lessons from Completed Projects. Operations Evaluation Department Report No. 13000. Washington, D.C.: The World Bank.

Wyeth, P. 1981. Economic Development in Kenyan Agriculture. In papers on the Kenyan Economy: Performance, Problems and Policies, edited by T. Killick, 299-310. Nairobi: Heinemann.

Zwanenberg, R. 1975. Colonial Capitalism and labour in Kenya, 1919-1939. Nairobi: African Literature Bureau.

Zwanenberg, R.M.A. & A. King. 1975. An Economic History of Kenya and Uganda 1800-1970. London: The Macmillan Press Ltd.

Abbreviations

AFC	Agricultural Finance Corporation
ALDEV	African Lands Development Programme
CAN	Calcium Ammonium Nitrate
CPE	Certificate of Primary Education
CPCS	Co-operative Production Credit Service
DAO	District Agricultural Officer
DDC	District Development Committee
DDO	District Development Officer
FAO	Food and Agricultural Organization of the United Nations
FSR	Farming Systems Research
FTC	Farmer Training Centre
GDP	Gross Domestic Product
GNP	Gross National Production
GOK	Government of Kenya
GTZ	German Technical Agency
HH	Household
HHH	Household Head
HHT	Household Type
HYV	High Yielding Variety
HV	High Value
IADP	Integrated Agricultural Development Programme
ICIPE	International Centre for Insects and Plant Epidemiology
ICRAF	International Council for Research in Agroforestry
IDRC	International Development Research Centre
IFAD	International Fund for Agricultural Development
ILO	International Labour Organization
IMF	International Monetary Fund
IRDP	Integrated Rural Development Programme/Project
IUCN	World Conservation Union
KARI	Kenya Agricultural Research Institute
KCC	Kenya Co-operative Creameries
KCPE	Kenya Certificate Primary Education
KEFINCO	Kenya Finland Corporation
KENGO	Kenya Energy Non-Governmental Organization
KGGCU	Kenya Grain Growers Co-operative Union
KJSE	Kenya Junior Secondary Examination
KNTC	Kenya National Trading Company
KSC	Kenya Seed Company
Kshs	Kenya Shillings
KPCU	Kenya Planters Co-operative Union
KTDA	Kenya Tea Development Authority
KWAP	Kenya Woodfuel and Agro-forestry Programme
KWDP	The Kenya Woodfuel Development Programme

MOE	Ministry of Energy
NCPB	National Cereals and Produce Board
NEP	National Extension Programme
NFP	National Food Policy
NGO	Non-Governmental Organization
NPK	Nitrogen Phosphate Potassium
ODA	Overseas Development Administration
ODI	Overseas Development Institute
SCCS	Savings Credit and Co-operative Society
SCIP	Smallholder Coffee Improvement Project
SPSCP	Smallholder Production Services and Credit Project
SRDP	Special Rural Development Programme
SAPs	Structural Adjustment Programmes
SS	Sub-System
TOT	Total
T&V	Training and Visit Extension System
YWCA	Young Women Christian Association
UNCED	United Nations Convention for Environment and Development
UNDP	United Nations Development Programme
USAID	United States Agency for International Development
WCU	World Conservation Union

Glossary

Busaa	Local alcoholic brew (beer)
Baba	Respective way of addressing a father
Baraza	Chief's meeting
Chang'aa	Illegally local distilled alcohol, kind of gin or whisky
District province	A geographical area which is a constituent administrative unit of a headed by a district officer
Location	A geographical area which is a constituent administrative unit of a district headed by a chief
Jembe	Hand hoe
Mama	Respective way of addressing a mother
Mzee	A term used for purposes of respect for a married man and an old man
Shamba	Land (farm) under the jurisdiction of a household utilized for agriculture, livestock, tree cultivation and home garden in the rural areas
Shambaman	A hired worker on a farm or a gardener
Sub-location	A geographical area which is a constituent administrative unit of a location headed by an assistant chief
Sukuma wiki	Vegetable. A kind of spinach or Kale. Means "pushing through the week"
Ugali or posho	A staple dish prepared from maize flour -bread-like- that is eaten in many parts of Kenya

Appendix I

Research methodology and time schedule

This research project was divided into three phases and a number of methods were used in each given phase to collect the required information.

Phase One: The Netherlands, 1991
The following activities were carried out in this phase:
(1) Literature search on farm management, agricultural development with emphasis on Africa, Kenya in particular, and gender issues in agriculture in many university libraries (Institute of African Studies - Leiden, University of Utrecht, University of Amsterdam and Institute of Social Sciences, The Hague). (2) Courses on women and development, and ecologically sustainable development at the I.S.S,- The Hague, and an international course on computerized regional database development at the University of Utrecht.
(3) Writing of research Proposal. (4) Preparation of questionnaires for the sample survey. (5) Training in handling and processing data from the Kakamega District database. (6) Training in Complete Statistical System (CSS).

Phase Two: Fieldwork in Kenya, 1992 to 1994.
Secondary and primary data collection was the main activity during this phase. This data collection consisted of:

Secondary Data
An analysis of secondary data from a large variety of sources of information (books, articles, maps, and reports) collected from the following sources:- (1) Kenya Agricultural Information Centre Nairobi. (2) Kenya Agricultural Research Institute Nairobi and Kakamega. (3) Text Book Centre, (4) Ministry of Economic Planning Nairobi. (5) Ministry of Agriculture and Livestock Development, Nairobi, Kakamega and Vihiga offices. (6) Food and Agriculture Organization Library, Nairobi. (7) University of Nairobi and Moi University Libraries. Also archival research on Vihiga district in general, and North Maragoli in particular was carried out in Nairobi and Kakamega.

Primary Data
Sources of primary data. Three main sources of primary data were used: (a) key informants, (b) household heads, and (c) women groups. Information from household heads was collected in three stages: (1) Census of 1000 farm households, (2) General (one time visit) survey of 180 farm household heads selected from 1000 farm household heads, and (3) In-depth (monthly visits) of 33 farm households selected from the 180 farm household heads. At all stages, the households were categorized into male-headed, female-managed and female-headed households.

(a) Key Informants: government officials and related organizations in the field of farming: (i) Agricultural officers at the headquarters, provincial, district and divisional levels. (ii) Extension officers, (iii) Livestock officers, (iv) Environmental officers, (v) District Development officers, (vi) District Planners, (vii) Co-operative officers, (viii) Agricultural Training Centres in the district, (ix) Chiefs and subchiefs, and (x) a few NGOs operating in the area of agriculture.

Techniques of data collection from key informants: conversational and structured interviews, structured questions, group discussions, seminars and workshops. Types of data collected: Information on government and other institutional policies as they affect farmers, especially women in decision-making and farm management practices. (1) Information on land tenure system and how the tenure system affects farm management practices. (2) Information on extension services - types of services provided; types of information commonly conveyed by the extension officers and relevance, whether based on cash crop production or on all the crops; Organization of farming demonstrations, number of times, those in attendance, and the effectiveness of the methods; integration of local farming systems knowledge; Experimental stations, types of experiments, whether specific experiments for each farming system are carried out and how the information is used; Any specific problems experienced by officers when they serve women farmers; Ratio of extension officers to the number of farmers and finally frequency of visits to both male and female farms. (3) Policies on financing agriculture: financing institutions; criteria for extending credit facilities to farmers; requirements for one to qualify for a loan; the issue of women and finance for farming. (4) Policies on agricultural training: requirements for one to qualify for a training place at an agricultural training centre on short courses for farmers (Bukura training centre), women who have benefitted from such training and the type of courses compared to men; Constraints women face in such training places. (5) Policies on land management in relation to cash crop farming; Maintaining soil fertility; Recommendations of land use policies; Types of conservation and special programmes for women. (6) Policies on agricultural co-operative movement: Types of co-operatives in the agricultural sector; Requirements necessary for individuals to join farmers' co-operatives ; Services from the co-operatives to the farmers; Special problems of women. (7) Agricultural pricing policies: Whether the pricing policies favour mostly the crops grown by men. (8) Other institutional policies, Non-governmental institutions and their role in agriculture. Criterion for their choice of certain agricultural projects and the areas. The priorities of NGOs in agriculture.

(b) Heads of Households: On the basis of a census of 1000 households, 180 heads of households were sampled, representing in equal measure three types of farm households, viz. male-headed households, female-managed households (defacto fem.h), and female-headed households (dejure fem.h) and also the wives of male-headed households. These heads of households were interviewed by means of a structured questionnaire in a general one-visit survey carried out at household level to investigate farm management.

(c) Women Groups: The role of women groups in agricultural production and in general rural development.

North Maragoli Location in Vihiga District was selected for this study. This location was selected out of 16 locations because it was representative of the district in terms of agro-ecological conditions and the pattern of population density. Vihiga District effectively comprises one agro-ecological zone based on the main crops grown, soils, rainfall and population density - viz. the Tea/Coffee Zone. North Maragoli is found in the Tea/Coffee zone, with an altitude of more than 1500 metres, high rainfall between 1500-2000 mm, high rainfall reliability, and a population density of between 740-1300 persons per sq km.

Sample Design and Sample Size

On the basis of a census of 1000 households in the mentioned locations, a selection of 180 heads of households was made. Stratified sampling was used as a means of selection. This method was adopted because we had to deal with different types of households within the given area. The respondents were, the farm household heads who were divided into three groups according to the type of farm household.

The selection of the farm household heads was based on the following criteria.
(a) Age: the farm household heads selected were aged between 30-70 years. It is believed that such types of households are bound to be more stable, have acquired their own land, be well established and independent of their parents, be under more pressure to utilize resources optimally because of family commitments and have children old enough to provide some family labour. Initially, the researcher had anticipated selecting farmers between the ages of 30 - 50 years to avoid including very old farmers who were not strong enough to work and who are sometimes being cared for by their relatives but, once in the field it was realized that the age limit had to be raised as most women in female-headed households were found to be 60 years or more.
(b) Farming system: the heads of households selected had farms comprising all four farming sub-systems, viz. the cash crop, food crop, livestock and tree production sub-systems, to provide the basis for the analysis of farm management and agricultural production in the three types of households.
(c) Farm size: the heads selected had farms ranging between 0.2 to 2.5 hectares. This allows us to assess the differences in farm management based on the farm size. Farm household heads selected were interviewed on the basis of a structured questionnaires.

Data collected from heads of households and wives in male-headed households and the techniques of measurement of variables where necessary:

In the general survey, the above mentioned household heads and wives in male-headed households were interviewed about their farming operations, labour inputs, farm incomes, yields,

Table 1
Sample Size

Location and Farming System	Male-headed and female in male-headed households	Female-managed hhs	Female-headed hhs	Total
Tea/coffee agro-ecological zone	60 (60)	60	60	240
Total sample population	120	60	60	240

(between brackets: number of wives interviewed)

investments and perceptions on proper farm management, existing constraints and the relevance of government and institutional policies using a structured questionnaire.

After the general survey, 33 farmers were selected for in-depth case study analysis based on the presence of the four farming sub-systems with emphasis on maize and tea production to verify the anomalies that exist in farm management practices from a gender perspective, taking account of different interests and decision-making patterns in the three types of households mentioned.

Farm size
The system of agriculture prevailing and the lack of knowledge of the exact size of their farms on the part of the respondents required the measurement of farm sizes for the in-depth case studies. The respondents who had land title deeds knew the area covered by the title, but the land covered by the legal title sometimes was not equal to the holding or to the land under cultivation. This forced the researcher to measure all the farm holdings and plots under different crops for the in-depth studies only. During the general survey, estimates of farm sizes were made from land title deeds where these were available. A rough estimate of farm size during this phase of the research was also obtained by questioning the heads of households about the size of their farms.

For the in-depth case studies actual measurements of farm size and the area under the cultivation of different crops were made using the following methods: Triangulation and the fitting of a polygon to the shape of the area; measurements of distances were made using a measuring tape and a road measuring wheel. Surveyor's ranging poles were also used to mark the start and end points. The use of a measuring chain would probably have been the most accurate method of recording distances but because it is time consuming, the researcher decided against it. The measurements were accurately done.

After the measurements had been completed, the researcher first computed the area of the whole holding after which the parcels under different crops were computed using the following formula. Area measurement by trigulation method.
(i) Draw out the farm holding and indicate clearly the parcels under different crops or uses.
(ii) Find the sum of the areas of triangles.
(iii) Area of each triangle was worked out using the following formula: square Root of S(s-a)(s-b)(s-c). Where a, b, and c are the sides of the triangle and
S (semi perimeter)= a+b+c divide by 2

Where the diagonal could not be measured because of obstruction, the diagonal was computed using the following formula:- Diagonal = a squared (one side of the triangle) + b squared (another side of the triangle) and then the square root of the total. Add together all the areas of the triangles. Once the areas have been established divide it by 10,000 to change it from metres to hectares.

This method was used in combination with pocket calculators. It was then possible for the enumerators to make quick calculations of areas during field visits. The enumerators had to be plot the distances onto squared paper. This way of calculating farm size was quite cumbersome, time consuming and if the plotting was delayed, the almost inevitable high error rate in indicating distances at their right sides would be detected the following day or after calculating the area forcing us to revisit the farm to take another measurement. To avoid these repeat measurements, the researcher made sure that drawing was going on concurrently as the measurements were being taken. After completing the measurements the researcher checked the drawings with the enumerators to make sure that all sides had been measured and indicated.

Another problem that the researcher experienced in measuring crop areas were the overlaps between planting and harvesting, and the use of the same piece of land several times a year. This problem was solved by the multi-visit, in-depth case studies involving interviews, measurements and monthly visits to the farmers throughout the agricultural year. To avoid the complication presented by the use of a single piece of land on more than one occasion during the agricultural year which would complicate statistics on total crop areas and land under cultivation, a specification of the main season crops and their production was made, and again for the short season crops and their production. The worst problem in measuring parcels under different crops was the intercropping of about four to five different crops on the same plot. In this case, the researcher only considered the two most dominant crops with the highest intensity or concentration. An additional problem was the opposition from some farmers and relatives of household heads who were suspicious and did not want us to measure farms.

Crop Yields
There are two types of crop yields that agriculturalists normally measure, viz. the biological yield and the economic yield. This researcher was concerned with the economic yield. Casley (1981) refers to economic yield as holder harvested production. Estimates of yield from objective measurements (involving cutting and harvesting of the crop by the surveyor) of sub-plot crop cuts is best left to agricultural research stations and specialized surveys that specifically require estimates of biological yields (Casley, 1981). Objective sub-plot harvesting requires a static enumerator and a highly clustered sample which would be a very inefficient design in this research. This researcher relied on holder-specific economic production estimates which were simpler and more acceptable to many farmers since they would not involve enumerators trampling over their crops every day. It was easy to measure the production of cash crops since production is measured and recorded on a daily basis at the buying centres or at the co-operative societies. The production and yield levels indicated to some extent the level of farm management. High yields were expected on farms that were well managed and where farmers invested in the crops, and carried out the farming activities on time. The problem of mixed and associate cropping was solved by this researcher allocating three-quarters of the entire area to the main crop and the remaining one-quarter to the minor or subsidiary crop as suggested by Casley (1981).

Livestock
Livestock information was also provided by the farmers. Information on livestock was accurate and straight-forward because many farmers owned on average of two cows and a few poultry. Here the researcher was interested in both the small stock such as sheep, goats, poultry farming and mainly cattle keeping. The farmers provided the following information on animal husbandry and elements of farm management: (a) Type of cattle kept and other animals, whether cross-breed or local, breeding strategy, calving rate; (b) The practice of zero grazing, the diversity and additional feedstuffs (for instance planted special grasses for cattle, use of stocks from grains, artificial feeds); (c) cattle dipping; (d) veterinary care; (e) use of animals,(for instance milk use,

dung as manure, draught power); (f) labour allocation for animal husbandry, and (g) problems encountered by women livestock farming.

Labour Measurements
Classification was made of farming operations that require labour. Measuring labour required calculating the time spent on each farming activity per day and per week. This measure provided a basis for analyzing discrepancies between the gender division of labour, and gave an indication of the importance of sufficient labour for farm management in the three types of households. Further information was gathered on timeliness (specifically for the case studies) of labour where preparation, planting, weeding, harvesting to post harvest activities were concerned. Specific problems that women experience associated with labour, migration and remittances; and finally the relationship between remittances and agricultural investment, and farm management and production.

Farm Inputs and Investments in Farming
All the inputs purchased, quantity, quality and the amount of money spent especially on new seeds, fertilizers, extra labour, pesticides, sprays, transportation were considered. This highlighted the level of investments and showed at what level female-headed and female-managed farms were investing and how this affected their farm management practices.

Farm Incomes
This was measured by the amount of money received from each crop minus the production costs. The farmers were asked to estimate the amount of money earned on each crop. The researcher cross check the data on cash crops from the co-operatives or the buying organizations. However, in a place like North Maragoli where the production from staple crops is not sufficient to feed the family, farmers do not sell maize but buy it instead and so here we have a straight forward example of getting incomes from mostly tea, coffee and sometimes bananas. Good management was reflected in higher earnings.
Information on extension services, training, land tenure, co-operative movements, credit and financing institutions, pricing policies and NGOs were also checked from the farmers as well as the way these policies and programmes influenced farm management and specific problems that confronted women in this area.

Non-agricultural activities and earnings
Information was collected on non-agricultural activities and how such activities influence farm management, for instance the time spent on these activities, money earned from such activities and whether some farmers neglected farm management if non-agricultural activities were more rewarding.

Household income
The total household income and its influence on farm management was gauged:
All the variables that did not have quantitative values were evaluated on a qualitative scale of very good, good, fair, poor, and very poor indicating the best options and the reverse is the case for poor management.

Methods for Data Analysis and Data Presentation
The statistical package for social sciences (SPSS) and Harvard Graphics have been used for data analyses and presentation. Descriptive statistics, multiple regression and graphs have been used in the presentation of data.

Schedule of work while in the field
February to April 1992, a census of household heads and 1000 households was made. 180 households were selected on the basis of the criteria discussed above. Types of information that were extracted during this exercise were: (a) household head, (b) age, (c) farm size, (d) crops grown, (e) livestock. The Central Bureau of Statistics of Kenya was consulted on information on households in North Maragoli.

May to early November 1992, general survey. Sampled household heads were interviewed using a structured questionnaire; observations were made on farm preparations, weeding, and crop care; and production records for both 1991 and 1992 were made. September to November 1992 a survey of women groups was carried out with the help of a Dutch student as a research assistant. In all these field studies and surveys questionnaires were used. During this study between four to six local research assistants were used at different times. From December 1992 a selection of household heads was made for in-depth analysis. In-depth case study analyses were carried out from January to December 1993. The first analysis of data and the writing of the rough draft took place early 1994.

Phase Three. The Netherlands August 1994 to 1995
Final analysis, writing and presentation of the Ph.D thesis.

Appendices

Appendix 2

Definitions of operational terms

Household: A household comprises of a person or group of persons, generally bound by ties of kinship, normally living together under one roof or several roofs within the same compound or homestead area and sharing a community of life by their dependence on a common holding as a source of income and food, which normally, but not necessarily, involves them in eating from a 'common pot'. It is therefore, a unit of production and consumption.

Holding: A unit of agricultural production comprising all the land used completely or partly for agricultural purposes and all livestock kept and operated under the management of an individual or group without regard to legal ownership.

Household head: The main person who manages and makes decisions with regard to the household in every day life. His/her decisions are essential for the sustenance and management of the household.

Male-headed household: A household where a husband is present and engages in the decision making in farm management. It is also referred to as a joint household where both wife and husband are present throughout the year.

Female-headed household: A household where the female head may be a widow, a divorcee, a single woman who has never married but has acquired her own land or a woman who has been abandoned or forsaken by the husband. Here, the woman is almost entirely responsible for the household and farm management, and makes all key agricultural decisions. Where the husband is dead or if the woman is divorced these events should have taken place at least a year ago for the female to be considered a head of household.

Female-managed household: A household where the female head manages the household and the farm in the absence of the man who has migrated out of the home to work elsewhere. He may send money/goods to the household. Once in a while he comes home, and he helps the wife in farm management decisions. Otherwise he communicates these decisions either in writing, or be sending messages through relatives. Thus, the husband has a full-time occupation other than farming but continues being a part-time farmer by contributing some labour himself and/or by hiring labour for crops cultivated on his plot (mainly the cash crop plot) (Safilios, 1988 a). There were cases where the man is in and out of the home very frequently. For this research the household was considered as female-managed if, between March and August when farming operations are intense, the man was away from the household for more than three months. However, if the man was away for less than three months within this same period then such a household was considered to be male-headed.

Farm Income: All the earnings (money) received by the farmers from the sale of their agricultural produce. Such households also usually rely on their own produce for part of their food needs, but this production will not be reported as income, even if the respondent knows the amount or its value.

Household Income: This includes farm income and other non-agricultural earnings.

Farm Management: The way the individual farmer organizes the factors of production - land, labour, and capital - on his/her farm, adapts farming practices to his/her particular environment, and disposes of his/her product in such a way as to yield him/her the largest net return, while still maintaining the integrity of his/her land and equipment (Clayton, 1983).

Farming Sustainability: The growing of crops and/or rearing of animals to meet the present requirements of the household while maintaining or enhancing resource productivity on a long term basis.

Land Degradation: The word degradation implies 'reduction to a lower rank.' The 'rank' conceived is in relation to actual or possible uses, and a reduction implies a problem for those who use the land. Therefore, when land is degraded, its productivity declines unless measures are taken to restore that productivity and check further losses. In either case, the yield of the labour in terms of production is adversely affected. Land degradation, therefore, directly consumes the product of labour, and consumes capital inputs into production. Other things being equal, the product of work on degraded land is less than it would be on the same land if there was no degradation (Blaikie and Brookfield, 1987).

Land Management: The application of known or discovered skills to land use in such a way as to avoid, minimize or repair degradation, and to ensure that the capacity of the land is maintained beyond the present crop or other activity, such land use skills include (1) soil conservation measures (terracing, construction of gabions, contour ploughing, planting grass and trees; (2) land quality in terms of fertility measures, for instance the use of chemical fertilizers, manure, and composite waste; (3) water management, for instance avoiding water logging; and (4) methods of preventing animal damage (Blaikie and Brookfield, 1987).

Appendices

Appendix 3

A Summary of Major Crop Statistics in the Case Studies

In the next few pages a comparison of the farm-household systems in the three types of households is presented in tables 6.1, 6.2 and 6.3. Table 6.1 contains the summary information on tea and the total household income of the cases discussed in this chapter.

Table 6.1
Summary of Annual Tea Production, Tea Income and Total Households Income (source: North Maragoli In-depth Case Studies, 1993).

Farmer's Name	Farm size under Tea in Ha	No. of Tea Bushes	Total Tea Production 1993, kgs	Tea Yield per Hectare	Total Tea Income in 1993 Kshs	Total HH Income in 1993 Kshs
Agesa-mh	0.3 ha	3500	2350 kgs	7,833 kg	29,166.00	36,166.00
Eboso-mh	0.5 ha	5000	3371 kgs	6,742 kg	36,529.00	47,029.00
Manase-mh	0.2 ha	2500	70 kgs	350 kg	180.00	7,000.00
Musasa-mh	0.2 ha	3000	1552 kgs	7,769 kg	8,577.00	192,000.00
Iminza-fmh	0.3 ha	2000	438 kgs	1,460 kg	2,216.00	27,000.00
Chagenda-fm	0.7 ha	4000	3614 kgs	5,162 kg	42,556.00	42,556.00
Mmbone-fm	0.2 ha	1500	1650 kgs	8,250 kg	19,298.00	71,298.00
Muhonja-fm	0.1 ha	1456	1030 kgs	10,300 kg	11,893.00	45,000.00
Ajema-fh	0.4 ha	2500	347 kgs	868 kg	4,839.00	80,000.00
Kaveza-fh	0.3 ha	3000	2502 kgs	8,340 kg	30,540.00	42,540.00
Keyonzo-fh	0.3 ha	2500	20 kgs		167.60	167.60

Many tea fields measure 0.3 hectares (45%), yet a big differentiation exists in the number of tea bushes, the production and total annual tea incomes as shown in table 6.1. A few examples will be given to clarify this. Tea bushes, tea production, and total annual tea incomes on 0.3 hectares plots vary between 2,000 and 3,500 bushes; 20 kgs and 2,350 kgs of green tea leaves; and Kshs 167 to Kshs 29,166 of income respectively. These differences on a plot of similar size highlight the influence of the specific situations of a household on farm management and agricultural production. Important aspects are use of fertilizer, weeding of the crop, pruning, and plucking characteristics.

For these selected cases the total annual tea income contributed 33% of the total annual household income. In the various households tea contributed 30% in male- headed, 47% in female-managed and 28% in female-headed households in 1993. This implies that though farming is the main preoccupation in North Maragoli, its contribution in terms of earnings is low since it does not provide even 50% of the total-household income. Farmers have to devise other strategies in order to accumulate money for household requirements.

Accessibility to tea income and the expenditure patterns between the husband and the wife to some extent influenced the management of the crop. Many male-headed households perform better than the other two types of households both in tea production and in devising other income-earning strategies. In fact, it is apparent that production on some farms is very high and as good as the national average (see the column on production per hectare). Comparing with

Table 6.2
Summary Maize Production (source: North Maragoli In-depth Case Studies, 1993).

Farmer's name	Plot size	Production in kgs	Average yield per hectare
Agesa - male-hh	0.1	270	
Eboso - male-hh	0.2	90	
Manase - male-hh	1.0	00	
Musasa - male-hh	0.6	2,250	
Total for mh hhs	1.9	2,610	1,374 kg/ha
Chagenda-female-mh	0.4	1,800	
Mmbone-female-mh	0.1	45	
Muhonja-female-mh	0.1	180	
Total for fm hhs	0.6	2,025	3,375 kg/ha
Ajema-female-hh	0.1	270	
Kaveza-female-hh	0.1	90	
Keyonzo-female-hh	0.5	90	
Total for fh hhs	0.7	450	643 kg/ha
Iminza-female-mhh	0.4	270	
Total for all hhs	3.6	5,355	1,488 kg/ha

average yield for Vihiga District (see 3.4.1), North Maragoli attained the highest tea yields and production, an indication that management in this area was superior to the other areas of Vihiga District. We also learn from table 6.1 that some farmers in North Maragoli produce more tea per hectare than the national yield and are on the same level as small-scale farmers on the eastern side of the Rift Valley (see section 4.5.1). General averages hide the inter- and intra-household variations in management and production, and these differences can only be brought to the surface when individual cases are exposed. This may be the only way to find out where the weaknesses are and how to deal with them. We were suspicious of Mmbone and Muhonja's exaggerated high yields and production. It seems that this emanated from faulty recordings at the tea buying/collecting centre. We can draw the conclusion that women need to own and to have access to production factors. They need both technical and financial support in order to improve

Table 6.3
Range of Activities in Relation to Agricultural Production and Household Incomes (source: North Maragoli In-depth Case Studies, 1993).

Farmer N	Food Crops				Cash Crops	Livestock			Others						
	Maize	Beans	Bans	Oths	Tea, Coffee	Fb	Zebu	Cross/Grade	1	2	3	4	5	6	7
Agesa	*	*	*	*	*		*		*	*					
Eboso	*	*	*	*	*		*		*	*		*	*		
Manase	*	*	*	*	*										*
Musasa	*	*	*	*	*		*		*		*			*	*
Iminza	*	*	*	*	*			*	*			*	*		*
Chagenda	*	*	*	*	*	*	*		*						
Mmbone	*	*	*	*	*		*		*		*				
Muhonja	*	*	*	*	*			*	*	*		*			*
Ajema	*	*	*	*	*			*	*	*					*
Kaveza	*	*	*	*	*	*			*						
Keyonzo	*	*	*	*	*		*		*						

Key :
Bans = Bananas; Fb = French Beans; Oths = Others
1 = Trees; 2 = Horticulture (for sale); 3 = Permanent Wage Employment;
4 = Trade; 5 = Renting in land; 6 = Rental Houses; 7 = Others

management and production. In fact, what they need most is an independent source of income. However, female-headed households seem to be in dire need as compared to the other two types of households. It was our assumption that widows could be more productive on their farms since with the death of their husbands they have the overall mandate in decision-making and management of the household resources. But this study reveals that this is not true for rural women in small-scale farming. The linkage and importance of a man in a household being linked with success is internalized. However, as explained elsewhere, the socio-economic conditions after the man's death are made very difficulty for the women by the local community which erodes her confidence and security.

The poor state of food production is clearly shown in Table 6.2. The household with the largest plot (1 ha) under maize produced zero in 1993. Female-managed household performed better in maize production; 3,375 kg/ha where male-heads produced 1,374 kg/ha and female-heads 643 kg/ha. The yield for all the farmers was 1,488 kg/ha. This still shows that yields are low in this area when compared to the minimum expected of 6,000 kg/ha. Poor husbandry and poverty contribute greatly to this situation. Many farm households are unable to invest in good maize husbandry practices because they lack the capital to do so. Maize is further green plucked before the harvest, largely reducing the expected production. Any little money that many households receive is used to buy food.

Table 6.3 presents the range of activities in each of the case study households. Male-headed households seem to have more activities than the other two types of households. It also appears that the households with more activities enjoyed higher total monthly and annual incomes.

Appendix 4

Table 8.4
A List of Women Groups in North Maragoli Location 1992

Name of Group	Year of Reg	No. Members	Projects and Activities of the Group
Mwenyelitsi	1986	32	Zero-grazing, Horticulture, Petty business, choir, Tree planting, Tailoring and Group farming
Chanderema	1977	60	Zero-grazing, Petty business, Horticulture, Group farming, Revolving fund, Nursery and Adult literacy classes
Vokoli	1986	52	Zero-grazing, Petty business, Horticulture, Tailoring, Poultry, Group farming and Revolving fund
Chegobero	1981	36	Pig keeping, Horticulture, Revolving funds and Group farming
Gurugwa	1988	55	Group farming, Horticulture, Petty business, Choir, Plot, Rental houses, Butchery, Hotel, Revolving fund
Mukuyu	1981	20	Poultry keeping, Horticulture, Petty business, Group farming and Revolving fund
Mudete	1979	27	Plot, Commercial building with office space, Tailoring Knitting, Making table clothes, Tailoring school, Petty business, Group farming and Revolving fund
Gaigedi	1985	50	Horticulture, Petty business, Group farming, Timber-sawing and Revolving fund
Lwenya	1987	40	Medical clinic, Traditional mid-wifery, Brick-making, Group farming and Revolving fund
Gavudia	1989	30	Poultry keeping, Making table clothes, Revolving fund, Petty business, Horticulture, choir and Group farming
Lumavo	1977	45	Horticulture, Petty business, Brick-making, Poultry-keeping, Group farming and Revolving fund
Givudimbuli	1989	25	Poultry keeping, Horticulture, Petty business, Group farming and Tree nursery
Wadigula	1989	50	Handicraft, Fish farming, Bee-keeping, Rabbit-rearing, Horticulture, Brick-making, Tree nursery, Revolving fund, Group farming and Timber sawing
Ambenge Youth Group	1988	31	Tie and dye, Printing, Knitting, Tailoring, Handicraft Hotel, Petty business, Tailoring school, Group farming
Mukunyo	1986	37	Horticulture, Petty business, Group farming and Revolving fund
C.P.K Lososi	1987	30	Handicraft, Horticulture, Petty business, Group farming and Revolving fund
Kisuru	1990	18	Group farming, Petty business
Masigolo	1990	30	Horticulture, Petty Business, Poultry keeping, Group farming, Revolving fund

Name	Year	Members	Activities
Kwenyelitsa Group	1991	36	Pottery, Petty business, Revolving fund, Group farming Horticulture
Kisangula	1981	40	Horticulture, Handicraft, Revolving fund, Group farming, Plot
Kivagala	1981	45	Group farming, Handicraft, Petty business, Revolving fund
Kigama	1989	25	Choir, Petty business, Group farming, Horticulture, Revolving fund
Inzale Self-Help Men Group	1990	32	Horticulture, Petty business, Group farming, Brick-making, Timber sawing
Mulina	1990	15	Horticulture, Revolving fund, Petty business, Group farming, Handicraft
Mahindi	1988	20	Handicraft, Petty business, Revolving fund, Adult literacy class, Horticulture, Group farming
Givungi	1986	22	Handicraft, Petty business, Revolving fund, Group farming, Horticulture
Akwigwira	1990	30	Horticulture, Petty business, Revolving fund, Group farming
Dabanga	1991	31	Horticulture, Group farming
Mkonyi	1990	26	Horticulture, Petty business, Mid-wifery, Group farming
Gaigedi Men's Group	1990	20	Brick-making, Timber sawing, Revolving fund
Avugwi	1982	60	Tree nursery, Petty business, Horticulture, Organic farming, Group farming, Plot, Zero-grazing, Processing of cooking oil, Revolving fund, Handicraft
Digula	1990	15	Handicraft, Horticulture, Group farming, Revolving fund
Gaigedi Youth Association	1992	25	Horticulture, Bee-keeping, Fish farming, Agro-forestry Group farming, saving scheme
Evojo	1983	32	Zero-grazing, Horticulture, Group farming, Cookery, Choir, Revolving fund
Mago	1990	15	Zero-grazing, Horticulture, Group farming, Handicraft, Making and selling energy saving stoves, Group farming Revolving fund
Wekhonye	1987	20	Zero-grazing, Choir, Petty business, Group farming, Revolving fund
Watanyalitsa	1991	15	A group of very old women known as 'help-age', soliciting for food aid and medical care

Curriculum Vitae

Basilida Anyona Mutoro was born on 4th April 1956 in a family of eleven children in Kakamega District, Kenya. Her primary education was undertaken in a local primary school- Shikoti Girls Primary Boarding School. After completing her Secondary and Advanced levels education at Mukumu (Kakamega District) and Ngandu (Nyeri District) girls high schools respectively, she started tertiary education at Kenyatta University College then a constituent college of Nairobi University. At this level she pursued a Bachelor of Education degree (B.ED) between 1978 and 1981 specializing in Geography, Philosophy and Religious Studies, and Education. An Upper Second Class Honours was obtained on completion. Between 1982 and 1985, Basilida Anyona Mutoro successfully pursued and obtained an M.A degree in geography at the Department of Geography, Nairobi University on a government scholarship. Her thesis was titled Selected Environmental and Socio-Economic Factors Affecting Agricultural Production in South Maragoli Location in the Western Province of Kenya. On successful completion of the M.A, she worked as a teacher trainer at Kaimosi Teachers College between 1985 and 1988. She joined the Department of Geography at Moi University August 1988 as a lecturer. She transferred from this Department to the School of Environmental Studies in 1991. Between 1991 and 1995 this study was undertaken, written and published for the award of the Ph.D. degree of the University of Utrecht in The Netherlands. The research she has carried out and her research interests are focused on rural development problems, integrating gender issues, environmental and resource management problems, socio-economic effects, and policy issues as they impact on the livelihood and agricultural production of rural households and development of rural areas at large.